Advanced Algebra

Advanced Algebra

Edited by
Derek Portman

Willford Press
www.willfordpress.com

Published by Willford Press,
118-35 Queens Blvd., Suite 400,
Forest Hills, NY 11375, USA

ISBN: 978-1-68285-634-5

Cataloging-in-Publication Data

Advanced algebra / edited by Derek Portman.
p. cm.
Includes bibliographical references and index.
ISBN 978-1-68285-634-5
1. Algebra. 2. Mathematical analysis. 3. Mathematics. I. Portman, Derek.
QA155 .A38 2019
512--dc23

For information on all Willford Press publications visit our website at www.willfordpress.com

Willford Press

Contents

Preface

The main aim of this book is to educate learners and enhance their research focus by presenting diverse topics covering this vast field. This is an advanced book which compiles significant studies by distinguished experts in the area of analysis. This book addresses successive solutions to the challenges arising in the area of application, along with it; the book provides scope for future developments.

Algebra is a sub-division of mathematics that involves elementary equation solving and study of abstractions. Advanced algebra or abstract algebra is the study of algebraic structures such as groups, rings, fields, vector spaces, lattices, etc. Algebraic structures that involve single binary operations are magma, group, semigroup, monoid and quasigroup. Fields, modules, vector spaces, lie algebra, lattice, Boolean algebra, etc. are all examples of algebraic structures that involve several operations. This book is a compilation of chapters that discuss the most vital concepts and emerging trends in the field of advanced algebra. Such selected concepts that redefine this discipline have been presented in this book. It will serve as a valuable source of reference for graduate and post graduate students as well as experts.

It was a great honour to edit this book, though there were challenges, as it involved a lot of communication and networking between me and the editorial team. However, the end result was this all-inclusive book covering diverse themes in the field.

Finally, it is important to acknowledge the efforts of the contributors for their excellent chapters, through which a wide variety of issues have been addressed. I would also like to thank my colleagues for their valuable feedback during the making of this book.

Editor

1

Upper and lower $\alpha(\mu X, \mu Y)$-continuous multifunctions

M. Akdağ[a], F. Erol[a*]

[a] *Cumhuriyet University Science Faculty Department of Mathematics 58140 SIVAS / TURKEY.*

Abstract. In this paper, a new class of multifunctions, called generalized $\alpha(\mu_X, \mu_Y)$-continuous multifunctions, has been defined and studied. Some characterizations and several properties concerning generalized $\alpha(\mu_X, \mu_Y)$-continuous multifunctions are obtained. The relationships between generalized $\alpha(\mu_X, \mu_Y)$-continuous multifunctions and some known concepts are also discussed.

Keywords: Generalized open sets, multifunction, generalized continuity.

1. Introduction

There are various types of functions which play an important role in the classical theory of set topology. A great deal of works on such functions has been extended to the setting of multifunctions. A multifunction is a set-valued function. The theory of multifunctions was first codified by Berge [10]. In the last three decades, the theory of multifunctions has advanced in a variety of ways and applications of this theory, can be found for example, in economic theory, noncooparative games, artificial intelligence, medicine, information sciences and decision theory (See [14] and references therein). Continuity is a basic concept for the study of general topological spaces. This concept has been extended to the setting of multifunctions and has been generalized by weaker forms of open sets such as α-open sets [18], semiopen sets [17], preopen sets [9], $\beta-$

*Corresponding author.
E-mail address: feerol@cumhuriyet.edu.tr (F. Erol).

open sets [16] and semi-preopen sets [13]. Multifunctions and of course continuous multifunctions stand among the most important and most researched points in the whole of the mathematical science. Many different forms of continuous multifunctions have been introduced over the years. Csaszar [1] introduced the notions of generalized topological spaces and neighborhood systems are contained in these classes, respectively. Specifically, he introduced the notions of continuous functions on generalized topological spaces and investigated the characterizations of generalized continuous functions. By using these consepts, Min [21] introduced the notions of $\alpha(g_X, g_Y)$-continuity, $\beta(g_X, g_Y)$-continuity, $pre\,(g_X, g_Y)$-continuity and $semi\,(g_X, g_Y)$-continuity of functions on generalized topological spaces. Kanibir and Reilly [8] extended these concepts to multifunctions. Also, Boonpok [12] studied the $\beta(\mu_X, \mu_Y)$-continuous multifunctions. Then Akdağ and Erol [15], studied the $pre\,(\mu_X, \mu_Y)$ -continuous multifunctions. In this paper our purpose is to define $\alpha(\mu_X, \mu_Y)$-continuous multifunctions and to obtain some characterizations and several properties concerning such multifunctions. Moreover, the relationships between generalized $\alpha(\mu_X, \mu_Y)$-continuous multifunctions and some known concepts are also discussed.

2. Preliminaries

Let X be a nonempty set and μ be a collection of subsets of X. Then μ is called a generaized topology (briefly GT) on X iff $\emptyset \in \mu$ and an arbitrary union of elements of μ belongs to μ [1]. A set with a GT μ is said to be generalized topological spaces (briefly GTS) and denoted by (X, μ). The elements of μ are called $\mu - open$ sets and their complements are called $\mu - closed$ sets. The generalized interior of a subset A of X denoted by $i_\mu(A)$ is the union of $\mu - open$ sets contained in A, and the generalized closure of A denoted by $c_\mu(A)$ is the intersection of $\mu - closed$ sets containing A. It is easy to verify that $c_\mu(A) = X - i_\mu(X - A)$ and $i_\mu(A) = X - c_\mu(X - A)$. Let μ be a GT on a set $X \neq \varnothing$. Clear that $X \in \mu$ must not hold; if all the same $X \in \mu$, then we say that the GT μ is strong [2]. In general M_μ denote the union of all elements of μ, of course $M_\mu \in \mu$ and μ is strong GT if and only if $M_\mu = X$ [2]. Let us now consider those GTS μ satisfy the following condition: if $M, M' \in \mu$, then $M \cap M' \in \mu$. We call such a GT *quasitopology (briefly QT)* [3], the QTS clearly are very near to the topologies.

A subset A of a generalized topological spaces (X, μ) is said to be $\mu regular - open$ [3] (resp. $\mu regular - closed$) if $A = i_\mu(c_\mu(A))$ (resp. $A = c\mu(i_\mu(A))$). A subset A of a generalized topological spaces (X, μ) is said to be $\mu - semiopen$ [3] (resp. $\mu - preopen$, $\mu - \alpha - open$ and $\mu - \beta - open$) if $A \subseteq c\mu(i_\mu(A))$ (resp. $A \subseteq i_\mu(c_\mu(A))$, $A \subseteq i_\mu(c_\mu(i_\mu(A)))$, $A \subseteq c_\mu(i_\mu(c_\mu(A)))$). The family of all $\mu - semiopen$ (resp. $\mu - preopen$, $\mu - \alpha - open$ and $\mu - \beta - open$) sets of X is denoted by $\sigma(\mu)$ (resp. $\pi(\mu)$, $\alpha(\mu)$ and $\beta(\mu)$). It is shown in [3], that $\alpha(\mu) = \pi(\mu) \cap \sigma(\mu)$ and it is obvious that $\sigma(\mu) \cup \pi(\mu) \subseteq \beta(\mu)$. The complement of a $\mu - semiopen$ (resp. $\mu - preopen$, $\mu - \alpha - open$ and $\mu - \beta - open$) set is said to be $\mu - semiclosed$ (resp. $\mu - preclosed$, $\mu - \alpha - closed$ and $\mu - \beta - closed$). The intersection of all $\mu - semiclosed$ (resp. $\mu - preclosed$, $\mu - \alpha - closed$ and $\mu - \beta - closed$) sets of X containing A is denoted by $c_\sigma(A)$, (resp. $c_\pi(A)$, $c_\alpha(A)$ and $c_\beta(A)$) are defined similarly. The union of all $\mu - \alpha - open$ ($\mu - \beta - open$) sets containing in A is denoted by $i_\alpha(A)$ ($i_\beta(A)$).

By a multifunction $F : X \longrightarrow Y$, we mean a point-to-set correspondence from X to Y, and we always assume that $F(x) \neq \emptyset$ for all $x \in X$. For a multifunction $F : X \longrightarrow Y$, following we shall denote the upper and lower inverse set of a set B of Y by $F^+(B)$ and $F^-(B)$ respectively, that is, $F^+(B) = \{x \in X : F(x) \subseteq B\}$ and $F^-(B) = \{x \in X :$

$F(x) \cap B \neq \emptyset\}$. In particular, $F^{-}(y) = \{x \in X : y \in F(x)\}$ for each point $y \in Y$. Also, $F(A) = \bigcup_{x \in X} F(x)$, for each $A \subseteq X$. Then F is said to be surjection if $F(X) = Y$, or equivalently, if for each $y \in Y$ there exists an $x \in X$ such that $y \in F(x)$. Throughout this paper (X, μ_X) and (Y, μ_Y) (or simply X and Y) always mean generalized topological spaces.

3. Upper and Lower $\alpha(\mu_X, \mu_Y)$-Continuous Multifunctions

Lemma 3.1 [21] Let A be a subset of a generalized topological space (X, μ_X). Then,
(i) $x \in c_{\alpha_X}(A)$ if and only if $A \cap U \neq \varnothing$ for each $U \in \alpha(\mu_X)$ containing x.
(ii) $c_{\alpha_X}(X - A) = X - i_{\alpha_X}(A)$.
(iii) $c_{\alpha_X}(A)$ is $\mu_X - \alpha-$ closed in X.

Definition 3.2 Let (X, μ_X) and (Y, μ_Y) be a generalized topological spaces. Then a multifunction $F : X \longrightarrow Y$ is said to be;
(i) upper $\alpha(\mu_X, \mu_Y)$ -continuous at a point $x \in X$ if for each μ_Y-open set V of Y containing $F(x)$, there exists $U \in \alpha(\mu_X)$ containing x such that $F(U) \subseteq V$.
(ii) lower $\alpha(\mu_X, \mu_Y)$ -continuous at a point $x \in X$ if for each μ_Y-open set V of Y such that $F(x) \cap V \neq \varnothing$, there exists $U \in \alpha(\mu_X)$ containing x such that $F(z) \cap V \neq \varnothing$ for every $z \in U$.
(iii) upper(lower) $\alpha(\mu_X, \mu_Y)-$ continuous if F has this property at each point of X.

Definition 3.3 [12] Let (X, μ_X) and (Y, μ_Y) be a generalized topological spaces. Then a multifunction $F : X \longrightarrow Y$ is said to be;
(i) upper $\beta(\mu_X, \mu_Y)$-continuous at a point $x \in X$ if for each μ_Y-open set V of Y containing $F(x)$, there exists $U \in \beta(\mu_X)$ containing x such that $F(U) \subseteq V$.
(ii) lower $\beta(\mu_X, \mu_Y)$-continuous at a point $x \in X$ if for each μ_Y-open set V of Y such that $F(x) \cap V \neq \varnothing$, there exists $U \in \beta(\mu_X)$ containing x such that $F(z) \cap V \neq \varnothing$ for every $z \in U$.
(iii) upper(lower) $\beta(\mu_X, \mu_Y)$- continuous if F has this property at each point of X.

Definition 3.4 [8] Let (X, μ_X) and (Y, μ_Y) be a generalized topological spaces. Then a multifunction $F : X \longrightarrow Y$ is said to be
(i) upper (μ_X, μ_Y)-continuous at a point $x \in X$ if for each μ_Y-open set V of Y containing $F(x)$, there exists $U \in (\mu_X)$ containing x such that $F(U) \subseteq V$.
(ii) lower (μ_X, μ_Y)-continuous at a point $x \in X$ if for each μ_Y-open set V of Y such that $F(x) \cap V \neq \varnothing$, there exists $U \in (\mu_X)$ containing x such that $F(z) \cap V \neq \varnothing$ for every $z \in U$.
(iii) upper(lower) (μ_X, μ_Y)- continuous if F has this property at each point of X.

Remark 1 *For a multifunction $F : X \longrightarrow Y$, following implications hold:*
upper(lower) (μ_X, μ_Y)-continuous $\Longrightarrow$ upper(lower) $\alpha(\mu_X, \mu_Y)$-continuous $\Longrightarrow$ upper(lower) $\beta(\mu_X, \mu_Y)$ -continuous.
The following examples shows that these implications are not reversible.

***Example* 3.5** Let $X = \{a, b, c, d\} = Y$ and $\mu_X = \{\emptyset, \{a\}, \{a, b, c\}\} = \mu_Y$. Consider a multifunction $F : (X, \mu_X) \longrightarrow (Y, \mu_Y)$ defined by $F(a) = \{a\} = F(b)$ and $F(c) = \{b\}$, $F(d) = \{d\}$. Then $F^{+}(\{a, b, c\}) = \{a, b, c\}$ and $F^{+}(\{a\}) = \{a, b\}$. Now $\{a, b\}$ is $\mu_X - \alpha-$open but not μ_X-open. Hence F is upper $\alpha(\mu_X, \mu_Y)$-continuous but not upper (μ_X, μ_Y)-continuous.

***Example* 3.6** Let $X = \{a, b, c\} = Y$ and $\mu_X = \{\emptyset, \{a\}, \{b\}, \{a, b\}\} = \mu_Y$. Consider a multifunction $F : (X, \mu_X) \longrightarrow (Y, \mu_Y)$ defined by $F(a) = \{b\} = F(b)$ and $F(c) = \{a\}$. Then F is upper $\beta(\mu_X, \mu_Y)$-continuous but not upper $\alpha(\mu_X, \mu_Y)$-continuous.

Theorem 3.7 For a multifunction $F : (X, \mu_X) \longrightarrow (Y, \mu_Y)$ the following are equivalent.
(*i*) F is upper $\alpha(\mu_X, \mu_Y)$-continuous,
(*ii*) $F^+(V)$ is a $\mu_X - \alpha$-open set in X for each μ_Y-open set V of Y.
(*iii*) $F^-(K)$ is a $\mu_X - \alpha$-closed set in X for each μ_Y-closed set K of Y.
(*iv*) $c_{\alpha_X}(F^-(B)) \subseteq F^-(c_Y(B))$ for each subset B of Y.
(*v*) $F^+(i_Y(B)) \subseteq i_{\alpha_X}(F^+(B))$ for each subset B of Y.

Proof. (*i*) $\Longrightarrow$ (*ii*) Let V be any μ_Y-open set of Y and $x \in F^+(V)$. Then $F(x) \subseteq V$. There exists $U \in \alpha(\mu_X)$ containing x such that $F(U) \subseteq V$. Thus $x \in U \subseteq F^+(V)$. This implies that $x \in i_{\alpha_X}(F^+(V))$. This shows that $F^+(V) \subseteq i_{\alpha_X}(F^+(V))$. We have $i_{\alpha_X}(F^+(V)) \subseteq F^+(V)$. Therefore, $i_{\alpha_X}(F^+(V)) = F^+(V)$ and so $F^+(V)$ is $\mu_X - \alpha$-open set in X.

(*ii*) $\Longrightarrow$ (*iii*) Let K be any μ_Y-closed set of Y. If we take, $V = Y - K$, then V is a μ_Y-open set in Y. By (*ii*) $F^+(V)$ is a μ_X- open set in X. So $X - F^+(V) = X - F^+(Y - K) = F^-(K)$ is a μ_X-closed set in X.

(*iii*) $\Longrightarrow$ (*iv*) Let B be any subset of Y. Since $c_Y(B)$ is a μ_Y-closed set in Y, by (*iii*), $F^-(c_Y(B))$ is a $\mu_X - \alpha$-closed set in X. Thus $c_{\alpha_X}(F^-(c_Y(B))) \subseteq F^-(c_Y(B))$. So $c_{\alpha_X}(F^-(B)) \subseteq F^-(c_Y(B))$.

(*iv*) $\Longleftrightarrow$ (*v*) It follows from [Lemma 3.2, [18]].

(*v*) $\Longrightarrow$ (*i*) Let $x \in X$ and V be any μ_Y-open set V of Y containing $F(x)$. Then by (*v*), $x \in F^+(V) = F^+(i_Y(V)) \subseteq i_{\alpha_X}(F^+(V))$. Therefore, there exists a $\mu_X - \alpha-$open set $U = F^+(V)$ of X containing x such that $F(U) \subseteq V$. This implies F is upper $\alpha(\mu_X, \mu_Y)$-continuous at x. ∎

Theorem 3.8 For a multifunction $F : (X, \mu_X) \longrightarrow (Y, \mu_Y)$, the following are equivalent.
(*i*) F is lower $\alpha(\mu_X, \mu_Y)$-continuous.
(*ii*) $F^-(V)$ is a $\mu_X - \alpha$-open set in X for each μ_Y-open V of Y.
(*iii*) $F^+(K)$ is a $\mu_X - \alpha$-closed set in X for each μ_Y-closed set K of Y.
(*iv*) $c_{\alpha_X}(F^+(B)) \subseteq F^+(c_Y(B))$ for each subset B of Y.
(*v*) $F^-(i_Y(B)) \subseteq i_{\alpha_X}(F^-(B))$ for each subset B of Y.
(*vi*) $F(c_{\alpha_X}(A)) \subseteq c_Y(F(A))$ for each subset A of X.

Proof. We prove only the implications (*iv*) $\Longrightarrow$ (*vi*) and (*vi*) $\Longrightarrow$ (*v*) with the proofs the other being similar to those of Theorem 3.7.

(*iv*) $\Longrightarrow$ (*vi*) Let A be any subset of X. By (*iv*), we have $c_{\alpha_X}(A) \subseteq c_{\alpha_X} F^+(F(A)) \subseteq F^+(c_Y F(A))$ and $F(c_{\alpha_X}(A)) \subseteq c_Y(F(A))$.

(*vi*) $\Longrightarrow$ (*v*) Let B be any subset of Y. By (*vi*), we have $F(c_{\alpha_X}(F^+(Y - B))) \subseteq c_Y(F(F^+(Y - B))) \subseteq c_Y(Y - B) = Y - i_Y(B)$ and $F(c_{\alpha_X}(F^+(Y - B))) = F(c_{\alpha_X}(X - F^-(B))) = F(X - i_{\alpha_X}(F^-(B)))$.This implies $F^-(i_Y(B)) \subseteq i_{\alpha_X}(F^-(B))$. ∎

Theorem 3.9 Let (X, μ_X), (Y, μ_Y) be two generalized topological spaces and the multifunctions $F_1 : X \longrightarrow Y$, $F_2 : X \longrightarrow Y$ be upper $\alpha(\mu_X, \mu_Y)$-continuous. Then $F_1 \cup F_2$ the combination of F_1 and F_2 is defined $(F_1 \cup F_2)(x) = F_1(x) \cup F_2(x)$ is upper $\alpha(\mu_X, \mu_Y)$-continuous.

Proof. Let V be any μ_Y-open set of Y. Since F_1 and F_2 are upper $\alpha(\mu_X, \mu_Y)$-continuous then $F_1^+(V)$ and $F_2^+(V)$ are $\mu_X - \alpha$-open set in X. So $F_1^+(V) \cup F_2^+(V) = (F_1 \cup F_2)^+(V)$ is $\mu_X - \alpha$-open set in X. Therefore $F_1 \cup F_2$ is upper $\alpha(\mu_X, \mu_Y)$-continuous. ∎

Theorem 3.10 Let (X, μ_X), (Y, μ_Y) be two generalized topological spaces and the multifunctions $F_1 : X \longrightarrow Y$, $F_2 : X \longrightarrow Y$ be lower $\alpha\,(\mu_X, \mu_Y)$-continuous. Then $F_1 \cup F_2$ is lower $\alpha\,(\mu_X, \mu_Y)$-continuous.

Proof. The proof is similar to Theorem 3.9. ■

Theorem 3.11 Let (X, μ_X), (Y, μ_Y) be generalized topological spaces and the multifunctions $F_1 : X \longrightarrow Y$, $F_2 : X \longrightarrow Y$ be upper $\alpha\,(\mu_X, \mu_Y)$ -continuous. Then $F_1 \times F_2$ the product of F_1 and F_2 is defined $(F_1 \times F_2)(x) = F_1(x) \times F_2(x)$ is upper $\alpha\,(\mu_{X \times X}, \mu_{Y \times Y})$-continuous.

Proof. Let V be any μ_Y-open subset of Y. Since F_1 and F_2 are upper $\alpha\,(\mu_X, \mu_Y)$-continuous then $F_1^+(V)$ and $F_2^+(V)$ are $\mu_X - \alpha$-open set in X. So $F_1^+(V) \times F_2^+(V) = (F_1 \times F_2)^+(V)$ is $\mu_X - \alpha$-open set in $X \times X$. Therefore $F_1 \times F_2$ is upper $\alpha\,(\mu_{X \times X}, \mu_{Y \times Y})$-continuous. ■

Theorem 3.12 Let (X, μ_X) and (Y, μ_Y) be generalized topological spaces and the multifunctions $F_1 : X \longrightarrow Y$ and $F_2 : X \longrightarrow Y$ be lower $\alpha\,(\mu_X, \mu_Y)$-continuous. Then $F_1 \times F_2$ is lower $\alpha\,(\mu_{X \times X}, \mu_{Y \times Y})$-continuous.

Proof. The proof is similar to Theorem 3.11. ■

Theorem 3.13 Let (X, μ_X), (Y, μ_Y) and (Z, μ_Z) be three generalized topological spaces. The multifunctions $F_1 : X \longrightarrow Y$ be upper $\alpha\,(\mu_X, \mu_Y)$-continuous and $F_2 : Y \longrightarrow Z$ be upper (μ_Y, μ_Z)-continuous. Then the multifunction $F_1 \circ F_2$ is defined $(F_1 \circ F_2)(x) = F_1(F_2(x))$ is upper $\alpha\,(\mu_X, \mu_Z)$-continuous.

Proof. Let W be any μ_Z-open subset of Z. Since F_2 is upper (μ_Y, μ_Z)-continuous then $F_2^+(W)$ is μ_Y-open set in Y. Also, since F_1 is upper $\alpha\,(\mu_X, \mu_Y)$-continuous then $F_1^+(F_2^+(W)) = (F_1 \circ F_2)^+(W)$ is $\alpha - \mu_X$-open set in X. Therefore, $F_1 \circ F_2$ is upper $\alpha\,(\mu_X, \mu_Z)$-continuous. ■

Theorem 3.14 Let (X, μ_X), (Y, μ_Y) and (Z, μ_Z) be three generalized topological spaces. The multifunctions $F_1 : X \longrightarrow Y$ be lower $\alpha\,(\mu_X, \mu_Y)$-continuous and $F_2 : Y \longrightarrow Z$ lower (μ_X, μ_Z)-continuous. Then the multifunction $F_1 \circ F_2$ is lower $\alpha\,(\mu_X, \mu_Z)$ -continuous.

Proof. The proof is similar to Theorem 3.13. ■

Theorem 3.15 Let (X, μ_X), (Y, μ_Y) be two generalized topological spaces and $F(X)$ is endowed with subspace topology. If the multifunction $F : X \longrightarrow Y$ is upper $\alpha\,(\mu_X, \mu_Y)$-continuous then the multifunction $F : X \longrightarrow F(X)$ is upper $\alpha\left(\mu_X, \mu_{F(X)}\right)$-continuous.

Proof. Since F is upper $\alpha\,(\mu_X, \mu_Y)$ -continuous, for every μ_Y-open set V of Y, $F^+(V \cap F(X)) = F^+(V) \cap F^+(F(X)) = F^+(V) \cap X = F^+(V)$ is $\alpha - \mu_X$-open set in X. Therefore $F : X \longrightarrow F(X)$ is upper $\alpha\left(\mu_X, \mu_{F(X)}\right)$ -continuous. ■

Theorem 3.16 Let (X, μ_X), (Y, μ_Y) be two generalized topological spaces and $F(X)$ is endowed with subspace topology. If the multifunction $F : X \longrightarrow Y$ is lower $\alpha\,(\mu_X, \mu_Y)$-continuous then the multifunction $F : X \longrightarrow F(X)$ is lower $\alpha\left(\mu_X, \mu_{F(X)}\right)$-continuous.

Proof. The proof is similar to Theorem 3.15. ■

Let $K \neq \emptyset$ be an index set and $X_k \neq \varnothing$ for $k \in K$ and $X = \prod\limits_{k \in K} X_k$ the cartesian product of the sets X_k. We denote by p_k the projection function $p_k : X \to X_k$ defined by $x_k = p_k(x)$, for each $x \in X$. Suppose that for $k \in K$, μ_k is a given generalized topological space on X_k. Let us consider all sets of the form $\prod_{k\in K} M_k$, where $M_k \in \mu_k$, with the exception of a finite number of indices k, $M_k = X_k = M_{\mu_k}$. We denote by $\wp$ the collection of all these sets. Clearly $\varnothing \in \wp$ so that we can define a GT $\mu = \mu(\wp)$ having $\wp$ for base. We call μ the product [11] of the GT's μ_k and denoted by P_{μ_k} $(k \in K)$.

Lemma 3.17 [20] Let $A = \prod_{k\in K} A_k \subseteq \prod_{k\in K} X_k$ and let K_0 be a finite subset of K. If $A_k \in (M_{\mu_k}, X_k)$ for each $k \in K - K_0$, then $iA = \prod_{k\in K} i_k A_k$.

Lemma 3.18 [11] Let $A = \prod_{k\in K} A_k \subseteq \prod_{k\in K} X_k$, then $cA = \prod_{k\in K} c_k A_k$.

Lemma 3.19 [11] If every μ_k is strong, then $\mu = \prod_{k\in K} \mu_k$ is strong and p_k is (μ, μ_k)-continuous for $k \in K$.

Lemma 3.20 [11] The projection p_k is (μ, μ_k)- open.

Lemma 3.21 Let $f : X \longrightarrow Y$ be (μ_X, μ_Y)-continuous and (μ_X, μ_Y)-open function. If A is a $\alpha - \mu_X$-open set in X, then $f(A)$ is $\alpha - \mu_Y$-open set in Y.

Proof. Let A be a $\alpha - \mu_X$-open set of X. Since f is (μ_X, μ_Y)-open and (μ_X, μ_Y)-continuous, then $f(i_{\mu_X}(A)) \subseteq i_{\mu_Y}(f(A))$ and $f(c_{\mu_X}(A)) \subseteq c_{\mu_Y}(f(A))$. Thus, $f(A) \subseteq f(i_{\mu_X}(c_{\mu_X}(i_{\mu_X}(A))))$ and $f(A) \subseteq f(i_{\mu_X}(c_{\mu_X}(i_{\mu_X}(A)))) \subseteq i_{\mu_Y}(c_{\mu_Y}(i_{\mu_Y}(f(A))))$. This show that, $f(A)$ is $\alpha - \mu_Y$-open in Y. ∎

Theorem 3.22 Let X be a strong generalized topological space, $F : X \longrightarrow Y$ be a multifunction and $G_F : X \to X \times Y$ be the graph multifunction of F defined by $G_F(x) = (x, F(x))$ for each $x \in X$. If G_F is upper $\alpha(\mu_X, \mu_{X\times Y})$-continuous then F is upper $\alpha(\mu_X, \mu_Y)$-continuous.

Proof. Let V be any μ_Y-open set of Y. Then $X \times V$ is $\mu_{X\times Y}$-open set of $X \times Y$. Since G_F is upper $\alpha(\mu_X, \mu_{X\times Y})$-continuous, then $G_F^+(X \times V) = F^+(V) \cap X = F^+(V)$ is a $\alpha - \mu_X$-open set in X. Thus F is upper $\alpha(\mu_X, \mu_Y)$-continuous. ∎

Theorem 3.23 Let X be a strong generalized topological space, $F : X \longrightarrow Y$ be a multifunction and $G_F : X \to X \times Y$ be the graph multifunction of F. If G_F is lower $\alpha(\mu_X, \mu_{X\times Y})$-continuous then F is lower $\alpha(\mu_X, \mu_Y)$-continuous.

Proof. The proof is similar to Theorem 3.22. ∎

Let $\{X_k : k \in K\}$ and $\{Y_k : k \in K\}$ be any two families of generalized topological spaces with the same index set K. For each $k \in K$, let $F_k : X_k \longrightarrow Y_k$ be a multifunction. The product space $\prod_{k\in K} X_k$ is denoted by $\prod X_k$ and the product multifunction $\prod F_k : \prod X_k \longrightarrow \prod Y_k$, defined by $F(x) = \prod_{k\in K} F_k(x_k)$ for each $x = \{x_k\} \in \prod X_k$, is simply denoted by $F : \prod X_k \longrightarrow \prod Y_k$.

Theorem 3.24 Let $Y = \prod_{k\in K} Y_k$. If a multifunction $F : X \longrightarrow \prod_{k\in K} Y_k$ is upper $\alpha(\mu_X, \mu_Y)$-continuous and every μ_{Yk} is strong, then $p_k \circ F : X \to Y_k$ is upper $\alpha(\mu_X, \mu_{Y_k})$ -continuous, where p_k is the projection of $\prod_{k\in K} Y_k$ onto Y_k, for each $k \in K$.

Proof. Let $k \in K$ and V_k be any μ_{Y_k}-open set of Y_k. By Lemma 3.19, p_k is (μ_X, μ_{Y_k}) -continuous, so $p_k^{-1}(V_k)$ is a generalized open set in Y. Since F is upper $\alpha(\mu_X, \mu_{Y_k})$

-continuous, then $F^{-}\left(p_k^{-1}(V_k)\right) = (p_k \circ F)^{-}(V_k)$ is a $\alpha - \mu_X$-open set in X. Therefore, $p_k \circ F$ is upper $\alpha(\mu_X, \mu_{Y_k})$-continuous. ■

Theorem 3.25 Let $Y = \prod_{k \in K} Y_k$. If a multifunction $F : X \longrightarrow \prod_{k \in K} Y_k$ is lower $\alpha(\mu_X, \mu_Y)$-continuous and every μ_{Y_k} is strong, then $p_k \circ F : X \to Y_k$ is lower $\alpha(\mu_X, \mu_{Y_k})$ -continuous.

Proof. The proof is similar to Theorem 3.24. ■

Theorem 3.26 Let $k \in K$, μ_{Y_k} be strong generalized topological spaces and $F_k : X_k \to Y_k$ be multifunctions. If the product multifunction $F : X \to Y$ is upper $\alpha(\mu_X, \mu_Y)$-continuous where $X = \prod X_k$, $Y = \prod Y_k$, then $F_k : X_k \to Y_k$ is upper $\alpha(\mu_{X_k}, \mu_{Y_k})$-continuous for each $k \in K$.

Proof. Let k_0 be an arbitrary fixed index in K and V_{k_0} be any $\mu_{Y_{k_0}}$-open set of Y_{k_0}. Then $\prod_{k \neq k_0} Y_k \times V_{k_0}$ is a μ_Y-open set in Y. Since F is upper $\alpha(\mu_X, \mu_Y)$-continuous, then $F^{-}\left(\prod_{k \neq k_0} Y_k \times V_{k_0}\right) = \prod_{k \neq k_0} X_k \times F_{k_0}^{-}(V_{k_0})$ is a $\alpha - \mu_X$-open set in X. By Lemma 3.21, $F_{k_0}^{-}(V_{k_0})$ is a $\mu_{Y_{k_0}}$-open set in X_{k_0}. This shows that F_{k_0} is upper $\alpha\left(\mu_{X_{k_0}}, \mu_{Y_{k_0}}\right)$-continuous. ■

Theorem 3.27 Let $k \in K$ and let X_k and μ_{Y_k} be strong generalized topological spaces and $F_k : X_k \to Y_k$ be multifunctions. If the product multifunction $F : X \to Y$ is lower $\alpha(\mu_X, \mu_Y)$-continuous where $X = \prod X_k$, $Y = \prod Y_k$, then $F_k : X_k \to Y_k$ is lower $\alpha(\mu_{X_k}, \mu_{Y_k})$-continuous for each $k \in K$.

Proof. The proof is similar to Theorem 3.26. ■

Definition 3.28 [7] A space X said to be μ-compact (resp. $\alpha - \mu$-compact) if every μ-open (resp. $\alpha - \mu-$open) cover of X has a finite subcover.

Theorem 3.29 Let a multifunction $F : X \longrightarrow Y$ be upper $\alpha(\mu_X, \mu_Y)$-continuous and X is $\alpha - \mu_X$-compact, then Y is μ_Y-compact.

Proof. Let χ be a cover of Y by μ_Y-open sets in Y. Since F is upper $\alpha(\mu_X, \mu_Y)$-continuous then $\{F^{+}(A) : A \in \chi\}$ is a $\alpha - \mu_X$- open cover of X. Also, since X is $\alpha - \mu_X$-compact, so the cover of X has a finite subcover $\{F^{+}(A) : A \in \chi'\}$, where χ' is a subfamily of χ. Then $Y \subset \bigcup_{A \in \chi'} F(F^{+}(A)) = \bigcup_{A \in \chi'} A$. Therefore Y is μ_Y-compact. ■

Theorem 3.30 Let (X, μ_X) be a generalized topological space and (Y, μ_Y) be a quasitopological space. If $F : X \longrightarrow Y$ be upper $\alpha(\mu_X, \mu_Y)$-continuous such that $F(x)$ is $\alpha - \mu_Y-$compact for each $x \in X$ and M is $\alpha - \mu_X-$compact set of X, then $F(M)$ is $\alpha - \mu_Y-$compact.

Proof. Let $\{V_i : i \in I\}$ be any cover of $F(M)$ by $\alpha - \mu_Y$-open sets. For each $x \in M$, $F(x)$ is $\alpha - \mu_Y$- compact and there exists a finite subset $I_0(x)$ of I such that $F(x) \subseteq \cup\{V_i : i \in I_0(x)\}$. Now set $V(x) = \cup\{V_i : i \in I_0(x)\}$. Then we have $F(x) \subseteq V(x)$ and $V(x)$ is $\alpha - \mu_Y$-open set of Y. Since F is upper $\alpha(\mu_X, \mu_Y)$ -continuous, there exists an $\alpha - \mu_X$-open set $U(x)$ containing x such that $F(U(x)) \subseteq V(x)$. The family $\{U(x) : x \in M\}$ is a cover of M by $\alpha - \mu_X$-open sets. Since M is $\alpha - \mu_X$-compact, there exists a finite number of points, say, $x_1, x_2, ..., x_n$ in M such that $M \subseteq \cup\{U(x_m) : x_m \in M,\ 1 \leqslant m \leqslant n\}$. Therefore, we obtain $F(M) \subseteq$

$\cup\{F(U(x_m)) : x_m \in M,\ 1 \leqslant m \leqslant n\} \subseteq \cup\{V_i : i \in i(x_m),\ x_m \in M,\ 1 \leqslant m \leqslant n\}$. This shows that $F(M)$ is $\alpha - \mu_Y$-compact. ■

Corollary 3.31 Let (X, μ_X) be a generalized topological space and (Y, μ_Y) be a quasitopological space. If $F : X \longrightarrow Y$ be upper $\alpha(\mu_X, \mu_Y)$ -continuous such that $F(x)$ is $\alpha - \mu_Y$-compact for each $x \in X$ and (X, μ_X) is $\alpha - \mu_X$-compact, then (Y, μ_Y) is $\alpha - \mu_Y$-compact.

Definition 3.32 [10] A space X is said to be μ_X-connected if there are no nonempty disjoint sets $U, V \subset \mu_X$ such that $U \cup V = X$.

Definition 3.33 A space X is said to be $\alpha - \mu_X$-connected if there are no nonempty disjoint α-open sets $U, V \subset \mu_X$ such that $U \cup V = X$.

Theorem 3.34 Let (X, μ_X), (Y, μ_Y) be generalized topological spaces and the multifunctions $F : X \longrightarrow Y$ be upper $\alpha(\mu_X, \mu_Y)$-continuous. If (X, μ_X) is $\alpha - \mu_X$-connected, then (Y, μ_Y) is μ_Y-connected.

Proof. Suppose there are two nonempty disjoint μ_Y-open subsets U, V of Y, such that $U \cup V = Y$. Since F is upper $\alpha(\mu_X, \mu_Y)$-continuous, so $F^+(U)$,$F^+(V)$ are $\alpha - \mu_X$-open subsets of X. Also $F^+(U) \cap F^+(V) = F^+(U \cap V) = F^+(\emptyset) = \emptyset$ and $F^+(U) \cup F^+(V) = F^+(U \cup V) = F^+(Y) = X$. So (X, μ_X) is α-disconnected. Therefore (Y, μ_Y) is μ_Y-connected. ■

Lemma 3.35 Let A and B be subsets of a GTS (X, μ).
  (a) If $A \in \sigma(\mu) \cup \pi(\mu)$ and $B \in \alpha(\mu)$, then $A \cap B \in \alpha(\mu)$.
  (b) If $A \subseteq B \subseteq X$, $A \in \alpha(B\mu)$ and $B \in \alpha(\mu)$, then $A \in \alpha(\mu)$.

Proof. Obvious. ■

Theorem 3.36 Let (X, μ_X), (Y, μ_Y) be generalized topological spaces and the multifunctions $F : X \longrightarrow Y$ be upper (resp. lower) $\alpha(\mu_X, \mu_Y)$-continuous. If $A \in \sigma(\mu) \cup \pi(\mu)$,then defined as $(F|_A)(x) = F(x)$ the restriction $F|_A : A \to Y$ is upper (resp. lower) $\alpha(\mu_A, \mu_Y)$-continuous.

Proof. We prove only the assertion for F upper $\alpha(\mu_A, \mu_Y)$-continuous, the proof for F lower $\alpha(\mu_A, \mu_Y)$-continuous being analogous. Let $x \in A$ and V be μ_Y-open set of Y such that $(F|_A)(x) \subseteq V$. Since F is upper $\alpha(\mu_X, \mu_Y)$-continuous and $(F|_A)(x) = F(x)$, there exists $U \in \alpha(\mu_X)$ containing x such that $F(U) \subseteq V$. Set $U_0 = U \cap A$, then by Lemma 5, we have $x \in U_0 \in \alpha(\mu_A)$ and $(F|_A)(U_0) \subseteq V$.This shows that $(F|_A) : A \to Y$ is upper $\alpha(\mu_A, \mu_Y)$-continuous. ■

Theorem 3.37 Let (X, μ_X), (Y, μ_Y) be generalized topological spaces and the multifunctions $F : X \longrightarrow Y$ is upper (resp. lower) $\alpha(\mu_X, \mu_Y)$-continuous. If for each $x \in X$, there exists $A \in \alpha(\mu_X)$ containing x, the restriction $(F|_A) : A \to Y$ is upper (resp. lower) $\alpha(\mu_A, \mu_Y)$-continuous.

Proof. The proof is similar to Theorem 3.36. ■

Corollary 3.38 Let (X, μ_X), (Y, μ_Y) be generalized topological spaces and $\{A_i : i \in I\}$ be an $\alpha - \mu_X$-open cover of X . A multifunctions $F : X \longrightarrow Y$ is upper (resp. lower) $\alpha(\mu_X, \mu_Y)$-continuous if and only if $(F|_A) : A_i \to Y$ is upper (resp. lower) $\alpha(\mu_X, \mu_Y)$-continuous for each $i \in I$.

References

[1] A. Csaszar, Generalized topology, generalized continuity, Acta Math. Hungar., 96, 351-357, 2002.
[2] A. Csaszar, Extremally disconnected generalized topologies, Annales Univ. Sci. Budapest., 47, 91-96, 2004.
[3] A. Csaszar, "$\delta-$and $\theta-$modifications of generalized topologies," Acta Mathematica Hungarica, vol. 120, pp. 274-279, 2008.
[4] A. Csaszar, "Product of generalized topologies," Acta Mathematica Hungarica, vol. 123, no:1-2, pp. 127-132, 2009.
[5] A. Csaszar, γ-connected sets, Acta Math. Hungar., 101, 273-279, 2003.
[6] A. Csaszar, "Further remarks on the formula for $\gamma-$interior," Acta Mathematica Hungarica, vol. 113, no: 4, pp. 325-332, 2006.
[7] A. Csaszar, "Generalized open sets in generalized topologies," Acta Mathematica Hungarica, vol. 106,no: 1-2 pp. 53-66, 2005.
[8] A. Kanibir and I. L. Reilly, "Generalized continuity for multifunctions, " Acta Mathematica Hungarica, vol. 122, no . 3, pp. 283-292, 2009.
[9] A. S. Mashour, M. E. Abd El-Monsef, and S. N. El-Deeb, "On precontinuous and weakprecontinuous functions, "Proceedings of the Mathematical and Physical Society of Egypt, pp. 47-53, 1982.
[10] C. Berge, Topological Spaces, Macmillian, New York, 1963. English translation by E. M. Patterson of Espaces Topologiques, Fonctions Multivoques, Dunod, Paris, 1959.
[11] C. Cao, J. Yang, W. Wang, B. Wang, Some generalized continuities functions on generalized topological spaces, Hacettepe Jou. of Math. and Stat., 42(2), 159-163, 2013.
[12] C. Boonpok, "On upper and Lower $\beta\,(\mu_X, \mu_Y)$- Continuous multifunctions", Int. J. of Math. and Math. Sci., 2012 Doi: 10. 1155/2012/931656.
[13] D. Andrijevic, "Semipreopen sets, " Matematicki Vesnik, vol. 38, no. 2, pp. 24-32, 1986.
[14] J. P. Aubin, H. Frankowska, Set-Valued Analysis, Birkhauser, Boston, 1990.
[15] M. Akdağ and F. Erol, Upper and Lower $Pre(\mu_X, \mu_Y)$ Continuous Multifunctions, Scientific Journal of Mathematics Research Oct. 2014, Vol. 4 Iss. 5, PP. 46-52.
[16] M. E. Abd El-Monsef, S. N. El-Deeb, and R. A. Mahmoud, "β-open sets and β-continuous mapping, "Bulletin of the Faculty of Science. Assiult Universty, vol. 12, no. 1, pp. 77-90, 1983.
[17] N. Levine, "Semi-open sets and semi-contiuity in topological spaces, " The American Mathematical Montly, vol. 70, pp. 36-41, 1963.
[18] O. Njastad, On some classes of nearly open sets, Pacific Journal of Math., vol.15, 961-870, 1965.
[19] R. Shen, Remarks on products of generalized topologies, Acta Math. Hungar., 124, 363-369, 2009.
[20] R. Shen, A note on generalized connectedness, Acta Math. Hungar., 122, 231-235, 2009.
[21] W. K. Min, Generalized continuous functions defined by generalized open sets on generalized topological spaces, Acta Math. Hun., 128, 299-306, 2010.

2

On duality of modular G-Riesz bases and G-Riesz bases in Hilbert C*-modules

M. Rashidi-Kouchi*

Young Researchers and Elite Club Kahnooj Branch, Islamic Azad University, Kerman, Iran.

Abstract. In this paper, we investigate duality of modular g-Riesz bases and g-Riesz bases in Hilbert C*-modules. First we give some characterization of g-Riesz bases in Hilbert C*-modules, by using properties of operator theory. Next, we characterize the duals of a given g-Riesz basis in Hilbert C*-module. In addition, we obtain sufficient and necessary condition for a dual of a g-Riesz basis to be again a g-Riesz basis. We find a situation for a g-Riesz basis to have unique dual g-Riesz basis. Also, we show that every modular g-Riesz basis is a g-Riesz basis in Hilbert C*-module but the opposite implication is not true.

Keywords: Modular G-Riesz basis, G-Riesz basis, dual G-Riesz basis, Hilbert C^*-module.

1. Introduction

Frames in Hilbert spaces were first introduced in 1952 by Duffin and Schaeffer [5] in the study of nonharmonic Fourier series. They were reintroduced and developed in 1986 by Daubechies, Grossmann and Meyer [4], and popularized from then on.

Let H be a Hilbert space, and J a set which is finite or countable. A sequence $\{f_j\}_{j\in J} \subseteq H$ is called a frame for H if there exist constants $C, D > 0$ such that

$$C\|f\|^2 \leqslant \sum_{j\in J} |\langle f, f_j\rangle|^2 \leqslant D\|f\|^2 \tag{1}$$

*Corresponding author.
E-mail address: m_rashidi@kahnoojiau.ac.ir (M. Rashidi-Kouchi).

for all $f \in H$. The constants C and D are called the frame bounds. We have a tight frame if $C = D$ and a Parseval frame if $C = D = 1$. We refer the reader to [2, 3] for more details.

In [15] Sun introduced a generalized notion of frames and suggested further generalizations, showing that many basic properties of frames can be derived within this more general framework.

Let U and V be two Hilbert spaces and $\{V_j : j \in J\}$ be a sequence of subspaces of V, where J is a subset of $\mathbb{Z}$. Let $L(U, V_j)$ be the collection of all bounded linear operators from U to V_j . We call a sequence $\{\Lambda_j \in L(U, V_j) : j \in J\}$ a generalized frame (or simply a g-frame) for U with respect to $\{V_j : j \in J\}$ if there are two positive constants C and D such that

$$C\|f\|^2 \leqslant \sum_{j\in J} \|\Lambda_j f\|^2 \leqslant D\|f\|^2 \tag{2}$$

for all $f \in U$. The constants C and D are called g-frame bounds. If $C = D$ we call have a tight g-frame and if $C = D = 1$ we have a Parseval g-frame.

The notions of frames and g-frames in Hilbert C^*-modules were introduced and investigated in [7, 10, 11, 16]. Frank and Larson [6, 7] defined the standard frames in Hilbert C^*-modules in 1998 and got a series of results for standard frames in finitely or countably generated Hilbert C^*-modules over unital C^*-algebras. Extending the results to this more general framework is not a routine generalization, as there are essential differences between Hilbert C^*-modules and Hilbert spaces. For example, any closed subspace in a Hilbert space has an orthogonal complement, but this fails in Hilbert C^*-module. Also there is no explicit analogue of the Riesz representation theorem of continuous functionals in Hilbert C^*-modules. We refer the readers to [13] for more details on Hilbert C^*-modules, and to [11] and [16], for a discussion of basic properties of g-frame in Hilbert C^*-modules.

Alijani and Dehghan in [1] studied dual g-frames in Hilbert C*-modules. They give some characterizations of dual g-frames for Hilbert spaces and Hilbert C*-modules. The main goal of this paper is to study duals of g-Riesz basis in Hilbert C*-modules.

This paper is organized as follows. In section 2 we review some basic properties of Hilbert C*-modules and g-Riesz bases in this space. In particular we characterize g-frames and g-Riesz bases in Hilbert C*-modules. In section 3 we study dual g-Riesz bases in Hilbert C^*-modules and characterize the duals of a given g-Riesz basis in Hilbert C*-module. We also obtain sufficient and necessary condition for a dual of a g-Riesz basis to be again a g-Riesz basis. We find a situation for a g-Riesz basis to have unique dual g-Riesz basis. Also, we show that every modular g-Riesz basis is a g-Riesz basis in Hilbert C*-module but the opposite implication is not true.

2. Preliminaries

In this section we review basic properties of g-frames in Hilbert C^*-modules. We also prove some results related to the notion of stability which is used in the next section. Our basic reference for Hilbert C^*-modules is [13]. For basic details on frames in Hilbert C^*-modules we refer the reader to [7].

Definition 2.1 Let A be a C^*-algebra with involution $*$. An inner product A-module (or pre Hilbert A-module) is a complex linear space H which is a left A-module with an inner product map $\langle .,. \rangle : H \times H \to A$ which satisfies the following properties:

1) $\langle \alpha f + \beta g, h\rangle = \alpha\langle f, h\rangle + \beta\langle g, h\rangle$ for all $f, g, h \in H$ and $\alpha, \beta \in \mathbb{C}$;
2) $\langle af, g\rangle = a\langle f, g\rangle$ for all $f, g \in H$ and $a \in A$;
3) $\langle f, g\rangle = \langle g, f\rangle^*$ for all $f, g \in H$;
4) $\langle f, f\rangle \geqslant 0$ for all $f \in H$ and $\langle f, f\rangle = 0$ iff $f = 0$.

For $f \in H$, we define a norm on H by $\|f\|_H = \|\langle f, f\rangle\|_A^{1/2}$. If H is complete in this norm, it is called a (left) Hilbert C^*-module over A or a (left) Hilbert A-module.

An element a of a C^*-algebra A is positive if $a^* = a$ and the spectrum of a is a subset of positive real number. In this case, we write $a \geqslant 0$. It is easy to see that $\langle f, f\rangle \geqslant 0$ for every $f \in H$, hence we define $|f| = \langle f, f\rangle^{1/2}$.

If H be a Hilbert C^*-module, and J a set which is finite or countable, a sequence $\{f_j\}_{j\in J} \subseteq H$ is called a frame for H if there exist constants $C, D > 0$ such that

$$C\langle f, f\rangle \leqslant \sum_{j\in J}\langle f, f_j\rangle\langle f_j, f\rangle \leqslant D\langle f, f\rangle \tag{3}$$

for all $f \in H$. The constants C and D are called the frame bounds. The notion of (standard) frames in Hilbert C^*-modules is first defined by Frank and Larson [7]. Basic properties of frames in Hilbert C^*-modules are discussed in [8–10].

A. Khosravi and B. Khosravi [11] defined g-frame in Hilbert C^*-modules. Let U and V be two Hilbert C^*-modules over the same C^*-algebra A and $\{V_j : j \in J\}$ be a sequence of subspaces of V, where J is a subset of $\mathbb{Z}$. Let $End_A^*(U, V_j)$ be the collection of all adjointable A-linear maps from U into V_j, i.e. $\langle Tf, g\rangle = \langle f, T^*g\rangle$ for all $f, g \in H$ and $T \in End_A^*(U, V_j)$. We call a sequence $\{\Lambda_j \in End_A^*(U, V_j) : j \in J\}$ a generalized frame (or simply a g-frame) for Hilbert C^*-module U with respect to $\{V_j : j \in J\}$ if there are two positive constants C and D such that

$$C\langle f, f\rangle \leqslant \sum_{j\in J}\langle \Lambda_j f, \Lambda_j f\rangle \leqslant D\langle f, f\rangle \tag{4}$$

for all $f \in U$. The constants C and D are called g-frame bounds. Those sequences which satisfy only the upper inequality in (2.2) are called g-Bessel sequences. A g-frame is tight, if $C = D$. If $C = D = 1$, it is called a Parseval g-frame.

Definition 2.2 [16] A g-frame $\{\Lambda_j \in End_A^*(U, V_j) : j \in J\}$ in Hilbert C*-module U with respect to $\{V_j : j \in J\}$ is called a g-Riesz basis if it satisfies:

(1) $\Lambda_j \neq 0$ for any $j \in J$;
(2) If an A-linear combination $\sum_{j\in K}\Lambda_j^* g_j$ is equal to zero, then every summand $\Lambda_j^* g_j$ is equal to zero, where $\{g_j\}_{j\in K} \in \bigoplus_{j\in K} V_j$ and $K \subseteq J$.

***Example* 2.3** Let H be an ordinary Hilbert space, then H is a Hilbert $\mathbb{C}$-module. Let $\{e_j : j \in J\}$ be an orthonormal basis for H, then $\{e_j : j \in J\}$ is a Parseval frame for Hilbert $\mathbb{C}$-module H.

***Example* 2.4** Let U be an ordinary Hilbert space, $J = N$ and $\{e_j\}_{j=1}^{\infty}$ be an orthonormal basis for Hilbert $\mathbb{C}$-module U. For j=1,2,... we let $V_j = \overline{span}\{e_1, e_2, ..., e_j\}$, and $\Lambda_j : U \to V_j$, $\Lambda_j f = \sum_{k=1}^{j}\langle f, \frac{e_j}{\sqrt{j}}\rangle e_k$.

We have $\sum_{j=1}^{\infty}\langle \Lambda_j f, \Lambda_j f\rangle = \sum_{j=1}^{\infty}|\langle f, e_j\rangle|^2 = \langle f, f\rangle$, which implies that $\{\Lambda_j\}_{j=1}^{\infty}$ is a g-Parseval frame for U with respect to $\{V_j : j \in J\}$.

Theorem 2.5 [16] Let $\Lambda_j \in End_A^*(U, V_j)$ for any $j \in J$ and $\sum_{j\in J}\langle \Lambda_j f, \Lambda_j f\rangle$ converge in norm for $f \in U$. Then $\{\Lambda_j : j \in J\}$ is a g-frame for U with respect to $\{V_j : j \in J\}$ if and only if there exist constants $C, D > 0$ such that

$$C\|f\|^2 \leqslant \left\|\sum_{j\in J}\langle \Lambda_j f, \Lambda_j f\rangle\right\| \leqslant D\|f\|^2, \quad f \in U.$$

Definition 2.6 Let $\{\Lambda_j \in End_A^*(U, V_j) : j \in J\}$ be a g-frame in Hilbert C*-module U with respect to $\{V_j : j \in J\}$ and $\{\Gamma_j \in End_A^*(U, V_j) : j \in J\}$ be a sequence of A-linear operators. Then $\{\Gamma_j : j \in J\}$ is called a dual sequence operator of $\{\Lambda_j : j \in J\}$ if

$$f = \sum_{j\in J}\Lambda_j^*\Gamma_j f$$

for all $f \in U$. The sequences $\{\Lambda_j : j \in J\}$ and $\{\Gamma_j : j \in J\}$ are called a dual g-frame when moreover $\{\Gamma_j : j \in J\}$ is a g-frame.

In [11] the authors defined the g-frame operator S in Hilbert C^*-module as follow

$$Sf = \sum_{j\in J}\Lambda_j^*\Lambda_j f, \quad f \in U,$$

and showed that S is invertible, positive, and self-adjoint. Since

$$\langle Sf, f\rangle = \langle\sum_{j\in J}\Lambda_j^*\Lambda_j f, f\rangle = \sum_{j\in J}\langle \Lambda_j f, \Lambda_j f\rangle,$$

it follows that

$$C\langle f, f\rangle \leqslant \langle Sf, f\rangle \leqslant D\langle f, f\rangle,$$

and the following reconstruction formula holds

$$f = SS^{-1}f = S^{-1}Sf = \sum_{j\in J}\Lambda_j^*\Lambda_j S^{-1}f = \sum_{j\in J}S^{-1}\Lambda_j^*\Lambda_j f,$$

for all $f \in U$. Let $\tilde{\Lambda}_j = \Lambda_j S^{-1}$, then

$$f = \sum_{j\in J}\Lambda_j^*\tilde{\Lambda}_j f = \sum_{j\in J}\tilde{\Lambda}_j\Lambda_j^* f.$$

The sequence $\{\tilde{\Lambda}_j : j \in J\}$ is also a g-frame for U with respect to $\{V_j : j \in J\}$(see [11]) which is called the canonical dual g-frame of $\{\Lambda_j : j \in J\}$.

Definition 2.7 Let $\{\Lambda_j : j \in J\}$ be a g-frame in Hilbert C*-module U with respect to $\{V_j : j \in J\}$, then the related analysis operator $T : U \to \bigoplus_{j\in J} V_j$ is defined by $Tf = \{\Lambda_j f : j \in J\}$, for all $f \in U$. We define the synthesis operator $F : \bigoplus_{j\in J} V_j \to U$

by $Ff = F(f_j) = \sum_{j\in J} \Lambda_j^* f_j$, for all $f = \{f_j\}_{j\in J} \in \bigoplus_{j\in J} V_j$, where

$$\bigoplus_{j\in J} V_j = \left\{ f = \{f_j\} : f_j \in V_j, \left\| \sum_{j\in J} |f_j|^2 \right\| < \infty \right\}.$$

It has been showed in [16] that if for any $f = \{f_j\}_{j\in J}$ and $g = \{g_j\}_{j\in J}$ in V_j the A-valued inner product is defined by $\langle f, g\rangle = \sum_{j\in J} \langle f_j, g_j\rangle$ and the norm is defined by $\|f\| = \|\langle f, f\rangle\|^{1/2}$, then $\bigoplus_{j\in J} V_j$ is a Hilbert A-module. Hence the above operators are definable. Moreover, since for any $g = \{g_j\}_{j\in J} \in \bigoplus_{j\in J} V_j$ and $f \in U$,

$$\begin{aligned}\langle Tf, g\rangle &= \sum_{j\in J} \langle \Lambda_j f, g_j\rangle = \sum_{j\in J} \langle f, \Lambda_j^* g_j\rangle \\ &= \langle f, \sum_{j\in J} \Lambda_j^* g_j\rangle = \langle f, Fg\rangle,\end{aligned}$$

it follows that T is adjointable and $T^* = F$. Also

$$T^*Tf = T^*(\Lambda_j f) = \sum_{j\in J} \Lambda_j^* \Lambda_j f = Sf,$$

for all $f \in U$. Let P_n be the projection on $\bigoplus_{j\in J} V_j$ that is $P_n : \bigoplus_{j\in J} V_j \to \bigoplus_{j\in J} V_j$ is defined by $P_n f = P_n(\{f_j\}_{j\in J}) = U_j$, for $f = \{f_j\}_{j\in J} \in \bigoplus_{j\in J} V_j$, and $U_j = f_n$ when $j = n$ and $U_j = 0$ when $j \neq n$.

Theorem 2.8 ([14]) Let $\{\Lambda_j \in End_A^*(U, V_j) : j \in J\}$ be a g-frame in Hilbert C*-module U with respect to $\{V_j\}_{j\in J}$, then $\{\Lambda_j\}_{j\in J}$ is a g-Riesz basis if and only if $\Lambda_n \neq 0$ and $P_n(RangT) \subseteq RangT$ for all $n \in J$, where T is the analysis operator of $\{\Lambda_j\}_{j\in J}$.

Corollary 2.9 A g-frame $\{\Lambda_j \in End_A^*(U, V_j) : j \in J\}$ in Hilbert C*-module U with respect to $\{V_j : j \in J\}$ is a g-Riesz basis if and only if

(1) $\Lambda_j \neq 0$ for any $j \in J$;
(2) If an A-linear combination $\sum_{j\in K} \Lambda_j^* g_j$is equal to zero, then every summand $\Lambda_j^* g_j$ is equal to zero, where $\{g_j\}_{j\in K} \in \bigoplus_{j\in K} V_j$ and $K \subseteq J$.

3. Dual of g-Riesz bases in Hilbert C*-modules

In this section, we study dual g-Riesz bases in Hilbert C^*-modules and characterize the duals of a given g-Riesz basis in Hilbert C*-module. We also obtain sufficient and necessary condition for a dual of a g-Riesz basis to be again a g-Riesz basis. We find a situation for a g-Riesz basis to have unique dual g-Riesz basis.

Proposition 3.1 Let $\{\Lambda_j \in End_A^*(U, V_j) : j \in J\}$ and $\{\Gamma_j \in End_A^*(U, V_j) : j \in J\}$ be two g-Bessel sequences in Hilbert C*-module U with respect to $\{V_j : j \in J\}$. If $f = \sum_{j\in J} \Lambda_j^* \Gamma_j f$ holds for any $f \in U$, then both $\{\Lambda_j : j \in J\}$ and $\{\Gamma_j : j \in J\}$ are g-frames in Hilbert C*-module U with respect to $\{V_j : j \in J\}$ and $f = \sum_{j\in J} \Gamma_j^* \Lambda_j f$.

Proof. Let us denote the g-Bessel bound of $\{\Gamma_j : j \in J\}$ by B_Γ. For all $f \in U$ we have

$$\begin{aligned}\|f\|^4 &= \|\langle \sum_{j\in J} \Lambda_j^* \Gamma_j f, f\rangle\|^2 = \|\sum_{j\in J}\langle \Gamma_j f, \Lambda_j f\rangle\|^2 \\ &\leqslant \|\sum_{j\in J}\langle \Gamma_j f, \Gamma_j f\rangle\|.\|\sum_{j\in J}\langle \Lambda_j f, \Lambda_j f\rangle\| \\ &\leqslant B_\Gamma \|f\|^2.\|\sum_{j\in J}\langle \Lambda_j f, \Lambda_j f\rangle\|.\end{aligned}$$

It follows that

$$B_\Gamma^{-1}\|f\|^2 \leqslant \|\sum_{j\in J}\langle \Lambda_j f, \Lambda_j f\rangle\|.$$

This implies that $\{\Lambda_j : j \in J\}$ is a g-frame in Hilbert C*-module. Similarly we can show that $\{\Gamma_j : j \in J\}$ is also a g-frame of U respect to $\{V_j : j \in J\}$. ■

Lemma 3.2 Let $\{\Lambda_j \in End_A^*(U, V_j) : j \in J\}$ be a g-frame in Hilbert C*-module U with respect to $\{V_j : j \in J\}$. Suppose that $\{\Gamma_j \in End_A^*(U, V_j) : j \in J\}$ and $\{\Theta_j \in End_A^*(U, V_j) : j \in J\}$ are dual g-frames of $\{\Lambda_j : j \in J\}$ with the property that either $RangT_\Gamma \subseteq RangT_\Theta$ or $RangT_\Theta \subseteq RangT_\Gamma$. Then $\Gamma_j = \Theta_j \ \forall j \in J$.

Proof. Suppose that $RangT_\Theta \subseteq RangT_\Gamma$. Then for each $f \in U$ there exists $g_f \in U$ such that $T_\Theta g_f = T_\Gamma f$.
Applying T_Λ^* on both sides, we arrive at

$$g_f = \sum_{j\in J}\Lambda_j^*\Theta_j g_f = T_\Lambda^* T_\Theta g_f = T_\Lambda^* T_\Gamma f = \sum_{j\in J}\Lambda_j^*\Gamma_j f = f$$

and so $T_\Gamma f = T_\Theta f, \forall f \in U$.
Equivalently $\Gamma_j f = \Theta_j f, \forall j \in J$. ■

Theorem 3.3 Let $\{\Lambda_j \in End_A^*(U, V_j) : j \in J\}$ be a g-frame in Hilbert C*-module U with respect to $\{V_j : j \in J\}$ with analysis operator T_Λ, then the following are equivalence:
(1) $\{\Lambda_j : j \in J\}$ has a unique dual g-frame;
(2) $RangT_\Lambda = \oplus_{j\in J}V_j$;
(3) If $\sum_{j\in J}\Lambda_j^* f_j = 0$ for some sequence $\{f_j\}_{j\in J} \in \bigoplus_{j\in J} V_j$, then $f_j = 0$ for each $j \in J$.
In case the equivalent conditions are satisfied, $\{\Lambda_j : j \in J\}$ is a g-Riesz basis.

Proof. (2)$\Rightarrow$ (1) Let $\{\tilde{\Lambda}_j : j \in J\}$ be the canonical dual g-frame of $\{\Lambda_j : j \in J\}$ with analysis operator $T_{\tilde{\Lambda}}$. Then $\tilde{\Lambda}_j = \Lambda_j S^{-1}$, where S is g-frame operator.
Let $\{\Gamma_j : j \in J\}$ be any dual g-frame of $\{\Lambda_j : j \in J\}$ with analysis operator T_Γ.
Then

$$RangT_\Gamma \subseteq \oplus_{j\in J}V_j = RangT_\Lambda = RangT_{\tilde{\Lambda}}.$$

By Lemma 3.2, $\Gamma_j = \tilde{\Lambda}_j$ for all $j \in J$.

(1)$\Rightarrow$ (2) Assume on the contrary that $RangT_\Lambda \neq \oplus_{j\in J}V_j$. We have

$$\bigoplus_{j\in J} V_j = RangT_\Lambda \bigoplus KerT_\Lambda^*. \tag{5}$$

Let P_Λ be orthogonal projection from $\bigoplus_{j\in J} V_j$ onto $RangT_\Lambda$, then

$$\bigoplus_{j\in J} V_j = P_\Lambda(\bigoplus_{j\in J} V_j) \bigoplus P_\Lambda^\perp(\bigoplus_{j\in J} V_j).$$

Therefore $P_\Lambda^\perp = KerT_\Lambda^* \neq \{0\}$.
Choose $f_{j_0} \in \bigoplus_{j\in J} V_j$ such that $P_\Lambda^\perp f_{j_0} \neq 0$ where $f_{j_0} = 1_{j_0}$ if $j = j_0$ and $f_{j_0} = 0$ if $j \neq j_0$ and 1_{j_0} is unital element of V_{j_0}. Define an operator $W : \bigoplus_{j\in J} V_j \to U$ by $W\{g_j\} = \Lambda_{j_0}^* g_{j_0}$.
Now, let $\{\widetilde{\Lambda}_j : j \in J\}$ be the canonical dual of $\{\Lambda_j : j \in J\}$ with upper bound $D_{\widetilde{\Lambda}}$ and $\Gamma_j = \widetilde{\Lambda}_j + \Pi_j \Pi W^*$ where $\Pi : \bigoplus_{j\in J} V_j \to KerT_\Lambda^*$ and $\Pi_j : \bigoplus_{j\in J} V_j \to V_j$ are projection operators.
We have

$$\begin{aligned}
\sum_{j\in J}\langle \Gamma_j f, \Gamma_j f\rangle &\leqslant 2\left(\sum_{j\in J}\langle \widetilde{\Lambda}_j f, \widetilde{\Lambda}_j f\rangle + \sum_{j\in J}\langle \Pi_j\Pi W^* f, \Pi_j \Pi W^* f\rangle\right)\\
&\leqslant 2\left(D_{\widetilde{\Lambda}}\langle f, f\rangle + \langle \Pi W^* f, \Pi W^* f\rangle\right)\\
&\leqslant (D_{\widetilde{\Lambda}} + \|\Pi W^*\|^2)\langle f, f\rangle,
\end{aligned}$$

which implies that $\{\Gamma_j : j \in J\}$ is a g-Bessel sequence.

Now for any $f \in U$,

$$\sum_{j\in J} \Lambda_j^* \Pi_j \Pi W^* f = T_\Lambda^*\{\Pi_j \Pi W^* f\} = 0.$$

This yields that $\sum_{j\in J} \Lambda_j^* \Gamma_j f = \sum_{j\in J} \Lambda_j^* \widetilde{\Lambda}_j f = f$ for all $f \in U$. By Proposition 3.1, $\{\Gamma_j : j \in J\}$ is a dual g-frame of $\{\Lambda_j : j \in J\}$ and is different from $\{\widetilde{\Lambda}_j : j \in J\}$, which contradicts with the uniqueness of dual g-frame of $\{\Lambda_j : j \in J\}$.
(2)$\Leftrightarrow$ (3) Obvious by (3.1). ■

Theorem 3.4 Suppose that $\{\Lambda_j \in End_A^*(U, V_j) : j \in J\}$ is a g-Riesz basis in Hilbert C*-module U with respect to $\{V_j : j \in J\}$ and $\{\Gamma_j \in End_A^*(U, V_j) : j \in J\}$ is a sequence of A-linear operators. Then the following are equivalence:

(1) $\{\Gamma_j : j \in J\}$ is a dual g-frame of $\{\Lambda_j : j \in J\}$;
(2) $\{\Gamma_j : j \in J\}$ is a dual g-Bessel sequence of $\{\Lambda_j : j \in J\}$;
(3) For each $j \in J$, $\Gamma_j = \Lambda_j S^{-1} + \Theta_j$, where S is the g-frame operator of $\{\Lambda_j : j \in J\}$ and $\{\Theta_j : j \in J\}$ is a dual g-Bessel sequence of U with respect to $\{V_j : j \in J\}$ satisfying $\Lambda_j^* \Theta_j f = 0$ for all $f \in U$ and $j \in J$.

Theorem 3.5 Let $\{\Lambda_j \in End_A^*(U, V_j) : j \in J\}$ be a g-Riesz basis in Hilbert C*-module U with respect to $\{V_j : j \in J\}$ and $\{\Gamma_j : j \in J\}$ a sequence of A-linear operators. Then $\{\Gamma_j : j \in J\}$ is a dual g-Riesz basis of $\{\Lambda_j : j \in J\}$ if and only if for each $j \in J$,

$\Gamma_j = \Lambda_j S^{-1} + \Theta_j$, where S is the g-frame operator of $\{\Lambda_j : j \in J\}$ and $\{\Theta_j : j \in J\}$ is a g-Bessel sequence of U with respect to $\{V_j : j \in J\}$ with the property that for each $j \in J$ there exists operator $F_j \in End_A^*(V_j, V_j)$ such that $\Theta_j = F_j\Lambda_j S^{-1}$ and $\Lambda_j^* F_j \Lambda_j f = 0$ holds for all $f \in U$.

Proof. $\Rightarrow$ Suppose that $\{\Gamma_j : j \in J\}$ is a dual g-Riesz basis of $\{\Lambda_j : j \in J\}$ and let $\Theta_j = \Gamma_j - \Lambda_j S^{-1}$. It is easy to see that $\{\Theta_j : j \in J\}$ is a g-Bessel sequence of U.
Now fix an $n \in J$. From $\sum_{j\in J} \Gamma_j \Lambda_j^*(\Gamma_n f) = \Gamma_n f$ we can infer that $\Gamma_n = \Gamma_n \Lambda_n^* \Gamma_n$, i.e.

$$\Lambda_n S^{-1} + \Theta_n = (\Lambda_n S^{-1} + \Theta_n)\Lambda_n^*(\Lambda_n S^{-1} + \Theta_n)$$

Consequently, we have

$$\begin{aligned}\Theta_n &= (\Lambda_n S^{-1}\Lambda_n^* + \Theta_n\Lambda_n^*)(\Lambda_n S^{-1} + \Theta_n) - \Lambda_n S^{-1}\\ &= \Lambda_n S^{-1}\Lambda_n^*\Lambda_n S^{-1} + \Lambda_n S^{-1}\Lambda_n^*\Theta_n + \Theta_n\Lambda_n^*\Lambda_n S^{-1} + \Theta_n\Lambda_n^*\Theta_n - \Lambda_n S^{-1}\\ &= \Lambda_n S^{-1}\Lambda_n^*\Theta_n + \Theta_n\Lambda_n^*\Lambda_n S^{-1} + \Theta_n\Lambda_n^*\Theta_n.\end{aligned}$$

We show that $\Lambda_n S^{-1}\Lambda_n^*\Theta_n + \Theta_n\Lambda_n^*\Lambda_n S^{-1} = 0$.
Note that

$$f = \sum_{j\in J} \Lambda_j^*\Gamma_j f = \sum_{j\in J} \Lambda_j^*(\Lambda_j S^{-1} + \Theta_j)f = \sum_{j\in J} \Lambda_j^*\Lambda_j S^{-1} f + \sum_{j\in J} \Lambda_j^*\Theta_j f = f + \sum_{j\in J} \Lambda_j^*\Theta_j f,$$

which implies that $\sum_{j\in J} \Lambda_j^*\Theta_j f = 0$ and $\Lambda_j^*\Theta_j f = 0$ for all $f \in U$ and $j \in J$.
Particularly, we have $\Lambda_n^*\Theta_n f = 0$ for all $f \in U$. This yields that $\Lambda_n S^{-1}\Lambda_n^*\Theta_n = 0$ and $\Theta_n\Lambda_n^*\Theta_n = 0$.
Therefore $\Theta_n = \Theta_n\Lambda_n^*\Lambda_n S^{-1}$. Suppose $F_n = \Theta_n\Lambda_n^*$, then $\Theta_n = F_n\Lambda_n S^{-1}$.
From $\Lambda_n^*\Theta_n = 0$, we have $\Lambda_n^*\Theta_n\Lambda_n^*\Lambda_n f = 0$ i.e.$F_n\Lambda_n^*\Lambda_n f = 0$.
$\Leftarrow$ Suppose that for each $j \in J$ there exists operator $F_j \in End_A^*(V_j, V_j)$ such that $\Theta_j = F_j\Lambda_j S^{-1}$ and $\Lambda_j^* F_j \Lambda_j f = 0$ holds for all $f \in U$. Then for all $f \in U$ we have

$$\sum_{j\in J} \Lambda_j^*\Gamma_j f = \sum_{j\in J} \Lambda_j^*\Lambda_j S^{-1} f = \sum_{j\in J} \Lambda_j^*\Theta_j f = f + \sum_{j\in J} \Lambda_j^* F_j \Lambda_j S^{-1} f = f.$$

Therefore $\{\Gamma_j : j \in J\}$ is a dual sequence of $\{\Lambda_j : j \in J\}$.
With similar proof of Theorem 3.3, $\{\Gamma_j : j \in J\}$ is a g-Bessel sequence and by Proposition 3.1, $\{\Gamma_j : j \in J\}$ is dual g-frame of $\{\Lambda_j : j \in J\}$.
To complete the proof, we need to show that $\{\Gamma_j : j \in J\}$ is a g-Riesz basis of U with respect to $\{V_j : j \in J\}$.
Let $\sum_{j\in J} \Gamma_j^* f_j = 0$, then we have

$$0 = \sum_{j\in J} (S^{-1}\Lambda_j^* + \Theta_j^*) f_j = \sum_{j\in J} (S^{-1}\Lambda_j^* + S^{-1}\Lambda_j^* F_j^*) f_j = \sum_{j\in J} S^{-1}\Lambda_j^*(I_j + F_j^*) f_j.$$

Since $\{S^{-1}\Lambda_j^* : j \in J\}$ is a g-Riesz basis then $S^{-1}\Lambda_j^*(I_j + F_j^*) f_j = 0$, i.e. $\Gamma_j^* f_j = 0$ for all $j \in J$.
We now show that $\Gamma_j \neq 0$ for all $j \in J$.
Assume on the contrary that $\Gamma_n = 0$ for some $n \in J$. Then $\Theta_n = -\Lambda_n S^{-1}$. It follows

that

$$0 = \Lambda_n^* F_n \Lambda_n f = \Lambda_n^* S \Theta_n f = -\Lambda_n^* \Lambda_n f$$

holds for all $f \in U$.
In particular, letting $f = S^{-1}\Lambda_n^* g$ for some $g \in U$, we have $-\Lambda_n^*\Lambda_n S^{-1}\Lambda_n^* g = -\Lambda_n^* g = 0$, therefore $\Lambda_n = 0$, a contradiction. This completes the proof. ■

Corollary 3.6 Suppose that $\{\Lambda_j \in End_A^*(U, V_j) : j \in J\}$ is a g-Riesz basis in Hilbert C*-module U with respect to $\{V_j : j \in J\}$ and Λ_j is surjective for any $j \in J$. Then $\{\Lambda_j : j \in J\}$ has a unique dual g-Riesz basis.

Proof. Let $f_j \in V_j$ for some $j \in J$, then there exists $f \in U$ such that $\Lambda_j f = f_j$. Therefore, we have

$$\Theta_j^* f_j = S^{-1}\Lambda_j^* F_j^* \Lambda_j f = S^{-1} 0 = 0.$$

■

Corollary 3.7 Suppose that $\{f_j : j \in J\}$ is a Riesz basis in Hilbert A-module H and operator $T_j : H \to A$ defined by $T_j f =< f, f_j >$ is surjective for any $j \in J$. Then $\{f_j : j \in J\}$ has a unique dual Riesz basis.

Definition 3.8 Let $\{\Lambda_j \in End_A^*(U, V_j) : j \in J\}$

(i) If the A-linear hull of $\bigcup_{j\in J} \Lambda_j^*(V_j)$ is dense in U, then $\{\Lambda_j : j \in J\}$ is g-complete.

(ii) If $\{\Lambda_j : j \in J\}$ is g-complete and there exist real constant A, B such that for any finite subset $S \subseteq J$ and $g_j \in V_j, j \in S$

$$A\Big\|\sum_{j\in S} |g_j|^2\Big\| \leqslant \Big\|\sum_{j\in S} \Lambda_j^* g_j\Big\|^2 \leqslant B\Big\|\sum_{j\in S} |g_j|^2\Big\|,$$

then $\{\Lambda_j : j \in J\}$ is a modular g-Riesz basis for U with respect to $\{V_j : j \in J\}$. A and B are called bounds of $\{\Lambda_j \in End_A^*(U, V_j) : j \in J\}$.

Theorem 3.9 ([12]) A sequence $\{\Lambda_j \in End_A^*(U, V_j) : j \in J\}$ is a modular g-Riesz basis if and only if the synthesis operator F is a homeomorphism.

Theorem 3.10 Let $\{\Lambda_j \in End_A^*(U, V_j) : j \in J\}$ then the following two statements are equivalent:

(1) The sequence $\{\Lambda_j : j \in J\}$ is a modular g-Riesz basis for Hilbert C*-module U with respect to $\{V_j : j \in J\}$ with bounds A and B;

(2)The sequence $\{\Lambda_j : j \in J\}$ is a g-frame for Hilbert C*-module U with respect to $\{V_j : j \in J\}$ with bounds A and B, and if an A-linear combination $\sum_{j\in S} \Lambda_j^* g_j = 0$ for $\{g_j\}_{j\in J} \in \oplus_{j\in J} V_j$, then $g_j = 0$ for all $j \in J$.

Proof. (1)→ (2) By Theorem 3.9 the operator $F : \oplus V_j \to U$ is a linear homeomorphism. Hence the operator F is onto and therefore by Theorem 3.2 in [16] $\{\Lambda_j : j \in J\}$ is a g-frame. Also, since F is injective

$$KerF = \Big\{\{g_j\}_{j\in J} \in \oplus_{j\in J} V_j : F(\{g_j\}_{j\in J}) = \sum_{j\in S} \Lambda_j^* g_j = 0\Big\} = \{0\}. \tag{6}$$

This implies the statement (2).

(2)→ (1)By Theorem 3.2 in [16] the operator F is injective and by (3.2) F is injective. Therefore F is homeomorphism and by Theorem ? $\{\Lambda_j : j \in J\}$ is a modular g-Riesz basis. ■

Corollary 3.11 Every modular g-Riesz basis is a g-Riesz basis.

Proof. By Definition 3.8 and Theorem 3.9. ■

Corollary 3.12 Let $\{\Lambda_j \in End_A^*(U, V_j) : j \in J\}$ then the following two statements are equivalent:

(1) The sequence $\{\Lambda_j : j \in J\}$ is a modular g-Riesz basis for Hilbert C*-module U with respect to $\{V_j : j \in J\}$.

(2)The sequence $\{\Lambda_j : j \in J\}$ has a unique dual modular g-Riesz basis.

Proof. (1)→ (2) Every modular g-Riesz basis is a g-Riesz basis and every g-Riesz basis is a g-frame. So every modular g-Riesz basis is a g-frame. Now by Theorem 3.3 and Theorem 3.10 $\{\Lambda_j : j \in J\}$ has a unique dual modular g-Riesz basis.

(2)→ (1)The proof by Theorem 3.3 and Theorem 3.10 is straightforward. ■

Next example shows in Hilbert C*-module setting, every Riesz basis is not a modular Riesz basis, so every g-Riesz basis is not a modular g-Riesz basis.

***Example* 3.13** Let $A = M_{2\times 2}(C)$ be the C*-algebra of all 2×2 complex matrices. Let $H = A$ and for any $A, B \in H$ define $\langle A, B\rangle = AB^*$. Then H is a Hilbert A-module.

Let $E_{i,j}$ be the matrix with 1 in the (i, j)th entry and 0 elsewhere, where $1 \leqslant i, j \leqslant 2$. Then $\Phi = \{E_{1,1}, E_{2,2}\}$ is a Riesz basis of H but is not a modular Riesz basis.

Acknowledgements

The author thanks the referee(s) for comments and suggestions which improved the quality of the paper.

References

[1] A. Alijan, M. A. Dehghan, g-frames and their duals for Hilbert C*-modules, Bull. Iran. Math. Soci., 38(3), (2012), 567-580.

[2] O. Christensen, An Introduction to Frames and Riesz Bases, Birkhauser, Boston, 2003.

[3] I. Daubechies, Ten Lectures on Wavelets, SIAM, Philadelphia, 1992.

[4] I. Daubechies, A. Grossmann,Y. Meyer, Painless nonorthogonal expansions, J. Math. Phys. 27 (1986), 1271-1283.

[5] R.J. Duffin, A.C. Schaeffer, A class of nonharmonic Fourier series, Trans. Amer. Math. Soc. 72 (1952), 341-366.

[6] M. Frank, D. R. Larson, A module frame concept for Hilbert C.-modules, in: Functional and Harmonic Analysis of Wavelets, San Antonio, TX, January 1999, Contemp. Math. 247, Amer. Math. Soc., Providence, RI 207-233, 2000.

[7] M. Frank, D.R. Larson, Frames in Hilbert C^*-modules and C^*-algebras, J. Operator Theory 48 (2002), 273-314.

[8] D. Han, W. Jing, D. Larson, R. Mohapatra, Riesz bases and their dual modular frames in Hilbert C^*-modules, J. Math. Anal. Appl. 343 (2008), 246-256.

[9] D. Han, W. Jing, R. Mohapatra, Perturbation of frames and Riesz bases in Hilbert C^*-modules, Linear Algebra Appl. 431 (2009), 746-759.

[10] A. Khosravi, B. Khosravi, Frames and bases in tensor products of Hilbert spaces and Hilbert C^*-modules, Proc. Indian Acad. Sci. Math. Sci. 117 (2007), 1-12.

[11] A. Khosravi, B. Khosravi, Fusion frames and g-frames in Hilbert C^*-modules, Int. J. Wavelets Multiresolut. Inf. Process. 6 (2008), 433-466.

[12] A. Khosravi, B. Khosravi, g-frames and modular Riesz bases in Hilbert C^*-modules, Int. J. Wavelets Multiresolut. Inf. Process. 10(2) (2012), 1250013 1-12.

[13] E.C. Lance, Hilbert C^*-Modules: A Toolkit for Operator Algebraists, London Math. Soc. Lecture Note Ser. 210, Cambridge Univ. Press, 1995.

[14] M. Rashidi-Kouchi, A. Nazari, M. Amini, On stability of g-frames and g-Riesz bases in Hilbert C*-modules, Int. J. Wavelets Multiresolut. Inf. Process. 12(6) (2014), 1450036 1-16.

[15] W. Sun, g-Frames and g-Riesz bases, J. Math. Anal. Appl. 322 (2006), 437-452.

[16] X.-C. Xiao, X.-M. Zeng, Some properties of g-frames in Hilbert C^*-modules J. Math. Anal. Appl. 363 (2010), 399-408.

3

Fixed Point Theorems for semi λ-subadmissible Contractions in b-Metric spaces

R.J. Shahkoohi[a], A. Razani[a*]

[a] *Department of Mathematics, Science and Research Branch, Islamic Azad University, Tehran, Iran.*

Abstract. Here, a new certain class of contractive mappings in the $b-$metric spaces is introduced. Some fixed point theorems are proved which generalize and modify the recent results in the literature. As an application, some results in the $b-$metric spaces endowed with a partial ordered are proved.

Keywords: Fixed point, b-metric.

1. Introduction

The existence of a fixed point is studied by many authors. The notion of b-metric space was first explained by Bakhtin in [2] and then widely utilized by Czerwik in [6] (this space is a metric type spaces defined by Khamsi and Hussain [18]). Since then, many researches deal with fixed point theory for single-valued and multi-valued mappings in b-metric spaces (see, [3, 6, 7] and references therein). Meanwhile, Samet *et al.* [30] presented the notions of α-ψ-contractive and α-admissible mappings and founded several fixed point theorems for such mappings outline under the complete metric spaces. Subsequently, Salimi *et al.* [28] and Hussain *et al.* [13] improved the concepts of α-ψ-contractive and α-admissible mappings and studied some fixed point theorems. In this paper, a new classes of contractive mappings is introduced in order to study some fixed point theorems in the $b-$metric spaces.

*Corresponding author.
E-mail address: razani@ipm.ir (A. Razani).

Definition 1.1 [6] Let X be a nonempty set and $s \geqslant 1$. A function $d : X \times X \to \mathbb{R}^+$ is a $b-$metric if and only if for all $x, y, z \in X$, the following conditions hold:

(b$_1$) $d(x, y) = 0$ iff $x = y$,
(b$_2$) $d(x, y) = d(y, x)$,
(b$_3$) $d(x, z) \leqslant s[d(x, y) + d(y, z)]$.

Then the tripled (X, d, s) is called a $b-$metric space.

Definition 1.2 [5] Let (X, d) be a $b-$metric space. A sequence $\{x_n\}$ in X is called:

(a) b-convergent if and only if there exists $x \in X$ such that $d(x_n, x) \to 0$, as $n \to +\infty$. In this case, we write $\lim\limits_{n\to\infty} x_n = x$.

(b) b-Cauchy if and only if $d(x_n, x_m) \to 0$, as $n, m \to +\infty$.

Proposition 1.3 [5, Remark 2.1] In a $b-$metric space (X, d) the following assertions hold:

p_1. A b-convergent sequence has a unique limit.
p_2. Each b-convergent sequence is b-Cauchy.
p_3. In general a $b-$metric is not continuous.

Lemma 1.4 [1] Let (X, d) be a $b-$metric space with $s \geqslant 1$. Suppose that $\{x_n\}$ and $\{y_n\}$ are b-convergent to x, y, respectively. Then

$$\frac{1}{s^2}d(x, y) \leqslant \liminf_{n\longrightarrow\infty} d(x_n, y_n) \leqslant \limsup_{n\longrightarrow\infty} d(x_n, y_n) \leqslant s^2 d(x, y).$$

In particular, if $x = y$ then $\lim\limits_{n\longrightarrow\infty} d(x_n, y_n) = 0$. Moreover, for each $z \in X$

$$\frac{1}{s}d(x, z) \leqslant \liminf_{n\longrightarrow\infty} d(x_n, z) \leqslant \limsup_{n\longrightarrow\infty} d(x_n, z) \leqslant sd(x, z).$$

For more details on b-metric spaces the reader can refer to [7]-[11].

Definition 1.5 [30] Let T be a self-mapping on X and $\alpha : X \times X \to [0, +\infty)$ be a function. T is an α-admissible mapping if

$$x, y \in X, \quad \alpha(x, y) \geqslant 1 \quad \Longrightarrow \quad \alpha(Tx, Ty) \geqslant 1.$$

Definition 1.6 [16] Let T be an α-admissible mapping. We say that T is a triangular α-admissible mapping if $\alpha(x, y) \geqslant 1$ and $\alpha(y, z) \geqslant 1$ implies that $\alpha(x, z) \geqslant 1$.

Lemma 1.7 [16] Let T be a triangular α-admissible mapping. Assume that there exists $x_0 \in X$ such that $\alpha(x_0, Tx_0) \geqslant 1$. Define sequence $\{x_n\}$ by $x_n = T^n x_0$. Then

$$\alpha(x_m, x_n) \geqslant 1 \text{ for all } m, n \in \mathbb{N} \text{ with } m < n.$$

Definition 1.8 [12] Let $\alpha : X \times X \to [0, \infty)$ and $T : X \to X$. We say that T is an α-continuous mapping if for given $x \in X$ and sequence $\{x_n\}$ with $x_n \to x$ and $\alpha(x_n, x_{n+1}) \geqslant 1$ for all $n \in \mathbb{N}$ one has $Tx_n \to Tx$.

Definition 1.9 Let T be a self-mapping on X and let $\lambda : X \to [0, +\infty)$ be a function. We say that T is a semi λ-subadmissible mapping if

$$x \in X, \quad \lambda(x) \leqslant 1 \quad \Longrightarrow \quad \lambda(Tx) \leqslant 1.$$

***Example* 1.10** Let $T : \mathbb{R} \to \mathbb{R}$ be defined by $Tx = x^3$. Suppose that $\lambda : \mathbb{R} \to \mathbb{R}^+$ is given by $\lambda(x) = e^x$ for all $x \in \mathbb{R}$. Then T is a semi λ-subadmissible mapping. Indeed, if $\lambda(x) = e^x \leqslant 1$ then $x \leqslant 0$ which implies that $Tx \leqslant 0$. Therefore $\lambda(Tx) = e^{Tx} \leqslant 1$.

Consistent with Khan *et al.* [17] we denote by Ψ the set of all function $\varphi : [0, +\infty) \to [0, +\infty)$ (which is called an altering distance function) if the following conditions hold:

- φ is continuous and non-decreasing.
- $\varphi(t) = 0$ if and only if $t = 0$.

Motivated by Kumam and Roldán [20] we introduce the following class of mappings which is suitable for our results.

Let Θ denote the set of all functions $\theta : R^{+^4} \to R^+$ satisfying:
(Θ_1) θ is continuous and increasing in all its variables;
(Θ_2) $\theta(t_1, t_2, t_3, t_4) = 0$ iff either $t_1 = 0$ or $t_4 = 0$.

2. Main Theorems

In this section we stat the Main results. The first theorem is based on [7, Theorem 4] and [27, Theorem 3].

Theorem 2.1 Let (X, d, s) be a complete b-metric space, T be a self-mapping on X and $\alpha : X \times X \to [0, \infty)$ and $\lambda : X \to [0, +\infty)$ be two functions. Suppose that the following assertions hold.

(i) There exists $x_0 \in X$ such that $\alpha(x_0, Tx_0) \geqslant 1$ and $\lambda(x_0) \leqslant 1$.
(ii) T is α-continuous, triangular α-admissible and semi λ-subadmissible mapping.
(iii) For all $x, y \in X$ with $\alpha(x, y) \geqslant 1$

$$\psi(sd(Tx, Ty)) \leqslant \lambda(x)\lambda(y)\Big[\psi(M(x, y)) - \varphi(M(x, y))\Big] + \theta\Big(d(x, Tx), d(y, Ty), d(x, Ty), d(y, Tx)\Big) \tag{1}$$

where $\psi, \varphi \in \Psi$, $\theta \in \Theta$ and

$$M(x, y) = \max\left\{d(x, y), d(x, Tx), d(y, Ty), \frac{d(x, Ty) + d(y, Tx)}{2s}\right\}.$$

Then T has a fixed point.

Proof. Let $x_0 \in X$ be such that $\alpha(x_0, Tx_0) \geqslant 1$ and $\lambda(x_0) \leqslant 1$. We define a sequence $\{x_n\}$ as follows

$$x_n = T^n x_0 = Tx_{n-1}$$

for all $n \in \mathbb{N}$. If $x_n = x_{n+1}$ for some $n \in \mathbb{N}$ then $x_n = Tx_n$ and so x_n is a fixed point of f. Hence we assume that $x_n \neq x_{n+1}$, for all $n \in \mathbb{N}$. Since T is a triangular α-admissible mapping then by Lemma 1.7

$$\alpha(x_m, x_n) \geqslant 1 \text{ for all } m, n \in \mathbb{N} \text{ with } m < n.$$

Also, since T is a semi λ-subadmissible mapping and $\lambda(x_0) \leqslant 1$ then $\lambda(x_1) = \lambda(Tx_0) \leqslant 1$. Again, since T is semi λ-subadmissible, then $\lambda(x_2) = \lambda(Tx_1) \leqslant 1$. Continuing this process

$\lambda(x_n) \leqslant 1$ for all $n \in \mathbb{N} \cup \{0\}$. Then by (iii),

$$\begin{aligned}\psi(d(x_n, x_{n+1})) &\leqslant \psi(sd(x_n, x_{n+1})) \\ &= \psi(sd(Tx_{n-1}, Tx_n)) \\ &\leqslant \lambda(x_{n-1})\lambda(x_n)\Big[\psi(M(x_{n-1}, x_n)) - \varphi(M(x_{n-1}, x_n))\Big] \\ &\quad +\theta\big(d(x_{n-1}, Tx_{n-1}), d(x_n, Tx_n), d(x_{n-1}, Tx_n), d(x_n, Tx_{n-1})\big) \\ &\leqslant \psi(M(x_{n-1}, x_n)) - \varphi(M(x_{n-1}, x_n)) \\ &\quad +\theta\big(d(x_{n-1}, Tx_{n-1}), d(x_n, Tx_n), d(x_{n-1}, Tx_n), d(x_n, Tx_{n-1})\big)\end{aligned} \tag{2}$$

where

$$\begin{aligned}M(x_{n-1}, x_n) &= \max\left\{d(x_{n-1}, x_n), d(x_{n-1}, Tx_{n-1}), d(x_n, Tx_n), \tfrac{d(x_{n-1}, Tx_n)+d(x_n, Tx_{n-1})}{2s}\right\} \\ &= \max\left\{d(x_{n-1}, x_n), d(x_n, x_{n+1}), \tfrac{d(x_{n-1}, x_{n+1})}{2s}\right\} \\ &\leqslant \max\left\{d(x_{n-1}, x_n), d(x_n, x_{n+1}), \tfrac{sd(x_{n-1}, x_n)+sd(x_n, x_{n+1})}{2s}\right\} \\ &= \max\left\{d(x_{n-1}, x_n), d(x_n, x_{n+1}), \tfrac{d(x_{n-1}, x_n)+d(x_n, x_{n+1})}{2}\right\} \\ &= \max\left\{d(x_{n-1}, x_n), d(x_n, x_{n+1})\right\}\end{aligned} \tag{3}$$

and

$$\begin{aligned}&\theta\big(d(x_{n-1}, Tx_{n-1}), d(x_n, Tx_n), d(x_{n-1}, Tx_n), d(x_n, Tx_{n-1})\big) \\ &= \theta\big(d(x_{n-1}, x_n), d(x_n, x_{n+1}), d(x_{n-1}, x_{n+1}), d(x_n, x_n)\big) \\ &= \theta\big(d(x_{n-1}, x_n), d(x_n, x_{n+1}), d(x_{n-1}, x_{n+1}), 0\big) = 0.\end{aligned} \tag{4}$$

By (2)-(4) and the properties of ψ and φ we obtain

$$\begin{aligned}\psi(d(x_n, x_{n+1})) &\leqslant \psi\Big(\max\Big\{d(x_{n-1}, x_n), d(x_n, x_{n+1})\Big\}\Big) - \varphi\Big(M(x_{n-1}, x_n)\Big) \\ &< \psi\Big(\max\Big\{d(x_{n-1}, x_n), d(x_n, x_{n+1})\Big\}\Big).\end{aligned} \tag{5}$$

Now if

$$\max\Big\{d(x_{n-1}, x_n), d(x_n, x_{n+1})\Big\} = d(x_n, x_{n+1}),$$

then by (5)

$$\begin{aligned}\psi(d(x_n, x_{n+1})) &\leqslant \psi(d(x_n, x_{n+1})) - \varphi(M(x_{n-1}, x_n)) \\ &< \psi(d(x_n, x_{n+1})),\end{aligned}$$

which is a contradiction. Hence

$$\max\Big\{d(x_{n-1}, x_n), d(x_n, x_{n+1})\Big\} = d(x_{n-1}, x_n).$$

Therefore

$$\psi(d(x_n, x_{n+1})) \leqslant \psi(d(x_n, x_{n-1})) - \varphi(M(x_{n-1}, x_n)) < \psi(d(x_n, x_{n-1})). \tag{6}$$

Since ψ is a non-decreasing mapping, then $\{d(x_n, x_{n+1}) : n \in \mathbb{N} \cup \{0\}\}$ is a non-increasing sequence of positive numbers. Then there exists $r \geqslant 0$ such that

$$\lim_{n\to\infty} d(x_n, x_{n+1}) = r.$$

Letting $n \to \infty$ in (6), we have

$$\psi(r) \leqslant \psi(r) - \varphi(\lim_{n\to\infty} M(x_{n-1}, x_n)) \leqslant \psi(r).$$

Therefore $\varphi(\lim_{n\to\infty} M(x_{n-1}, x_n)) = 0$ and hence $r = 0$, *i.e.*,

$$\lim_{n\to\infty} d(x_n, x_{n+1}) = 0. \tag{7}$$

Now, we show that $\{x_n\}$ is a $b-$Cauchy sequence in X. Assume the contrary, that $\{x_n\}$ is not a $b-$Cauchy sequence. Then there exists $\varepsilon > 0$ and two subsequences $\{x_{m_i}\}$ and $\{x_{n_i}\}$ of $\{x_n\}$ such that n_i is the smallest index for which

$$n_i > m_i > i \text{ dna } d(x_{m_i}, x_{n_i}) \geqslant \varepsilon. \tag{8}$$

That is

$$d(x_{m_i}, x_{n_i-1}) < \varepsilon. \tag{9}$$

By using (8), (9) and the triangular inequality

$$\begin{aligned}\varepsilon &\leqslant d(x_{m_i}, x_{n_i})\\ &\leqslant sd(x_{m_i}, x_{m_i-1}) + sd(x_{m_i-1}, x_{n_i})\\ &\leqslant sd(x_{m_i}, x_{m_i-1}) + s^2d(x_{m_i-1}, x_{n_i-1}) + s^2d(x_{n_i-1}, x_{n_i}).\end{aligned}$$

Now, using (7) and taking the upper limit as $i \to \infty$

$$\frac{\varepsilon}{s^2} \leqslant \limsup_{i\longrightarrow\infty} d(x_{m_i-1}, x_{n_i-1}).$$

On the other hand

$$d(x_{m_i-1}, x_{n_i-1}) \leqslant sd(x_{m_i-1}, x_{m_i}) + sd(x_{m_i}, x_{n_i-1}).$$

Using (7), (9) and taking the upper limit as $i \to \infty$

$$\limsup_{i\longrightarrow\infty} d(x_{m_i-1}, x_{n_i-1}) \leqslant \varepsilon s.$$

Hence

$$\frac{\varepsilon}{s^2} \leqslant \limsup_{i\longrightarrow\infty} d(x_{m_i-1}, x_{n_i-1}) \leqslant \varepsilon s. \tag{10}$$

Again using the triangular inequality

$$d(x_{m_i-1}, x_{n_i}) \leqslant sd(x_{m_i-1}, x_{n_i-1}) + sd(x_{n_i-1}, x_{n_i}), \tag{11}$$

$$\varepsilon \leqslant d(x_{m_i}, x_{n_i}) \leqslant sd(x_{m_i}, x_{m_i-1}) + sd(x_{m_i-1}, x_{n_i}) \tag{12}$$

and

$$\varepsilon \leqslant d(x_{m_i}, x_{n_i}) \leqslant sd(x_{m_i}, x_{n_i-1}) + sd(x_{n_i-1}, x_{n_i}). \tag{13}$$

Using (7) and (10) and taking the upper limit as $i \to \infty$ in (11) and (12) we get

$$\frac{\varepsilon}{s} \leqslant \limsup_{i\longrightarrow\infty} d(x_{m_i-1}, x_{n_i}) \leqslant \varepsilon s^2. \tag{14}$$

Again using (7) and (9) and taking the upper limit as $i \to \infty$ in (13)

$$\frac{\varepsilon}{s} \leqslant \limsup_{i\longrightarrow\infty} d(x_{m_i}, x_{n_i-1}) \leqslant \varepsilon. \tag{15}$$

Since $\alpha(x_{m_i-1}, x_{n_i-1}) \geqslant 1$, $\lambda(x_{m_i-1}) \leqslant 1$ and $\lambda(x_{n_i-1}) \leqslant 1$ then from (iii) we have

$$\begin{aligned}
&\psi(sd(x_{m_i}, x_{n_i})) = \psi(sd(Tx_{m_i-1}, Tx_{n_i-1}))\\
&\leqslant \lambda(x_{m_i-1})\lambda(x_{n_i-1})\Big[\psi(M(x_{m_i-1}, x_{n_i-1})) - \varphi(M(x_{m_i-1}, x_{n_i-1}))\Big]\\
&+\theta\Big(d(x_{m_i-1}, Tx_{m_i-1}), d(x_{n_i-1}, Tx_{n_i-1}), d(x_{m_i-1}, Tx_{n_i-1}), d(x_{n_i-1}, Tx_{m_i-1})\Big)\\
&\leqslant \psi(M(x_{m_i-1}, x_{n_i-1})) - \varphi(M(x_{m_i-1}, x_{n_i-1}))\\
&+\theta\Big(d(x_{m_i-1}, Tx_{m_i-1}), d(x_{n_i-1}, Tx_{n_i-1}), d(x_{m_i-1}, Tx_{n_i-1}), d(x_{n_i-1}, Tx_{m_i-1})\Big),
\end{aligned} \tag{16}$$

where

$$\begin{aligned}
M(x_{m_i-1}, x_{n_i-1}) &= \max\Big\{d(x_{m_i-1}, x_{n_i-1}), d(x_{m_i-1}, Tx_{m_i-1}), d(x_{n_i-1}, Tx_{n_i-1}),\\
&\qquad \tfrac{d(x_{m_i-1}, Tx_{n_i-1}) + d(Tx_{m_i-1}, x_{n_i-1})}{2s}\Big\}\\
&= \max\Big\{d(x_{m_i-1}, x_{n_i-1}), d(x_{m_i-1}, x_{m_i}), d(x_{n_i-1}, x_{n_i}),\\
&\qquad \tfrac{d(x_{m_i-1}, x_{n_i}) + d(x_{m_i}, x_{n_i-1})}{2s}\Big\},
\end{aligned} \tag{17}$$

and

$$\begin{aligned}
&\theta\Big(d(x_{m_i-1}, Tx_{m_i-1}), d(x_{n_i-1}, Tx_{n_i-1}), d(x_{m_i-1}, Tx_{n_i-1}), d(x_{n_i-1}, Tx_{m_i-1})\Big)\\
&= \theta\Big(d(x_{m_i-1}, x_{m_i}), d(x_{n_i-1}, x_{n_i}), d(x_{m_i-1}, x_{n_i}), d(x_{n_i-1}, x_{m_i})\Big).
\end{aligned} \tag{18}$$

Taking the upper limit as $i \to \infty$ in (17) and (18) and using (7), (10), (14) and (15) we get

$$\frac{\varepsilon}{s^2} = \min\left\{\frac{\varepsilon}{s^2}, \frac{\frac{\varepsilon}{s}+\frac{\varepsilon}{s}}{2s}\right\} \leqslant \limsup_{i\longrightarrow\infty} M(x_{m_i-1}, x_{n_i-1})$$
$$= \max\{\limsup_{i\longrightarrow\infty} d(x_{m_i-1}, x_{n_i-1}), 0, 0,$$
$$\frac{\limsup_{i\longrightarrow\infty} d(x_{m_i-1}, x_{n_i}) + \limsup_{i\longrightarrow\infty} d(x_{m_i}, x_{n_i-1})}{2s}\}$$
$$\leqslant \max\left\{\varepsilon s, \frac{\varepsilon s^2+\varepsilon}{2s}\right\} = \varepsilon s.$$

So

$$\frac{\varepsilon}{s^2} \leqslant \limsup_{i\longrightarrow\infty} M(x_{m_i-1}, x_{n_i-1}) \leqslant \varepsilon s, \tag{19}$$

and

$$\begin{aligned}&\limsup_{i\to\infty}\theta\Big(d(x_{m_i-1}, Tx_{m_i-1}), d(x_{n_i-1}, Tx_{n_i-1}), d(x_{m_i-1}, Tx_{n_i-1}), d(x_{n_i-1}, Tx_{m_i-1})\Big)\\ &= \limsup_{i\to\infty}\theta\Big(d(x_{m_i-1}, x_{m_i}), d(x_{n_i-1}, x_{n_i}), d(x_{m_i-1}, x_{n_i}), d(x_{n_i-1}, x_{m_i})\Big) = 0.\end{aligned} \tag{20}$$

Similarly

$$\frac{\varepsilon}{s^2} \leqslant \liminf_{i\longrightarrow\infty} M(x_{m_i-1}, x_{n_i-1}) \leqslant \varepsilon s. \tag{21}$$

Now, taking the upper limit as $i \to \infty$ in (16) and using (8), (19) and (20) we have

$$\psi(\varepsilon s) \leqslant \psi(s\limsup_{i\longrightarrow\infty} d(x_{m_i}, x_{n_i}))$$
$$\leqslant \psi(\limsup_{i\longrightarrow\infty} M(x_{m_i-1}, x_{n_i-1})) - \liminf_{n\longrightarrow\infty}\varphi(M(x_{m_i-1}, x_{n_i-1}))$$
$$\leqslant \psi(\varepsilon s) - \varphi(\liminf_{i\longrightarrow\infty} M(x_{m_i-1}, x_{n_i-1})),$$

which implies

$$\varphi(\liminf_{i\longrightarrow\infty} M(x_{m_i-1}, x_{n_i-1})) = 0,$$

so $\liminf_{i\longrightarrow\infty} M(x_{m_i-1}, x_{n_i-1}) = 0$, which is a contradiction with (21). So $\{x_{n+1}\}$ is a $b-$Cauchy sequence in X. Since X is a complete $b-$metric space, there exists $x^* \in X$ such that $x_n \to x^*$ as $n \to \infty$. Also, from (ii) we know T is an $\alpha-$continuous mapping. Hence $Tx_n \to Tx^*$ as $n \to \infty$. Then

$$d(x^*, Tx^*) \leqslant sd(x^*, Tx_n) + sd(Tx_n, Tx^*).$$

Letting $n \to \infty$ in the above inequality

$$d(x^*, Tx^*) \leqslant s \lim_{n\to\infty} d(x^*, Tx_n) + s \lim_{n\to\infty} d(Tx_n, Tx^*) = 0.$$

So $Tx^* = x^*$. ■

For self-mappings that are not continuous or $\alpha-$continuous we have the following result.

Theorem 2.2 Let (X, d, s) be a complete b-metric space, T be a self-mapping on X and $\alpha : X \times X \to [0, \infty)$ and $\lambda : X \to [0, +\infty)$ be two functions. Suppose that the following assertions hold.

(i) There exists $x_0 \in X$ such that $\alpha(x_0, Tx_0) \geqslant 1$ and $\lambda(x_0) \leqslant 1$.
(ii) T is a triangular α-admissible and semi λ-subadmissible mapping.
(iii) For all $x, y \in X$ with $\alpha(x, y) \geqslant 1$

$$\psi(sd(Tx, Ty)) \leqslant \lambda(x)\lambda(y)\Big[\psi(M(x, y)) - \varphi(M(x, y))\Big] + \theta\Big(d(x, Tx), d(y, Ty), d(x, Ty), d(y, Tx)\Big),$$

where $\psi, \varphi \in \Psi$, $\theta \in \Theta$ and

$$M(x, y) = \max\left\{d(x, y), d(x, Tx), d(y, Ty), \frac{d(x, Ty) + d(y, Tx)}{2s}\right\}.$$

(v) If $\{x_n\}$ be a sequence such that $\alpha(x_n, x_{n+1}) \geqslant 1$, $\lambda(x_n) \leqslant 1$ for all $n \in \mathbb{N} \cup \{0\}$ and $x_n \to x$ as $n \to \infty$, then $\alpha(x_n, x) \geqslant 1$ for all $n \in \mathbb{N} \cup \{0\}$ and $\lambda(x) \leqslant 1$.

Then T has a fixed point.

Proof. Let $x_0 \in X$ be such that $\alpha(x_0, Tx_0) \geqslant 1$ and $\lambda(x_0) \leqslant 1$. Define a sequence $\{x_n\}$ in X by $x_n = T^n x_0 = Tx_{n-1}$ for all $n \in \mathbb{N}$. Following the proof of the Theorem 2.1, we obtain that $\{x_n\}$ is a $b-$Cauchy sequence such that $\alpha(x_n, x_{n+1}) \geqslant 1$ and $\lambda(x_n) \leqslant 1$ for all $n \in \mathbb{N} \cup \{0\}$. Since X is complete, there exists $x^* \in X$ such that the sequence $\{x_n\}$ b-converges to x^*. Using the assumption (v), we have $\alpha(x_n, x^*) \geqslant 1$ for all $n \in \mathbb{N} \cup \{0\}$ and $\lambda(x^*) \leqslant 1$. By (iii)

$$\begin{aligned}
\psi(sd(x_{n+1}, Tx^*)) &= \psi(sd(Tx_n, Tx^*)) \\
&\leqslant \lambda(x_n)\lambda(x^*)\Big[\psi(M(x_n, x^*)) - \varphi(M(x_n, x^*))\Big] \\
&\quad +\theta\Big(d(x_n, Tx_n), d(x^*, Tx^*), d(x_n, Tx^*), d(x^*, Tx_n)\Big) \\
&\leqslant \psi(M(x_n, x^*)) - \varphi(M(x_n, x^*)) \\
&\quad +\theta\Big(d(x_n, Tx_n), d(x^*, Tx^*), d(x_n, Tx^*), d(x^*, Tx_n)\Big),
\end{aligned} \tag{22}$$

where

$$\begin{aligned}
M(x_n, x^*) &= \max\left\{d(x_n, x^*), d(x_n, Tx_n), d(x^*, Tx^*), \tfrac{d(x_n, Tx^*) + d(Tx_n, x^*)}{2s}\right\} \\
&= \max\left\{d(x_n, x^*), d(x_n, x_{n+1}), d(x^*, Tx^*), \tfrac{d(x_n, Tx^*) + d(x_{n+1}, x^*)}{2s}\right\}
\end{aligned} \tag{23}$$

and

$$\begin{aligned}&\theta\Big(d(x_n,Tx_n),d(x^*,Tx^*),d(x_n,Tx^*),d(x^*,Tx_n)\Big)\\&=\theta\Big(d(x_n,x_{n+1}),d(x^*,Tx^*),d(x_n,Tx^*),d(x^*,x_{n+1})\Big).\end{aligned}\tag{24}$$

Letting $n\to\infty$ in (23) and (24) and using lemma 1.4, we get

$$\begin{aligned}\tfrac{d(x^*,Tx^*)}{2s^2}&=\min\left\{d(x^*,Tx^*),\tfrac{d(x^*,Tx^*)}{2s^2}\right\}\leqslant\limsup_{n\longrightarrow\infty}M(x_n,x^*)\\&\leqslant\max\left\{d(x^*,Tx^*),\tfrac{sd(x^*,Tx^*)}{2s}\right\}=d(x^*,Tx^*),\end{aligned}\tag{25}$$

and

$$\theta\Big(d(x_n,Tx_n),d(x^*,Tx^*),d(x_n,Tx^*),d(x^*,Tx_n)\Big)\to0 \text{ as } n\to\infty.$$

Similarly

$$\frac{d(x^*,Tx^*)}{2s^2}\leqslant\liminf_{n\longrightarrow\infty}M(x_n,x^*)\leqslant d(x^*,Tx^*).\tag{26}$$

Again, taking the upper limit as $i\to\infty$ in (22) and using lemma 1.4 and (25) we get

$$\begin{aligned}\psi(d(x^*,Tx^*)=\psi(s\frac{1}{s}d(x^*,Tx^*))&\leqslant\psi(s\limsup_{n\longrightarrow\infty}d(x_{n+1},Tx^*))\\&\leqslant\psi(\limsup_{n\longrightarrow\infty}M(x_n,x^*))-\liminf_{n\longrightarrow\infty}\varphi(M(x_n,x^*))\\&\leqslant\psi(d(x^*,Tx^*))-\varphi(\liminf_{n\longrightarrow\infty}M(x_n,x^*)).\end{aligned}$$

Hence, $\varphi(\liminf_{n\longrightarrow\infty}M(x_n,x^*))=0$. Then, $\liminf_{n\longrightarrow\infty}M(x_n,x^*)=0$ which is a contradiction. So, $x^*=Tx^*$. ■

***Example* 2.3** Let $X=\mathbb{R}$ be endowed with the $b-$metric

$$d(x,y)=\begin{cases}(|x|+|y|)^2, \text{ if } x\neq y\\ 0 \qquad\qquad \text{ if } x=y\end{cases}$$

for all $x,y\in X$. Define $T:X\to X$, $\alpha:X\times X\to[0,\infty)$ and $\lambda:X\to[0,\infty)$ by

$$Tx=\begin{cases}2x^3+\sin x, & \text{if } x\in(-\infty,0)\\ \frac{1}{8}x^2, & \text{if } x\in[0,1)\\ \frac{1}{8}x, & \text{if } x\in[1,2)\\ \frac{1}{4} & \text{if } x\in[2,+\infty)\end{cases}\qquad \alpha(x,y)=\begin{cases}2, \text{ if } x,y\in[0,+\infty)\\ 0, \text{ otherwise}\end{cases}$$

and $\lambda(x) = \begin{cases} 1, & \text{if } x \in [0,+\infty) \\ \\ 2x^2+3, & \text{otherwise.} \end{cases}$

Also, define $\psi, \varphi : [0,\infty) \to [0,+\infty)$ and $\theta : [0,+\infty)^4 \to [0,+\infty)$ by $\psi(t) = t$, $\varphi(t) = \frac{3}{4}t$ and $\theta(t_1,t_2,t_3,t_4) = \min\{t_1,t_2,t_3,t_4\}$. Clearly (X,d,s) with $s=2$ is a complete $b-$metric space, $\psi,\varphi \in \Psi$ and $\theta \in \Theta$. Let $\alpha(x,y) \geqslant 1$, then $x,y \in [0,+\infty)$. On the other hand, $Tw \in [0,+\infty)$ for all $w \in [0,+\infty)$. Then $\alpha(Tx,Ty) \geqslant 1$. That is, T is an α-admissible mapping. Let $\alpha(x,y) \geqslant 1$ and $\alpha(y,z) \geqslant 1$. So $x,y,z \in [0,+\infty)$ *i.e.*, $\alpha(x,z) \geqslant 1$. Hence T is a triangular α-admissible mapping. Also, let $\lambda(x) \leqslant 1$. Thus $x \in [0,+\infty)$. That is, $\lambda(Tx) \leqslant 1$. Thus T is a semi λ-subadmissible mapping. Let $\{x_n\}$ be a sequence in X such that $\alpha(x_n,x_{n+1}) \geqslant 1$ and $\lambda(x_n) \leqslant 1$ with $x_n \to x$ as $n \to \infty$. Then, $x_n \in [0,+\infty)$ for all $n \in \mathbb{N}$. Also $[0,+\infty)$ is a closed set. Then $x \in [0,+\infty)$. That is $\alpha(x_n,x) \geqslant 1$ for all $n \in \mathbb{N} \cup \{0\}$ and $\lambda(x) \leqslant 1$. Clearly $\alpha(0,T0) \geqslant 1$ and $\lambda(0) \leqslant 1$.

Let $\alpha(x,y) \geqslant 1$. So $x,y \in [0,+\infty)$.

Now we consider the following cases:

- Let $x,y \in [0,1)$ then

$$\begin{aligned}
\psi(2d(Tx,Ty)) &= 2d(Tx,Ty) = 2(\tfrac{1}{8}x^2 + \tfrac{1}{8}y^2)^2 \\
&= \tfrac{1}{32}(x^2+y^2)^2 \\
&\leqslant \tfrac{1}{4}(x+y)^2 \\
&= \tfrac{1}{4}d(x,y) \\
&\leqslant \tfrac{1}{4}M(x,y) \\
&= \quad (M(x,y)) - \varphi(M(x,y)) \\
&= \lambda(x)\lambda(y)\Big[\psi(M(x,y)) - \varphi(M(x,y))\Big] \\
&\quad + \theta(d(x,Tx), d(y,Ty), d(x,Ty), d(y,Tx)).
\end{aligned}$$

- Let $x,y \in [1,2)$ then

$$\begin{aligned}
\psi(2d(Tx,Ty)) = 2d(Tx,Ty) &= 2(\tfrac{1}{8}x + \tfrac{1}{8}y)^2 \\
&= \tfrac{1}{32}(x+y)^2 \\
&\leqslant \tfrac{1}{4}(x+y)^2 \\
&= \tfrac{1}{4}d(x,y) \\
&\leqslant \tfrac{1}{4}M(x,y) \\
&= \quad (M(x,y)) - \varphi(M(x,y)) \\
&\leqslant \lambda(x)\lambda(y)\Big[\psi(M(x,y)) - \varphi(M(x,y))\Big] \\
&\quad + \theta(d(x,Tx), d(y,Ty), d(x,Ty), d(y,Tx)).
\end{aligned}$$

- Let $x,y \in [2,\infty)$ then

$$\begin{aligned}
\psi(2d(Tx,Ty)) &= 2d(Tx,Ty) = 2(\tfrac{1}{4} + \tfrac{1}{4})^2 \\
&= \tfrac{1}{2} \leqslant 1 \\
&= \tfrac{1}{4}(1+1)^2 \\
&\leqslant \tfrac{1}{4}(x+y)^2 \\
&= \tfrac{1}{4}d(x,y) \\
&\leqslant \tfrac{1}{4}M(x,y) \\
&= \quad (M(x,y)) - \varphi(M(x,y)) \\
&\leqslant \lambda(x)\lambda(y)\Big[\psi(M(x,y)) - \varphi(M(x,y))\Big] \\
&\quad + \theta(d(x,Tx), d(y,Ty), d(x,Ty), d(y,Tx)).
\end{aligned}$$

- Let $x \in [0,1)$ and $y \in [1,2)$ then

$$\begin{aligned}
\psi(2d(Tx,Ty)) &= 2d(Tx,Ty) = 2(\tfrac{1}{8}x^2 + \tfrac{1}{8}y)^2 \\
&\leqslant 2(\tfrac{1}{8}x + \tfrac{1}{8}y)^2 \\
&= \tfrac{1}{32}(x^2 + y^2)^2 \\
&\leqslant \tfrac{1}{4}(x+y)^2 \\
&= \tfrac{1}{4}d(x,y) \leqslant \tfrac{1}{4}M(x,y) \\
&= \quad (M(x,y)) - \varphi(M(x,y)) \\
&= \lambda(x)\lambda(y)\Big[\psi(M(x,y)) - \varphi(M(x,y))\Big] \\
&+\theta(d(x,Tx), d(y,Ty), d(x,Ty), d(y,Tx)).
\end{aligned}$$

- Let $x \in [0,1)$ and $y \in [2,\infty)$ then

$$\begin{aligned}
\psi(2d(Tx,Ty)) &= 2d(Tx,Ty) = 2t(\tfrac{1}{8}x^2 + \tfrac{1}{4})^2 \\
&\leqslant 2(\tfrac{1}{8}x + \tfrac{1}{8}y)^2 \\
&= \tfrac{1}{32}(x+y)^2 \\
&\leqslant \tfrac{1}{4}(x+y)^2 \\
&= \tfrac{1}{4}d(x,y) \\
&\leqslant \tfrac{1}{4}M(x,y) \\
&= \quad (M(x,y)) - \varphi(M(x,y)) \\
&= \lambda(x)\lambda(y)\Big[\psi(M(x,y)) - \varphi(M(x,y))\Big] \\
&+\theta(d(x,Tx), d(y,Ty), d(x,Ty), d(y,Tx)).
\end{aligned}$$

- Let $x \in [1,2)$ and $y \in [2,\infty)$ then

$$\begin{aligned}
\psi(2d(Tx,Ty)) &= 2d(Tx,Ty) = 2(\tfrac{1}{8}x + \tfrac{1}{4})^2 \\
&\leqslant 2(\tfrac{1}{8}x + \tfrac{1}{8}y)^2 \\
&= \tfrac{1}{32}(x+y)^2 \\
&\leqslant \tfrac{1}{4}(x+y)^2 \\
&= \tfrac{1}{4}d(x,y) \\
&\leqslant \tfrac{1}{4}M(x,y) \\
&= \quad (M(x,y)) - \varphi(M(x,y)) \\
&\leqslant \lambda(x)\lambda(y)\Big[\psi(M(x,y)) - \varphi(M(x,y))\Big] \\
&+\theta(d(x,Tx), d(y,Ty), d(x,Ty), d(y,Tx)).
\end{aligned}$$

Therefore $\alpha(x,y) \geqslant 1$ implies

$$\psi(2d(Tx,Ty)) \leqslant \lambda(x)\lambda(y)\Big[\psi(M(x,y)) - \varphi(M(x,y))\Big] + \theta(d(x,Tx), d(y,Ty), d(x,Ty), d(y,Tx))$$

Hence, all conditions of Theorem 2.2 holds and T has a fixed point. Here, $x = 0$ is a fixed point of T.

Corollary 2.4 Let (X,d,s) be a complete b-metric space, T be a self-mapping on X and $\alpha : X \times X \to [0,\infty)$ and $\lambda : X \to [0,+\infty)$ be two functions. Suppose that the following assertions hold.

(i) There exists $x_0 \in X$ such that $\alpha(x_0, Tx_0) \geqslant 1$ and $\lambda(x_0) \leqslant 1$.
(ii) T is a triangular α-admissible and semi λ-subadmissible mapping.
(iii) For all $x, y \in X$

$$\psi(s\alpha(x,y)d(Tx,Ty)) \leqslant \lambda(x)\lambda(y)\Big[\psi(M(x,y)) - \varphi(M(x,y))\Big] + \theta\Big(d(x,Tx), d(y,Ty), d(x,Ty), d(y,Tx)\Big) \tag{27}$$

where $\psi, \varphi \in \Psi$, $\theta \in \Theta$ and

$$M(x,y) = \max\left\{d(x,y), d(x,Tx), d(y,Ty), \frac{d(x,Ty)+d(y,Tx)}{2s}\right\}.$$

(v) If $\{x_n\}$ is a sequence such that $\alpha(x_n, x_{n+1}) \geqslant 1$, $\lambda(x_n) \leqslant 1$ for all $n \in \mathbb{N} \cup \{0\}$ and $x_n \to x$ as $n \to \infty$ then $\alpha(x_n, x) \geqslant 1$ for all $n \in \mathbb{N} \cup \{0\}$ and $\lambda(x) \leqslant 1$.

Then T has a fixed point.

Proof. Let $\alpha(x, y) \geqslant 1$. Since ψ is increasing then from (iii)

$$\begin{aligned} \psi(sd(Tx,Ty)) &\leqslant \psi(s\alpha(x,y)d(Tx,Ty)) \\ &\leqslant \lambda(x)\lambda(y)\Big[\psi(M(x,y)) - \varphi(M(x,y))\Big] \\ &+\theta\Big(d(x,Tx), d(y,Ty), d(x,Ty), d(y,Tx)\Big). \end{aligned}$$

Therefore all conditions of Theorem 2.2 holds and T has a fixed point. ■

If in Corollary 2.4 we take $\alpha(x, y) = 1$ for all $x, y \in X$, then we have the following corollary.

Corollary 2.5 Let (X, d, s) be a complete b-metric space and T be a self-mapping on X and $\lambda : X \to [0, +\infty)$ be a function. Suppose that the following assertions hold.

(i) there exists $x_0 \in X$ such that $\lambda(x_0) \leqslant 1$,
(ii) T is a semi λ-subadmissible mapping,
(iii) for all $x, y \in X$ we have

$$\psi(sd(Tx,Ty)) \leqslant \lambda(x)\lambda(y)\Big[\psi(M(x,y)) - \varphi(M(x,y))\Big] + \theta\Big(d(x,Tx), d(y,Ty), d(x,Ty), d(y,Tx)\Big) \tag{28}$$

where, $\psi, \varphi \in \Psi$, $\theta \in \Theta$ and

$$M(x,y) = \max\left\{d(x,y), d(x,Tx), d(y,Ty), \frac{d(x,Ty)+d(y,Tx)}{2s}\right\},$$

(v) if $\{x_n\}$ be a sequence such that $\lambda(x_n) \leqslant 1$ for all $n \in \mathbb{N} \cup \{0\}$ and $x_n \to x$ as $n \to \infty$ then $\lambda(x) \leqslant 1$.

Then T has a fixed point.

3. Some results in $b-$metric spaces endowed with a graph

In this section, we show that many fixed point results in $b-$metric spaces endowed with a graph G (see [4]) can be deduced easily from our presented theorems.
As in [14], let (E, d, s) be a $b-$metric space and Δ denotes the diagonal of the Cartesian product of $X \times X$. Consider a directed graph G such that the set $V(G)$ of its vertices coincides with X and the set $E(G)$ of its edges contains all loops, that is $E(G) \supseteq \Delta$. We assume that G has no parallel edges, so we can identify G with the pair $(V(G), E(G))$. Moreover, we may treat G as a weighted graph, see [15, P.309], by assigning to each edge the distance between its vertices. If x and y are vertices in a graph G then a path in G from x to y of length N ($N \in \mathbb{N}$) is a sequence $\{x_i\}_{i=0}^{N}$ of $N+1$ vertices such that $x_0 = x$, $x_N = y$ and $(x_{i-1}, x_i) \in E(G)$ for $i = 1, \ldots, N$.

Definition 3.1 [14] Let (X, d) be a metric space endowed with a graph G. We say that a self-mapping $T : X \to X$ is a Banach G-contraction or simply a G-contraction if T preserves the edges of G that is,

$$\text{for all } x,\ y \in X, \quad (x, y) \in E(G) \Longrightarrow (Tx, Ty) \in E(G)$$

and T decreases the weights of the edges of G in the following way:

$$\exists\, \alpha \in (0,1) \text{ such that for all } x, y \in X, \quad (x, y) \in E(G) \Longrightarrow d(Tx, Ty) \leq \alpha d(x, y).$$

Definition 3.2 [14] A mapping $T : X \to X$ is called G-continuous if given $x \in X$ and sequence $\{x_n\}$

$$x_n \to x \,\text{as}\, n \to \infty \text{ and } (x_n, x_{n+1}) \in E(G) \,\text{for all}\, n \in \mathbb{N} \text{ imply } Tx_n \to Tx.$$

Theorem 3.3 Let (X, d, s) be a complete b-metric space endowed with a graph G and T be a self-mapping on X. Suppose that the following assertions hold.

(i) there exists $x_0 \in X$ such that $(x_0, Tx_0) \in E(G)$ and $\lambda(x_0) \leqslant 1$,
(ii) T is G-continuous and semi λ-subadmissible mapping,
(iii) $\forall x, y \in X[(x, y) \in E(G) \Rightarrow (T(x), T(y)) \in E(G)]$
(iv) $\forall x, y, z \in X[(x, y) \in E(G) \,\text{and}\, (y, z) \in E(G) \Rightarrow (x, z) \in E(G)]$
(v) for all $x, y \in X$ with $(x, y) \in E(G)$ we have,

$$\psi(sd(Tx, Ty)) \leqslant \lambda(x)\lambda(y)\Big[\psi(M(x, y)) - \varphi(M(x, y))\Big] + \theta\Big(d(x, Tx), d(y, Ty), d(x, Ty), d(y, Tx)\Big)$$

where, $\psi, \varphi \in \Psi$, $\theta \in \Theta$ and

$$M(x, y) = \max\left\{d(x, y), d(x, Tx), d(y, Ty), \frac{d(x, Ty) + d(y, Tx)}{2s}\right\}.$$

Then T has a fixed point.

Proof. Define $\alpha : X^2 \to [0, +\infty)$ by

$$\alpha(x, y) = \begin{cases} 2, \text{ if } (x, y) \in E(G) \\ \frac{1}{2}, \text{ otherwise.} \end{cases}$$

First we show that T is a triangular α-admissible mapping. Let $\alpha(x,y) \geqslant 1$ then $(x,y) \in E(G)$. From (iii) $(Tx,Ty) \in E(G)$. That is $\alpha(Tx,Ty) \geqslant 1$. Also let $\alpha(x,y) \geqslant 1$ and $\alpha(y,z) \geqslant 1$. So $(x,y) \in E(G)$ and $(y,z) \in E(G)$. From (iv) we get $(x,z) \in E(G)$, i.e. $\alpha(x,z) \geqslant 1$. Thus T is a triangular α-admissible mapping. Let T be G-continuous. So

$$x_n \to x \text{ as } n \to \infty \text{ and } (x_n, x_{n+1}) \in E(G) \text{ for all } n \in \mathbb{N} \text{ imply } Tx_n \to Tx.$$

That is,

$$x_n \to x \text{ as } n \to \infty \text{ and } \alpha(x_n, x_{n+1}) \geqslant 1 \text{ for all } n \in \mathbb{N} \text{ imply } Tx_n \to Tx$$

which implies that T is α-continuous. From (i) there exists $x_0 \in X$ such that $(x_0, Tx_0) \in E(G)$. That is $\alpha(x_0, Tx_0) \geqslant 1$. Let $\alpha(x,y) \geqslant 1$ then $(x,y) \in E(G)$. Now from (v) we have

$$\psi(sd(Tx,Ty)) \leqslant \lambda(x)\lambda(y)\Big[\psi(M(x,y)) - \varphi(M(x,y))\Big] + \theta\Big(d(x,Tx), d(y,Ty), d(x,Ty), d(y,Tx)\Big)$$

Hence all conditions of Theorem 2.1 are satisfied and T has a fixed point. ■

In Theorem 3.3 we take $\theta(t_1, t_2, t_3, t_4) = \min\{t_1, t_2, t_3, t_4\}$.

Corollary 3.4 Let (X, d, s) be a complete b-metric space endowed with a graph G and T be a self-mapping on X. Suppose that the following assertions hold.

(i) there exists $x_0 \in X$ such that $(x_0, Tx_0) \in E(G)$ and $\lambda(x_0) \leqslant 1$,
(ii) T is G-continuous and semi λ-subadmissible mapping,
(iii) $\forall x, y \in X[(x,y) \in E(G) \Rightarrow (T(x), T(y)) \in E(G)]$
(iv) $\forall x, y, z \in X[(x,y) \in E(G) \text{ and } (y,z) \in E(G) \Rightarrow (x,z) \in E(G)]$
(v) for all $x, y \in X$ with $(x,y) \in E(G)$ we have,

$$\psi(sd(Tx,Ty)) \leqslant \lambda(x)\lambda(y)\ \ \psi\Big[M(x,y)) - \varphi(M(x,y))\ \ +\Big]L\min\{d(x,Tx), d(y,Ty), d(x,Ty), d(y,Tx)\}$$

where, $\psi, \varphi \in \Psi$, $L \geqslant 0$ and

$$M(x,y) = \max\left\{d(x,y), d(x,Tx), d(y,Ty), \frac{d(x,Ty) + d(y,Tx)}{2s}\right\}.$$

Then T has a fixed point.

Theorem 3.5 Let (X, d, s) be a complete b-metric space endowed with a graph G and T be a self-mapping on X. Suppose that the following assertions hold.

(i) there exists $x_0 \in X$ such that $(x_0, Tx_0) \in E(G)$ and $\lambda(x_0) \leqslant 1$,
(ii) T is semi λ-subadmissible mapping,
(iii) $\forall x, y \in X[(x,y) \in E(G) \Rightarrow (T(x), T(y)) \in E(G)]$
(iv) $\forall x, y, z \in X[(x,y) \in E(G) \text{ and } (y,z) \in E(G) \Rightarrow (x,z) \in E(G)]$
(v) for all $x, y \in X$ with $(x,y) \in E(G)$ we have,

$$\psi(sd(Tx,Ty)) \leqslant \lambda(x)\lambda(y)\Big[\psi(M(x,y)) - \varphi(M(x,y))\Big] + \theta\Big(d(x,Tx), d(y,Ty), d(x,Ty), d(y,Tx)\Big) \tag{29}$$

where, $(\psi, \varphi \in \Psi)$, $\theta \in \Theta$ and

$$M(x,y) = \max\left\{d(x,y), d(x,Tx), d(y,Ty), \frac{d(x,Ty)+d(y,Tx)}{2s}\right\}.$$

(vi) if $\{x_n\}$ be a sequence in X such that $(x_n, x_{n+1}) \in E(G)$, $\lambda(x_n) \leqslant 1$ for all $n \in \mathbb{N}\cup\{0\}$ and $x_n \to x$ as $n \to \infty$ then $(x_n, x) \in E(G)$ for all $n \in \mathbb{N}\cup\{0\}$ and $\lambda(x) \leqslant 1$.

Then T has a fixed point.

Proof. Define the mapping $\alpha : X^2 \to [0, +\infty)$ as in the proof of Theorem 3.3. Similar to the proof of Theorem 3.3 we can prove that the conditions (i)-(iii) of Theorem 2.2 are satisfied. Let $\{x_n\}$ be a sequence in X such that $\alpha(x_n, x_{n+1}) \geqslant 1$ and $\lambda(x_n) \leqslant 1$ for all $n \in \mathbb{N}\cup\{0\}$ and $x_n \to x$ as $n \to \infty$. Then $(x_n, x_{n+1}) \in E(G)$ and $\lambda(x_n) \leqslant 1$ for all $n \in \mathbb{N}\cup\{0\}$. From (vi) we get $(x_n, x) \in E(G)$ and $\lambda(x) \leqslant 1$. That is $\alpha(x_n, x) \geqslant 1$ for all $n \in \mathbb{N}\cup\{0\}$ and $\lambda(x) \leqslant 1$. Therefore all conditions of Theorem 2.2 holds and T has a fixed point. ∎

Corollary 3.6 Let (X, d, s) be a complete b-metric space endowed with a graph G and T be a self-mapping on X. Suppose that the following assertions hold.

(i) there exists $x_0 \in X$ such that $(x_0, Tx_0) \in E(G)$ and $\lambda(x_0) \leqslant 1$,
(ii) T is semi λ-subadmissible mapping,
(iii) $\forall x, y \in X[(x,y) \in E(G) \Rightarrow (T(x), T(y)) \in E(G)]$
(iv) $\forall x, y, z \in X[(x,y) \in E(G) \text{and} (y,z) \in E(G) \Rightarrow (x, z \in E(G)]$
(v) for all $x, y \in X$ with $(x,y) \in E(G)$ we have,

$$\psi(sd(Tx,Ty)) \leqslant \lambda(x)\lambda(y)\left[\psi(M(x,y)) - \varphi(M(x,y))\right] + L\min\{d(x,Tx), d(y,Ty), d(x,Ty), d(y,Tx)\}$$

where, $(\psi, \varphi \in \Psi)$, $L \geqslant 0$ and

$$M(x,y) = \max\left\{d(x,y), d(x,Tx), d(y,Ty), \frac{d(x,Ty)+d(y,Tx)}{2s}\right\}.$$

(vi) if $\{x_n\}$ be a sequence in X such that $(x_n, x_{n+1}) \in E(G)$, $\lambda(x_n) \leqslant 1$ for all $n \in \mathbb{N}\cup\{0\}$ and $x_n \to x$ as $n \to \infty$, then $(x_n, x) \in E(G)$ for all $n \in \mathbb{N}\cup\{0\}$ and $\lambda(x) \leqslant 1$.

Then T has a fixed point.

4. Some results in $b-$metric spaces endowed with a partial ordered

The existence of fixed points in partially ordered sets has been considered by many authors (such as [19], [21–26] and [29] etc.). Later on, some generalizations of [26] are given in [27]. Several applications of these results to matrix equations are presented in [26].

Let X be a nonempty set. If (X, d, s) is a $b-$metric space and $(X, \preceq)$ be a partially ordered set, then $(X, d, s, \preceq)$ is called an ordered $b-$metric space. Two elements $x, y \in X$ are called comparable if $x \preceq y$ or $y \preceq x$ hold. A mapping $T : X \to X$ is said to be non-decreasing if $x \preceq y$ implies $Tx \preceq Ty$ for all $x, y \in X$.

In this section, we will show that many fixed point results in partially ordered $b-$metric spaces can be deduced easily from our obtained results.

Theorem 4.1 Let $(X, d, s, \preceq)$ be a complete ordered b-metric space and T be a self-mapping on X. Suppose that the following assertions hold.

(i) there exists $x_0 \in X$ such that $x_0 \preceq Tx_0$ and $\lambda(x_0) \leqslant 1$,
(ii) T is continuous and semi λ-subadmissible mapping,
(iii) T is an increasing mapping,
(v) for all $x, y \in X$ with $x \preceq y$ we have,

$$\psi(sd(Tx,Ty)) \leqslant \lambda(x)\lambda(y)\Big[\psi(M(x,y))-\varphi(M(x,y))\Big]+\theta\Big(d(x,Tx),d(y,Ty),d(x,Ty),d(y,Tx)\Big)$$

where, $\psi, \varphi \in \Psi$, $\theta \in \Theta$ and

$$M(x,y) = \max\left\{d(x,y), d(x,Tx), d(y,Ty), \frac{d(x,Ty)+d(y,Tx)}{2s}\right\}.$$

Then T has a fixed point.

Proof. Define $\alpha : X^2 \to [0, +\infty)$ by

$$\alpha(x,y) = \begin{cases} 2, & \text{if } x \preceq y \\ \frac{1}{2}, & \text{otherwise} \end{cases}$$

First, we prove that T is a triangular α-admissible mapping. Let $\alpha(x,y) \geqslant 1$, then $x \preceq y$. Since T is increasing, then we have $Tx \preceq Ty$. That is, $\alpha(Tx,Ty) \geqslant 1$. Suppose that $\alpha(x,y) \geqslant 1$ and $\alpha(y,z) \geqslant 1$. Then $x \preceq y$ and $y \preceq z$. Hence $x \preceq z$ i.e., $\alpha(x,z) \geqslant 1$. Therefore, T is a triangular α-admissible mapping. Since T is continuous then it is α-continuous too. From (i) there exists $x_0 \in X$ such that $\alpha(x_0, Tx_0) \geqslant 1$. That is, $\alpha(x_0, Tx_0) \geqslant 1$. Let $\alpha(x,y) \geqslant 1$, then $x \preceq y$. Now, from (v) we have

$$\psi(sd(Tx,Ty)) \leqslant \lambda(x)\lambda(y)\Big[\psi(M(x,y))-\varphi(M(x,y))\Big]+\theta\Big(d(x,Tx),d(y,Ty),d(x,Ty),d(y,Tx)\Big).$$

Hence, all conditions of Theorem 2.1 are satisfied and T has a fixed point. ∎

If in Theorem 3.3 we take $\theta(t_1,t_2,t_3,t_4) = L\psi(\min\{t_1,t_4\})$ where $L \geqslant 0$, then we have the following Corollary.

Corollary 4.2 Let $(X, d, s, \preceq)$ be a complete ordered b-metric space and T be a self-mapping on X. Suppose that the following assertions hold.

(i) there exists $x_0 \in X$ such that $x_0 \preceq Tx_0$ and $\lambda(x_0) \leqslant 1$,
(ii) T is continuous and semi λ-subadmissible mapping,
(iii) T is an increasing mapping,
(v) for all $x, y \in X$ with $x \preceq y$ we have,

$$\psi(sd(Tx,Ty)) \leqslant \lambda(x)\lambda(y)\Big[\psi(M(x,y)) - \varphi(M(x,y))\Big] + L\psi(\min\{d(x,Tx), d(y,Tx)\})$$

where, $\psi, \varphi \in \Psi$, $L \geqslant 0$ and

$$M(x,y) = \max\left\{d(x,y), d(x,Tx), d(y,Ty), \frac{d(x,Ty)+d(y,Tx)}{2s}\right\}.$$

Then T has a fixed point.

If in Corollary 3.3 we take $\lambda(x) = 1$ for all $x \in X$, then we have the following Corollary.

Corollary 4.3 [27, Theorem 3] Let $(X, d, s, \preceq)$ be a complete ordered b-metric space and T be a self-mapping on X. Suppose that the following assertions hold.

(i) there exists $x_0 \in X$ such that $x_0 \preceq Tx_0$,
(ii) T is continuous,
(iii) T is an increasing mapping,
(v) for all $x, y \in X$ with $x \preceq y$ we have,

$$\psi(sd(Tx, Ty)) \leqslant \psi(M(x, y)) - \varphi(M(x, y)) + +L\psi(\min\{d(x, Tx), d(y, Tx)\})$$

where, $\psi, \varphi \in \Psi$, $L \geqslant 0$ and

$$M(x, y) = \max\left\{d(x, y), d(x, Tx), d(y, Ty), \frac{d(x, Ty) + d(y, Tx)}{2s}\right\}.$$

Then T has a fixed point.

Theorem 4.4 Let $(X, d, s, \preceq)$ be a complete partially ordered b-metric space and let T be a self-mapping on X. Suppose that the following assertions hold.

(i) there exists $x_0 \in X$ such that $x_0 \preceq Tx_0$ and $\lambda(x_0) \leqslant 1$,
(ii) T is a semi λ-subadmissible mapping,
(iii) T is an increasing mapping,
(iv) for all $x, y \in X$ with $x \preceq y$ we have,

$$\psi(sd(Tx, Ty)) \leqslant \lambda(x)\lambda(y)\left[\psi(M(x, y)) - \varphi(M(x, y))\right] + \theta\Big(d(x, Tx), d(y, Ty), d(x, Ty), d(y, Tx)\Big) \tag{30}$$

where, $(\psi, \varphi \in \Psi)$, $\theta \in \Theta$ and

$$M(x, y) = \max\left\{d(x, y), d(x, Tx), d(y, Ty), \frac{d(x, Ty) + d(y, Tx)}{2s}\right\}.$$

(v) if $\{x_n\}$ be an increasing sequence in X such that $\lambda(x_n) \leqslant 1$ for all $n \in \mathbb{N} \cup \{0\}$ and $x_n \to x$ as $n \to \infty$ then $x_n \preceq x$ for all $n \in \mathbb{N} \cup \{0\}$ and $\lambda(x) \leqslant 1$.

Then T has a fixed point.

Proof. Define the mapping $\alpha : X^2 \to [0, +\infty)$ as in the proof of Theorem 3.3. Analogous to the proof of Theorem 3.3 we can prove all the conditions (i)-(iii) of Theorem 2.2 are satisfied. Let $\{x_n\}$ be a sequence in X such that $\alpha(x_n, x_{n+1}) \geqslant 1$ and $\lambda(x_n) \leqslant 1$ for all $n \in \mathbb{N} \cup \{0\}$ and $x_n \to x$ as $n \to \infty$. Then $x_n \preceq x_{n+1}$ and $\lambda(x_n) \leqslant 1$ for all $n \in \mathbb{N} \cup \{0\}$. From (v) we get, $x_n \preceq x$ and $\lambda(x) \leqslant 1$. That is, $\alpha(x_n, x) \geqslant 1$ for all $n \in \mathbb{N} \cup \{0\}$ and $\lambda(x) \leqslant 1$. Therefore all conditions of Theorem 2.2 holds and T has a fixed point. ∎

Corollary 4.5 Let $(X, d, s, \preceq)$ be a complete partially ordered b-metric space and T be a self-mapping on X. Suppose that the following assertions hold.

(i) there exists $x_0 \in X$ such that, $x_0 \preceq Tx_0$ and $\lambda(x_0) \leqslant 1$,
(ii) T is a semi λ-subadmissible mapping,
(iii) T is an increasing mapping,

(iv) for all $x, y \in X$ with $x \preceq y$ we have,

$$\psi(sd(Tx,Ty)) \leqslant \lambda(x)\lambda(y)\Big[\psi(M(x,y)) - \varphi(M(x,y))\Big] + L\psi(\min\{d(x,Tx), d(y,Tx)\}) \tag{31}$$

where, $\psi, \varphi \in \Psi$, $\theta \in \Theta$ and

$$M(x,y) = \max\left\{d(x,y), d(x,Tx), d(y,Ty), \frac{d(x,Ty)+d(y,Tx)}{2s}\right\}.$$

(v) if $\{x_n\}$ be an increasing sequence in X such that $\lambda(x_n) \leqslant 1$ for all $n \in \mathbb{N} \cup \{0\}$ and $x_n \to x$ as $n \to \infty$, then $x_n \preceq x$ for all $n \in \mathbb{N} \cup \{0\}$ and $\lambda(x) \leqslant 1$.

Then T has a fixed point.

Corollary 4.6 [27, Theorem 4] Let $(X, d, s, \preceq)$ be a complete partially ordered b-metric space and T be a self-mapping on X. Suppose that the following assertions hold.

(i) there exists $x_0 \in X$ such that $x_0 \preceq Tx_0$,
(iii) T is an increasing mapping,
(iv) for all $x, y \in X$ with $x \preceq y$ we have,

$$\psi(sd(Tx,Ty)) \leqslant \psi(M(x,y)) - \varphi(M(x,y)) + L\psi(\min\{d(x,Tx), d(y,Tx)\}) \tag{32}$$

where, $(\psi, \varphi \in \Psi)$, $L \geqslant 0$ and

$$M(x,y) = \max\left\{d(x,y), d(x,Tx), d(y,Ty), \frac{d(x,Ty)+d(y,Tx)}{2s}\right\}.$$

(v) if $\{x_n\}$ be an increasing sequence in X such that $x_n \to x$ as $n \to \infty$ then $x_n \preceq x$ for all $n \in \mathbb{N} \cup \{0\}$.

Then T has a fixed point.

5. Some integral type contractions

Let Φ denotes the set of all functions $\phi : [0, +\infty) \to [0, +\infty)$ satisfying the following properties:

- every $\phi \in \Phi$ is a Lebesgue integrable function on each compact subset of $[0, +\infty)$,
- for any $\phi \in \Phi$ and any $\epsilon > 0$, $\int_0^\epsilon \phi(\tau)d\tau > 0$.

Note that if we take $\psi(t) = \int_0^t \phi(\tau)d\tau$ where $\phi \in \Phi$ then $\psi \in \Psi$.

Also note that if $\psi \in \Psi$ and $\theta \in \Theta$ then $\psi\theta \in \Theta$.

If in Theorem 2.1 we take $\psi(t) = \int_0^t \phi(\tau)d\tau$, $\varphi(t) = (1-r)\int_0^t \phi(\tau)d\tau$ for all $t \in [0, \infty)$ where $0 \leqslant r < 1$ and replace θ by $\psi\theta$ then we have the following theorem.

Theorem 5.1 Let (X, d, s) be a complete b-metric space, T be a self-mapping on X and $\alpha : X \times X \to [0, \infty)$ and $\lambda : X \to [0, +\infty)$ be two functions. Suppose that the following assertions hold.

(i) there exists $x_0 \in X$ such that $\alpha(x_0, Tx_0) \geqslant 1$ and $\lambda(x_0) \leqslant 1$,
(ii) T is α-continuous, triangular α-admissible and semi λ-subadmissible mapping,

(iii) for all $x, y \in X$ with $\alpha(x, y) \geqslant 1$ we have

$$\int_0^{d(Tx,Ty)} \phi(\tau)d\tau \leqslant \frac{r\lambda(x)\lambda(y)}{s} \int_0^{M(x,y)} \phi(\tau)d\tau + \int_0^{\theta\left(d(x,Tx),d(y,Ty),d(x,Ty),d(y,Tx)\right)} \phi(\tau)d\tau \tag{33}$$

where, $0 \leqslant r < 1$, $\phi \in \Phi$, $\theta \in \Theta$ and

$$M(x,y) = \max\left\{d(x,y), d(x,Tx), d(y,Ty), \frac{d(x,Ty)+d(y,Tx)}{2s}\right\}.$$

Then T has a fixed point.

Theorem 5.2 Let (X, d, s) be a complete b-metric space, T be a self-mapping on X and $\alpha : X \times X \to [0, \infty)$ and $\lambda : X \to [0, +\infty)$ be two functions. Suppose that the following assertions hold.

(i) there exists $x_0 \in X$ such that, $\alpha(x_0, Tx_0) \geqslant 1$ and $\lambda(x_0) \leqslant 1$,
(ii) T is a triangular α-admissible and semi λ-subadmissible mapping,
(iii) for all $x, y \in X$ with $\alpha(x, y) \geqslant 1$ we have

$$\int_0^{d(Tx,Ty)} \phi(\tau)d\tau \leqslant \frac{r\lambda(x)\lambda(y)}{s} \int_0^{M(x,y)} \phi(\tau)d\tau + \int_0^{\theta\left(d(x,Tx),d(y,Ty),d(x,Ty),d(y,Tx)\right)} \phi(\tau)d\tau \tag{34}$$

where, $0 \leqslant r < 1$, $\phi \in \Phi$, $\theta \in \Theta$ and

$$M(x,y) = \max\left\{d(x,y), d(x,Tx), d(y,Ty), \frac{d(x,Ty)+d(y,Tx)}{2s}\right\},$$

(v) if $\{x_n\}$ be a sequence such that $\alpha(x_n, x_{n+1}) \geqslant 1$, $\lambda(x_n) \leqslant 1$ for all $n \in \mathbb{N} \cup \{0\}$ and $x_n \to x$ as $n \to \infty$, then $\alpha(x_n, x) \geqslant 1$ for all $n \in \mathbb{N} \cup \{0\}$ and $\lambda(x) \leqslant 1$.

Then T has a fixed point.

Theorem 5.3 Let (X, d, s) be a complete b-metric space endowed with a graph G and T be a self-mapping on X. Suppose that the following assertions hold.

(i) there exists $x_0 \in X$ such that, $(x_0, Tx_0) \in E(G)$ and $\lambda(x_0) \leqslant 1$,
(ii) T is G-continuous and semi λ-subadmissible mapping,
(iii) $\forall x, y \in X[(x, y) \in E(G) \Rightarrow (T(x), T(y)) \in E(G)]$
(iv) $\forall x, y, z \in X[(x, y) \in E(G) \text{ and } (y, z) \in E(G) \Rightarrow (x, z) \in E(G)]$
(v) for all $x, y \in X$ with $(x, y) \in E(G)$ we have,

$$\int_0^{d(Tx,Ty)} \phi(\tau)d\tau \leqslant \frac{r\lambda(x)\lambda(y)}{s} \int_0^{M(x,y)} \phi(\tau)d\tau + \int_0^{\theta\left(d(x,Tx),d(y,Ty),d(x,Ty),d(y,Tx)\right)} \phi(\tau)d\tau \tag{35}$$

where, $0 \leqslant r < 1$, $\phi \in \Phi$, $\theta \in \Theta$ and

$$M(x,y) = \max\left\{d(x,y), d(x,Tx), d(y,Ty), \frac{d(x,Ty)+d(y,Tx)}{2s}\right\}.$$

Then T has a fixed point.

Theorem 5.4 Let (X, d, s) be a complete b-metric space endowed with a graph G and T be a self-mapping on X. Suppose that the following assertions hold.

(i) there exists $x_0 \in X$ such that $(x_0, Tx_0) \in E(G)$ and $\lambda(x_0) \leqslant 1$,
(ii) T is semi λ-subadmissible mapping,
(iii) $\forall x, y \in X[(x, y) \in E(G) \Rightarrow (T(x), T(y)) \in E(G)]$
(iv) $\forall x, y, z \in X[(x, y) \in E(G) \text{ and } (y, z) \in E(G) \Rightarrow (x, z) \in E(G)]$
(v) for all $x, y \in X$ with $(x, y) \in E(G)$ we have,

$$\int_0^{d(Tx,Ty)} \phi(\tau)d\tau \leqslant \frac{r\lambda(x)\lambda(y)}{s} \int_0^{M(x,y)} \phi(\tau)d\tau + \int_0^{\theta\left(d(x,Tx),d(y,Ty),d(x,Ty),d(y,Tx)\right)} \phi(\tau)d\tau \tag{36}$$

where, $0 \leqslant r < 1$, $\phi \in \Phi$, $\theta \in \Theta$ and

$$M(x, y) = \max\left\{d(x, y), d(x, Tx), d(y, Ty), \frac{d(x, Ty) + d(y, Tx)}{2s}\right\}.$$

(vi) if $\{x_n\}$ be a sequence in X such that $(x_n, x_{n+1}) \in E(G)$, $\lambda(x_n) \leqslant 1$ for all $n \in \mathbb{N} \cup \{0\}$ and $x_n \to x$ as $n \to \infty$ then $(x_n, x) \in E(G)$ for all $n \in \mathbb{N} \cup \{0\}$ and $\lambda(x) \leqslant 1$.

Then T has a fixed point.

Theorem 5.5 Let $(X, d, s, \preceq)$ be a complete ordered b-metric space and T be a self-mapping on X. Suppose that the following assertions hold.

(i) there exists $x_0 \in X$ such that $x_0 \preceq Tx_0$ and $\lambda(x_0) \leqslant 1$,
(ii) T is continuous and semi λ-subadmissible mapping,
(iii) T is an increasing mapping,
(v) for all $x, y \in X$ with $x \preceq y$ we have

$$\int_0^{d(Tx,Ty)} \phi(\tau)d\tau \leqslant \frac{r\lambda(x)\lambda(y)}{s} \int_0^{M(x,y)} \phi(\tau)d\tau + \int_0^{\theta\left(d(x,Tx),d(y,Ty),d(x,Ty),d(y,Tx)\right)} \phi(\tau)d\tau \tag{37}$$

where, $0 \leqslant r < 1$, $\phi \in \Phi$, $\theta \in \Theta$ and

$$M(x, y) = \max\left\{d(x, y), d(x, Tx), d(y, Ty), \frac{d(x, Ty) + d(y, Tx)}{2s}\right\}.$$

Then T has a fixed point.

References

[1] A. Aghajani, M. Abbas and J.R. Roshan, *Common fixed point of generalized weak contractive mappings in partially ordered b−metric spaces*, Math. Slovaca, **4** (2014), 941-960.
[2] I.A. Bakhtin, *The contraction mapping principle in quasimetric spaces*, (Russian), Func. An., Gos. Ped. Inst. Unianowsk **30** (1989), 26-37.
[3] V. Berinde, *Generalized contractions in quasimetric spaces*, Seminar on Fixed Point Theory, Preprint no. **3** (1993), 3-9.
[4] F. Bojor, *Fixed point theorems for Reich type contraction on metric spaces with a graph*, Nonlinear Anal. **75** (2012), 3895-3901.
[5] M. Boriceanu, *Strict fixed point theorems for multivalued operators in b−metric spaces*, Int. J. Modern Math., **4**(3) (2009), 285-301.
[6] S. Czerwick, *Contraction mappings in b-metric spaces*, Acta Mathematica et Informatica Universitatis Ostraviensis, **1** (1993), 5-11.

[7] N. Hussain, V. Parvaneh, J.R. Roshan and Z Kadelburg, *Fixed points of cyclic weakly (ψ, φ, L, A, B)-contractive mappings in ordered b-metric spaces with applications*, Fixed Point Theory Appl. **256** (2013).
[8] J.R. Roshan, V. Parvaneh and I. Altun, *Some coincidence point results in ordered b-metric spaces and applications in a system of integral equations*, Applied Mathematics and Computation **226** (2014), 725-737.
[9] Z. Mustafa, J.R. Roshan, V. Parvaneh and Z. Kadelburg, *Fixed point theorems for weakly T-Chatterjea and weakly T-Kannan contractions in b-metric spaces*, Journal of Inequalities and Applications **46** (2014).
[10] N. Hussain, J.R. Roshan, V. Parvaneh and M. Abbas, *Common fixed point results for weak contractive mappings in ordered b-dislocated metric spaces with applications*, Journal of Inequalities and Applications **486** (2013).
[11] Z. Mustafa, J.R. Roshan, V. Parvaneh and Z. Kadelburg, *Some common fixed point results in ordered partial b-metric spaces*, Journal of Inequalities and Applications **562** (2013).
[12] N. Hussain, M. A. Kutbi and P. Salimi, *Fixed point theory in α-complete metric spaces with applications*, Abstract and Applied Analysis **280817** (2014), 11 pages.
[13] N. Hussain, P. Salimi and A. Latif, *Fixed point results for single and set-valued α-η-ψ-contractive mappings*, Fixed Point Theory and Appl. **212** (2013).
[14] J. Jachymski, *The contraction principle for mappings on a metric space with a graph*, Proc. Amer. Math. Soc. **136** (2008), 1359–1373.
[15] R. Johnsonbaugh, Discrete Mathematics, Prentice-Hall, Inc., New Jersey, 1997.
[16] E. Karapınar, P. Kumam and P. Salimi, *On α-ψ-Meir-Keeler contractive mappings*, Fixed Point Theory and Appl. **94** (2013).
[17] M.S. Khan, M. Swaleh and S. Sessa, *Fixed point theorems by altering distancces between the points*, Bull. Aust. Math. Soc. **30** (1984), 1-9.
[18] M.A. Khamsi, N. Hussain, *KKM mappings in metric type spaces*, Nonlinear Anal. **73** (2010), 3123-3129.
[19] P. Kumam, H. Rahimi and G. Soleimani Rad, *The existence of fixed and periodic point theorems in cone metric type spaces*, J. Nonlinear Sci. Appl. **7 (4)** (2014), 255-263.
[20] P. Kumam and A. Roldán, *On existence and uniqueness of g-best proximity points under $(\varphi, \theta, \alpha, g)$-contractivity conditions and consequences*, Abstract and Applied Analysis, **Article ID 234027** (2014), 14 pages.
[21] J.J. Nieto and R. Rodríguez-López, *Contractive mapping theorems in partially ordered sets and applications to ordinary differential equations*, Order **22** (2005), 223-239.
[22] J.J. Nieto and R. Rodríguez-López, *Existence and uniqueness of fixed point in partially ordered sets and applications to ordinary differential equations*, Acta Math. Sin., (English Ser.), **23** (2007) 2205-2212.
[23] H. Rahimi and G. Soleimani Rad, Fixed point theory in various spaces, Lambert Academic Publishing, Germany, 2013.
[24] H. Rahimi and G. Soleimani Rad, *Some fixed point results in metric type space*, J. Basic. Appl. Sci. Res. **2 (9)** (2012), 9301-9308.
[25] H. Rahimi, P. Vetro, G. Soleimani Rad, *Some common fixed point results for weakly compatible mappings in cone metric type space*, Miskolc Math. Notes. **14 (1)** (2013), 233-243.
[26] A.C.M. Ran and M.C. Reurings, *A fixed point theorem in partially ordered sets and some applications to matrix equations*, Proc. Amer. Math. Soc. **132** (2004), 1435-1443.
[27] J.R. Roshan, V. Parvaneh, S. Sedghi, N. Shobkolaei and W. Shatanawi, *Common Fixed Points of Almost Generalized $(\psi, \varphi)_s-$Contractive Mappings in Ordered b−Metric Spaces*, Fixed Point Theory and Appl. **159** (2013).
[28] P. Salimi, A. Latif and N. Hussain, *Modified α-ψ-contractive mappings with applications*, Fixed Point Theory and Appl. **151** (2013).
[29] P. Salimi and P. Vetro, *A result of Suzuki type in partial G-metric spaces*, Acta Mathematica Scientia **34B(2)** (2014), 1-11.
[30] B. Samet, C. Vetro and P. Vetro, *Fixed point theorems for α-ψ-contractive type mappings*, Nonlinear Anal. **75** (2012), 2154-2165.

4

On the convergence of the homotopy analysis method to solve the system of partial differential equations

A. Fallahzadeh[a*], M. A. Fariborzi Araghi[b] and V. Fallahzadeh[c]

[a,b]*Department of Mathematics, Islamic Azad University, Central Tehran Branch, PO. Code 14168-94351, Iran* ;

[c]*Department of Mathematics, Islamic Azad University, Arac Branch, Iran.*

Abstract. One of the efficient and powerful schemes to solve linear and nonlinear equations is homotopy analysis method (HAM). In this work, we obtain the approximate solution of a system of partial differential equations (PDEs) by means of HAM. For this purpose, we develop the concept of HAM for a system of PDEs as a matrix form. Then, we prove the convergence theorem and apply the proposed method to find the approximate solution of some systems of PDEs. Also, we show the region of convergence by plotting the H-surface.

Keywords: Homotopy analysis method, System of partial differential equations, H-surface.

1. Introduction

Homotopy analysis method (HAM) was introduced by Liao in [8]. HAM is an efficient method for solving different kinds of partial differential equations. In recent years this method has been used to solve the various types of PDEs [1, 2, 5–7, 15]. The system of PDEs arise in mathematics, engineering and physical sciences. In [11], Sami Bataineh et al. used the HAM for solving the system of PDEs analytically, and in [3, 4], the homotopy perturbation method (HPM) was applied for solving the system of PDEs.

In this work, we introduce a development of the homotopy analysis method based on the matrix form of HAM and apply it to find the approximate solution of a given system of linear or nonlinear partial differential equations. Then, we prove the convergence of the

*Corresponding author.
E-mail address: amir_falah6@yahoo.com (A. Fallahzadeh).

proposed method and apply the method to find the solution of some systems of PDEs. Also, we introduce the H-surface to determine the region of convergence.
In section 2, we present the main idea of the work and introduce the HAM via matrix form. In section 3, we prove the convergence theorem of this method for a general form of system of partial differential equations. Section 4 contains some linear and nonlinear systems of PDEs which are solved based on the HAM in matrix form and the regions of convergence are shown.

2. Main idea

In this section, we develop the idea of HAM to solve a system of linear or nonlinear differential equations. We consider the following system,

$$N[U(X,t)] = 0, \tag{1}$$

$$N = \begin{pmatrix} N_1[U(X,t)] \\ \vdots \\ N_n[U(X,t)] \end{pmatrix} \quad , \quad U(X,t) = \begin{pmatrix} U_1(X,t) \\ \vdots \\ U_n(X,t) \end{pmatrix}$$

where N is the matrix of nonlinear operators, $X = (x, y, z)$ is the vector of variables and U is the vector of unknown functions. At first, we construct the zero-order deformation system as follows.

$$(I - Q)L[\phi(X,t;Q) - U^{(0)}(X,t)] = QHN[\phi(X,t;Q)], \tag{2}$$

where I is the identity matrix, $L = \begin{pmatrix} L_1 & & 0 \\ & \ddots & \\ 0 & & L_n \end{pmatrix}$ is an auxiliary linear operator matrix, $H = \begin{pmatrix} h_1 & & 0 \\ & \ddots & \\ 0 & & h_n \end{pmatrix}$ is an auxiliary parameter matrix, $\phi(X,t;Q)$ is the vector of unknown functions, $U^{(0)}(X,t)$ is the vector of initial guess and $Q = \begin{pmatrix} q_1 & & 0 \\ & \ddots & \\ 0 & & q_n \end{pmatrix}$, $0 \leqslant q_i \leqslant 1$, $1 \leqslant i \leqslant n$, is a diagonal matrix which denotes the embedding parameter matrix. It is obvious, when the q_i's, $1 \leqslant i \leqslant n$, increase from 0 to 1 or in other word, the embedding parameter matrix changes from $Q = \overline{0}$ to $Q = I$, the solution of system of equations (2) changes from $\phi(X,t;\overline{0}) = U^{(0)}(X,t)$ to $\phi(X,t;I) = U(X,t)$. Therefore, $\phi(X,t)$ varies from the initial guess $U^{(0)}(X,t)$ to the exact solution $U(X,t)$ of the system.

We consider $\phi(X,t;Q)$ in the following expansion in matrix form,

$$\phi(X,t;Q) = U^{(0)}(X,t) + \sum_{m=1}^{+\infty} Q^m U^{(m)}(X,t), \tag{3}$$

where

$$U^{(m)}(X,t) = \frac{1}{m!}\begin{pmatrix} \frac{\partial^m \phi_1(X,t,q_1)}{\partial q_1^m}|_{q_1=0} \\ \vdots \\ \frac{\partial^m \phi_n(X,t,q_n)}{\partial q_n^m}|_{q_n=0}. \end{pmatrix}. \tag{4}$$

The convergence of the vector series (3) depends upon the auxiliary parameter matrix H, if it is convergent at $Q = I$, we have

$$U(X,t) = U^{(0)}(X,t) + \sum_{m=1}^{+\infty} U^{(m)}(X,t). \tag{5}$$

Now, we define the vector,

$$\vec{U}_k = \{U^{(0)}(X,t), \ldots, U^{(k)}(X,t)\}, \tag{6}$$

where,

$$U^{(i)} = \begin{pmatrix} U_1^{(i)}(X,t) \\ \vdots \\ U_n^{(i)}(X,t) \end{pmatrix} \quad , \quad i = 0, \ldots, k. \tag{7}$$

Differentiating the zero-order system (2) m times with respect to the diagonal elements of the embedding parameter matrix Q and setting $Q = \bar{0}$ and finally dividing them by $m!$, we have the so-called mth-order deformation system as follows,

$$L[U^{(m)}(X,t) - \chi_m U^{(m-1)}(X,t)] = HR_m(\vec{U}_{m-1}), \tag{8}$$

where,

$$R_m(\vec{U}_{m-1}) = \frac{1}{(m-1)!}\begin{pmatrix} \frac{\partial^{m-1} N_1[\phi(X,t,Q)]}{\partial q_1^{m-1}}|_{Q=\bar{0}} \\ \vdots \\ \frac{\partial^{m-1} N_n[\phi(X,t,Q)]}{\partial q_n^{m-1}}|_{Q=\bar{0}} \end{pmatrix} \quad , \quad \chi_m = \begin{cases} \bar{0}, & m \leqslant 1, \\ I, & m > 1. \end{cases} \tag{9}$$

It should be emphasized that $U^{(m)}(X,t)$ for $m \geqslant 1$ is convergent by the linear system (8) with boundry conditions that comes from the original system.

3. Convergence of the HAM for system of PDEs

In this case, we consider the following general form of the system of partial differential equations,

$$ADS + BT + B'T' = C, \tag{10}$$

where A is the 3×3 coefficients matrix, D is the 3×3 differential operator matrix that

$$[D]_{ij} = \frac{\partial \alpha_{ij}^{(1)}}{\partial x^{\alpha_{ij}^{(1)}}} + \frac{\partial \alpha_{ij}^{(2)}}{\partial y^{\alpha_{ij}^{(2)}}} + \frac{\partial \alpha_{ij}^{(3)}}{\partial z^{\alpha_{ij}^{(3)}}} + \frac{\partial \alpha_{ij}^{(4)}}{\partial t^{\alpha_{ij}^{(4)}}}$$

S is unknowns vector $\left(u\ v\ w\right)^T$. B and B' are the 3×9 and 3×7 coefficients matrices of functions respectively and T and T' are the following vectors:

$$T = \left(u\ v\ w\ u^2\ v^2\ w^2\ uv\ uw\ vw\right)^T$$

$$T' = \left(uv^2\ uw^2\ vu^2\ vw^2\ wu^2\ wv^2\ uvw\right)^T$$

and $C = \left(c_1(x,y,z,t)\ c_2(x,y,z,t)\ c_3(x,y,z,t)\right)^T$ is the 3×1 vector of known functions.

In this part, we prove a theorem for convergence of the HAM for system of partial differential equations (10).

Theorem 3.1 If the series solution (5) of system (10) and also the series $\sum_{m=0}^{+\infty} DS_m$ where $S_m = \left(u_m\ v_m\ w_m\right)^T$ are convergent then the series (5) converges to the exact solution of the system (10).

Proof. Let:

$$S = \sum_{m=0}^{+\infty} S_m,$$

where

$$\lim_{m \to +\infty} S_m = \vec{0}. \tag{11}$$

We write

$$\sum_{m=1}^{n} [S_m - \chi_m S_{m-1}] = S_1 + (S_2 - S_1) + (S_3 - S_2) + \cdots + (S_n - S_{n-1}) = S_n.$$

Using (11), we have,

$$\sum_{m=1}^{+\infty} [u_m - \chi_m S_{m-1}] = \lim_{n \to +\infty} S_n = \vec{0}.$$

According to the definition of the operator L, we can write

$$\sum_{m=1}^{+\infty} L[S_m - \chi_m S_{m-1}] = L \sum_{m=1}^{+\infty} [S_m - \chi_m S_{m-1}] = \overrightarrow{0}.$$

From above expression and equation (8), we obtain

$$\sum_{m=1}^{+\infty} L[S_m - \chi_m S_{m-1}] = H \sum_{m=1}^{+\infty} [R_m(\overrightarrow{S}_{m-1})].$$

Since $H \neq \overline{0}$, we have

$$\sum_{m=1}^{+\infty} [R_m(\overrightarrow{S}_{m-1})] = \overrightarrow{0}. \tag{12}$$

From (12), it holds

$$\sum_{m=1}^{+\infty} R_m(\overrightarrow{S}_{m-1}) = \sum_{m=1}^{+\infty} ADS_{m-1}$$
$$+B \sum_{m=1}^{+\infty} \begin{pmatrix} \left(u_{m-1}\ v_{m-1}\ w_{m-1}\right)^T \\ \sum_{i=0}^{m-1} \left(u_{m-1}u_{m-1-i}\ v_{m-1}v_{m-1-i}\ w_{m-1}w_{m-1-i}\ u_{m-1}v_{m-1-i}\ u_{m-1}w_{m-1-i}\ v_{m-1}w_{m-1-i}\right)^T \end{pmatrix}$$
$$+B' \sum_{m=1}^{+\infty} \sum_{i=0}^{m-1} \sum_{k=0}^{m-i-1} \begin{pmatrix} \left(u_i v_k v_{m-i-k-1}\ u_i w_k w_{m-i-k-1}\ v_i u_k u_{m-i-k-1}\right)^T \\ \left(v_i w_k w_{m-i-k-1}\ w_i u_k u_{m-i-k-1}\ w_i v_k v_{m-i-k-1}\ u_i v_k w_{m-i-k-1}\right)^T \end{pmatrix}$$
$$-(I - \chi_m)C = \overrightarrow{0}.$$

We consider the following element from previous matrix equation. The similar manipulations can be done for other elements.

$$\begin{aligned}
&\sum_{m=1}^{+\infty}\sum_{i=0}^{m-1}\sum_{k=0}^{m-i-1} u_i v_k w_{m-i-k-1} \\
=&\sum_{i=0}^{+\infty}\sum_{m=i+1}^{m-1}\sum_{k=0}^{m-i-1} u_i v_k w_{m-i-k-1} \\
=&\sum_{i=0}^{+\infty} u_i \sum_{m=i+1}^{+\infty}\sum_{k=0}^{m-i-1} v_k w_{m-i-k-1} \\
=&\sum_{i=0}^{+\infty} u_i \sum_{m=1}^{+\infty}\sum_{k=0}^{m-1} v_k w_{m-k-1} \\
=&\sum_{i=0}^{+\infty} u_i \sum_{k=0}^{+\infty}\sum_{m=k+1}^{+\infty} v_k w_{m-k-1} \\
=&\sum_{i=0}^{+\infty} u_i \sum_{k=0}^{+\infty} v_k \sum_{m=0}^{+\infty} w_m.
\end{aligned}$$

In addition, we have,

$$\sum_{m=1}^{+\infty} ADS_{m-1} = AD\sum_{m=0}^{+\infty} S_m,$$

$$\sum_{m=1}^{+\infty}(I-\chi_m)C = C.$$

Therefore

$$ADS + BT + B'T' - C = \overrightarrow{0}. \tag{13}$$

■

4. Test Examples

In this part, we consider three sample systems of PDEs and apply the matrix form of HAM mentioned in previous section to solve these systems. The results in the tables have been provided by MAPLE.

***Example* 4.1** We consider the following linear system of PDEs:

$$\begin{aligned} u_{tt} + v_x + 2u = 0,\\ u_{xx} + v_t + 2u = 0, \end{aligned} \tag{14}$$

with the initial conditions:

$$\begin{aligned} u(x,0) &= \sin(x), \\ u_t(x,0) &= \cos(x), \\ v(x,0) &= \cos(x). \end{aligned} \tag{15}$$

The exact solution of this system is:

$$S(x,t) = \begin{pmatrix} \sin(x+t) \\ \cos(x+t) \end{pmatrix}. \tag{16}$$

We see:

$$\begin{aligned} N[S(x,t)] &= \begin{pmatrix} N_1[S(x,t)] \\ N_2[S(x,t)] \end{pmatrix} = \begin{pmatrix} u_{tt} + v_x + 2u \\ u_{xx} + v_t + 2u \end{pmatrix}, \\ S(x,t) &= \begin{pmatrix} u(x,t) \\ v(x,t) \end{pmatrix}. \end{aligned} \tag{17}$$

To solve the system (14) by means of HAM, we have:

$$N[\phi(x,t,Q)] = \begin{pmatrix} N_1[\phi(x,t,Q)] \\ N_2[\phi(x,t,Q)] \end{pmatrix} = \begin{pmatrix} \frac{\partial^2 \phi_1(x,t,q_1)}{\partial t^2} + \frac{\partial \phi_2(x,t,q_2)}{\partial x} + 2\phi_1(x,t,q_1) \\ \frac{\partial^2 \phi_1(x,t,q_1)}{\partial x^2} + \frac{\partial \phi_2(x,t,q_2)}{\partial t} + 2\phi_1(x,t,q_1) \end{pmatrix}, \tag{18}$$

where $Q = \left(\begin{smallmatrix} q_1 & 0 \\ 0 & q_2 \end{smallmatrix}\right)$ and the linear matrix operator

$$L[\phi(x,t,Q)] = \begin{pmatrix} \frac{\partial^2}{\partial t^2} & 0 \\ 0 & \frac{\partial}{\partial t} \end{pmatrix} \begin{pmatrix} \phi_1(x,t,q_1) \\ \phi_2(x,t,q_2) \end{pmatrix}, \tag{19}$$

with the property

$$L \begin{pmatrix} c_1(x) + c_2(x)t \\ c_3(x) \end{pmatrix} = 0, \tag{20}$$

where $c_1(x), c_2(x)$ and $c_3(x)$ are the integration constants. By using (8) under the initial conditions, we have:

$$R_m(\vec{u}_{m-1}) = \begin{pmatrix} \frac{\partial^2 u_{m-1}(x,t)}{\partial t^2} + \frac{\partial v_{m-1}(x,t)}{\partial x} + 2u_{m-1}(x,t) \\ \frac{\partial^2 u_{m-1}(x,t)}{\partial x^2} + \frac{\partial v_{m-1}(x,t)}{\partial t} + 2u_{m-1}(x,t) \end{pmatrix}. \tag{21}$$

The solution of the mth-order deformation system (8) for $m \geqslant 1$ becomes:

$$
\begin{aligned}
S_m(x,t) &= \chi_m S_{m-1}(x,t) + HL^{-1}R_m(\overrightarrow{S}_{m-1}) \\
&= \chi_m S_{m-1}(x,t) + \begin{pmatrix} h_1 & 0 \\ 0 & h_2 \end{pmatrix} \begin{pmatrix} \int\int dt & 0 \\ 0 & \int dt \end{pmatrix} \\
&\times \begin{pmatrix} \frac{\partial^2 u_{m-1}(x,t)}{\partial t^2} + \frac{\partial v_{m-1}(x,t)}{\partial x} + 2u_{m-1}(x,t) \\ \frac{\partial^2 u_{m-1}(x,t)}{\partial x^2} + \frac{\partial v_{m-1}(x,t)}{\partial t} + 2u_{m-1}(x,t) \end{pmatrix}.
\end{aligned} \tag{22}
$$

We choose the initial approximation as:

$$
S_0(x,t) = \begin{pmatrix} u_0 \\ v_0 \end{pmatrix} = \begin{pmatrix} \sin(x) + t\cos(x) \\ \cos(x) \end{pmatrix}. \tag{23}
$$

Applying (22) for $m \geqslant 1$, we have,

$$
S_1 = \begin{pmatrix} \frac{1}{6}h_1 t^2(3\sin(x) + 2t\cos(x)) \\ \frac{1}{2}h_2 t(2\sin(x) + t\cos(x)) \end{pmatrix}
$$

$$
S_2 = \begin{pmatrix} \frac{1}{120}(60\sin(x) + 40t\cos(x) + 60h_1\sin(x) + 40h_1 t\cos(x) + \\ 20h_2 t\cos(x) - 5h_2 t^2\sin(x) + 10h_1 t^2\sin(x) + 4h_1 t^3\cos(x)) \\ \\ \frac{1}{12}h_2 t(12\sin(x) + 6t\cos(x) + 12h_2\sin(x) + 6h_2 t\cos(x) + \\ 2h_1 t^2\sin(x) + h_1 t^3\cos(x)) \end{pmatrix}
$$

$$\vdots$$

In general, for $H = -I$ we have,

$$
\begin{aligned}
S_0 + S_1 + S_2 + \cdots &= \begin{pmatrix} (1 - \frac{t^2}{2!} + \frac{t^4}{4!} - \cdots)\sin(x) + (t - \frac{t^3}{3!} + \frac{t^5}{5!} - \cdots)\cos(x) \\ (1 - \frac{t^2}{2!} + \frac{t^4}{4!} - \cdots)\cos(x) - (t - \frac{t^3}{3!} + \frac{t^5}{5!} - \cdots)\sin(x) \end{pmatrix} \\
&= \begin{pmatrix} \sin(x+t) \\ \cos(x+t) \end{pmatrix}.
\end{aligned}
$$

We can see the convergence of this method at two points $(x_1, t_1) = (0.5, 1)$ and $(x_2, t_2) = (3, 1.5)$ for $H = -I$ in Table 1.

Table 1

	$(0.5, 1)$	$(3, 1.5)$
n=2	$\begin{pmatrix} 1.0202604 \\ 0.11240188 \end{pmatrix}$	$\begin{pmatrix} -1.1665823 \\ -0.42620403 \end{pmatrix}$
n=4	$\begin{pmatrix} 0.99747196 \\ 0.07079598 \end{pmatrix}$	$\begin{pmatrix} -0.97955171 \\ -0.21567996 \end{pmatrix}$
n=6	$\begin{pmatrix} 0.99749480 \\ 0.07073681 \end{pmatrix}$	$\begin{pmatrix} -0.97751894 \\ -0.21079367 \end{pmatrix}$
n=8	$\begin{pmatrix} 0.99749498 \\ 0.07073720 \end{pmatrix}$	$\begin{pmatrix} -0.97752994 \\ -0.21079568 \end{pmatrix}$
n=10	$\begin{pmatrix} 0.99749498 \\ 0.07073720 \end{pmatrix}$	$\begin{pmatrix} -0.97752999 \\ -0.21079578 \end{pmatrix}$

We present the H-surfaces of the example 4.1 which is a two-dimensional system to see the convergent region. In [8] the concept of h-curve was discussed to show the region of convergence in the HAM to solve a given linear or nonlinear equation, where h is the auxilary parameter. These curves convert to surfaces for a set of equations like a system of PDEs. By ploting the H-surface, it is easy to discover the valid region of the HAM and the optimal values of the entries of matrix H, which corresponds to the hyperplane parallel to the hyperplane that is made by the $h_1, \ldots, h_n$.
In figures 1, 2 and 3, we plot the 5-approximation of u, v, v_t, v_x, u_{tt} and u_{xx} when $x = 0.5$ and $t = 1$. In this case, when $H = -I$ we obtain the Taylor series of the exact solution of the system. In this figures, we can see the region of convergence of the u, v, v_t, v_x, u_{tt} and u_{xx} is $[-2, 1] \times [-2, 1]$.

***Example* 4.2** We consider the following linear system of PDEs with variable coefficients:

$$\begin{aligned} u_t + v_{zz} - w_{xx} - u &= 0, \\ u_{yy} + v_{tt} - e^x w_{xx} - v &= 0, \\ e^y u_{yy} + v_{xx} - w_{ttt} - w &= 0, \end{aligned} \tag{24}$$

with the initial conditions:

$$\begin{aligned} u(x, y, z, 0) &= y + z, \\ v(x, y, z, 0) &= y + x, \\ v_t(x, y, z, 0) &= y + x, \\ w(x, y, z, 0) &= x + y, \\ w_t(x, y, z, 0) &= x + y, \\ w_{tt}(x, y, z, 0) &= x + y. \end{aligned} \tag{25}$$

The exact solution of this system is:

$$S(x, y, z, t) = \begin{pmatrix} (y+z)e^t \\ (x+z)e^t \\ (x+y)e^t \end{pmatrix}. \tag{26}$$

Similar to the example 4.1, for $m \geqslant 1$ we have:

$$\begin{aligned} S_m(x, y, z, t) &= \chi_m S_{m-1}(x, y, z, t) + HL^{-1}R_m(\overrightarrow{S}_{m-1}) = \chi_m S_{m-1}(x, y, z, t) \\ &+ \begin{pmatrix} h_1 & 0 & 0 \\ 0 & h_2 & 0 \\ 0 & 0 & h_3 \end{pmatrix} \times \begin{pmatrix} \int dt & 0 & 0 \\ 0 & \int\int dt & 0 \\ 0 & 0 & \int\int\int dt \end{pmatrix} \\ &\times \begin{pmatrix} \frac{\partial u_{m-1}(x,y,z,t)}{\partial t} + \frac{\partial^2 v_{m-1}(x,y,z,t)}{\partial z^2} - \frac{\partial^2 w_{m-1}(x,y,z,t)}{\partial x^2} - u_{m-1}(x, y, z, t) \\ \frac{\partial^2 u_{m-1}(x,y,z,t)}{\partial y^2} + \frac{\partial^2 v_{m-1}(x,y,z,t)}{\partial t^2} - e^x \frac{\partial^2 w_{m-1}(x,y,z,t)}{\partial x^2} - v_{m-1}(x, y, z, t) \\ e^y \frac{\partial^2 u_{m-1}(x,y,z,t)}{\partial y^2} + \frac{\partial^2 v_{m-1}(x,y,z,t)}{\partial x^2} - \frac{\partial^3 w_{m-1}(x,y,z,t)}{\partial t^3} - w_{m-1}(x, y, z, t) \end{pmatrix}. \end{aligned} \tag{27}$$

where

$$L = \begin{pmatrix} \frac{\partial}{\partial t} & 0 & 0 \\ 0 & \frac{\partial^2}{\partial t^2} & 0 \\ 0 & 0 & \frac{\partial^3}{\partial t^3} \end{pmatrix}, \tag{28}$$

with the property

$$L \begin{pmatrix} c_1(x,y,z) \\ c_2(x,y,z) + c_3(x,y,z)t \\ c_4(x,y,z) + c_5(x,y,z)t + c_6(x,y,z)t^2 \end{pmatrix} = 0, \tag{29}$$

where $c_1(x,y,z), \ldots, c_6(x,y,z)$ are integration constant.

We choose the initial approximation as:

$$S_0(x,y,z,t) = \begin{pmatrix} u_0 \\ v_0 \\ w_0 \end{pmatrix} = \begin{pmatrix} y+z \\ (x+z)(1+t) \\ (x+y)(1+t+\frac{1}{2}t^2) \end{pmatrix}. \tag{30}$$

Applying (27) for $m \geqslant 1$ and $H = -I$ we have,

$$S_1 = \begin{pmatrix} t(y+z) \\ (\frac{t^2}{2!} + \frac{t^3}{3!})(x+z) \\ (\frac{t^3}{3!} + \frac{t^4}{4!} + \frac{t^5}{5!})(x+y) \end{pmatrix}$$

$$S_2 = \begin{pmatrix} \frac{t^2}{2!}(y+z) \\ (\frac{t^4}{4!} + \frac{t^5}{5!})(x+z) \\ (\frac{t^6}{6!} + \frac{t^7}{7!} + \frac{t^8}{8!})(x+y) \end{pmatrix}$$

$$\vdots$$

In general, we have,

$$S_0 + S_1 + S_2 + \cdots = \begin{pmatrix} (1+t+\frac{t^2}{2}+\cdots)(x+z) \\ (1+t+\frac{t^2}{2}+\cdots)(y+z) \\ (1+t+\frac{t^2}{2}+\cdots)(x+z) \end{pmatrix} = \begin{pmatrix} e^t(y+z) \\ e^t(x+z) \\ e^t(x+y) \end{pmatrix}.$$

We can see the convergence of this method at two points $(x_1, y_1, z_1, t_1) = (1, 2, 3, 0.5)$ and $(x_2, y_2, z_2, t_2) = (2, 3, 1, 3)$ for $H = -I$ in Table 2.

Table 2

	$(1,2,3,0.5)$	$(2,3,1,3)$
n=2	$\begin{pmatrix}8.1250000\\6.5947917\\4.9461639\end{pmatrix}$	$\begin{pmatrix}34.00000000\\55.20000000\\100.0457589\end{pmatrix}$
n=4	$\begin{pmatrix}8.2421875\\6.5948851\\4.9461639\end{pmatrix}$	$\begin{pmatrix}65.50000000\\60.19017857\\100.4276173\end{pmatrix}$
n=6	$\begin{pmatrix}8.2435981\\6.5948851\\4.9461639\end{pmatrix}$	$\begin{pmatrix}77.65000000\\60.25640578\\100.4276846\end{pmatrix}$
n=8	$\begin{pmatrix}8.2436064\\6.5948851\\4.9461639\end{pmatrix}$	$\begin{pmatrix}80.03660714\\60.25661055\\100.4276846\end{pmatrix}$
n=10	$\begin{pmatrix}8.2436064\\6.5948851\\4.9461639\end{pmatrix}$	$\begin{pmatrix}80.31866071\\60.25661077\\100.4276846\end{pmatrix}$

***Example* 4.3** We consider the nonlinear system of ordinary differential equations as follows:

$$\begin{aligned} u_{tt} - 2uv^2 &= 0, \\ v_t - uv &= 0, \end{aligned} \tag{31}$$

with the initial conditions:

$$\begin{aligned} u(0) &= 0, \\ u_t(0) &= 1, \\ v(0) &= 1. \end{aligned} \tag{32}$$

The exact solution of this system is:

$$S(t) = \begin{pmatrix} \tan(t) \\ \sec(t) \end{pmatrix}. \tag{33}$$

In this example, we have:

$$N[\phi(t,Q)] = \begin{pmatrix} N_1[\phi(t,Q)] \\ N_2[\phi(t,Q)] \end{pmatrix} = \begin{pmatrix} \frac{\partial^2\phi_1(t,q_1)}{\partial t^2} - 2\phi_1(t,q_1)\phi_2^2(t,q_2) \\ \frac{\partial\phi_2(t,q_2)}{\partial t} - \phi_1(t,q_1)\phi_2(t,q_2) \end{pmatrix}, \tag{34}$$

where $Q = \begin{pmatrix} q_1 & 0 \\ 0 & q_2 \end{pmatrix}$ and the linear matrix operator

$$L[\phi(t,Q)] = \begin{pmatrix} \frac{\partial^2}{\partial t^2} & 0 \\ 0 & \frac{\partial}{\partial t} \end{pmatrix} \begin{pmatrix} \phi_1(x,t,q_1) \\ \phi_2(x,t,q_2) \end{pmatrix}, \tag{35}$$

with the property

$$L\begin{pmatrix} c_1 + c_2 t \\ c_3 \end{pmatrix} = 0, \tag{36}$$

where c_1, c_2 and c_3 are the integration constant. The solution of the mth-order deformation system (8) for $m \geqslant 1$ becomes:

$$S_m(t) = \chi_m S_{m-1}(t) + HL^{-1}R_m(\vec{S}_{m-1}) = \chi_m S_{m-1}(t) + \begin{pmatrix} h_1 & 0 \\ 0 & h_2 \end{pmatrix} \begin{pmatrix} \int\int dt\, dt & 0 \\ 0 & \int dt \end{pmatrix} \times$$

$$\begin{pmatrix} \frac{\partial^2 S_{m-1}(t)}{\partial t^2} - 2\sum_{m=1}^{+\infty}\sum_{i=0}^{m-1}\sum_{k=0}^{m-i-1} u_i(t)v_k(t)v_{m-i-k-1}(t) \\ \frac{\partial u_{m-1}(t)}{\partial t} - \sum_{m=1}^{+\infty}\sum_{i=0}^{m-1} u_i(t)v_{m-i-1}(t) \end{pmatrix}. \tag{37}$$

We choose the initial approximation as:

$$S_0(t) = \begin{pmatrix} u_0 \\ v_0 \end{pmatrix} = \begin{pmatrix} t \\ 1 \end{pmatrix}. \tag{38}$$

Applying (37), we can see the results of the method at two points $t_1 = 0.5$ and $t_2 = 1$ for $H = -I$ in Table 3.

Table 3

	$t_1 = 0.5$	$t_2 = 1$
n=2	$\begin{pmatrix} 0.545833334 \\ 1.138020833 \end{pmatrix}$	$\begin{pmatrix} 1.46666667 \\ 1.708333333 \end{pmatrix}$
n=4	$\begin{pmatrix} 0.5462976742 \\ 1.139478798 \end{pmatrix}$	$\begin{pmatrix} 1.542504409 \\ 1.827405754 \end{pmatrix}$
n=6	$\begin{pmatrix} 0.5463024405 \\ 1.139493772 \end{pmatrix}$	$\begin{pmatrix} 1.554959773 \\ 1.846970484 \end{pmatrix}$
n=8	$\begin{pmatrix} 0.5463024894 \\ 1.139493926 \end{pmatrix}$	$\begin{pmatrix} 1.557005635 \\ 1.850184116 \end{pmatrix}$
n=10	$\begin{pmatrix} 0.5463024899 \\ 1.139493927 \end{pmatrix}$	$\begin{pmatrix} 1.557341679 \\ 1.850711974 \end{pmatrix}$

5. Conclusion

In this work, the homotopy analysis method was introduced in matrix form and applied to obtain the approximate solution of a linear or nonlinear system of partial differential equations. For this purpose, a convergence theorem was proved and some sample systems of PDEs were solved and the convergence of the HAM was discussed in each system. Also, the H-surface was introuced to illustrate the region of convergence in the HAM. Therefore, the HAM is able to solve the system of PDE via the matrix form and the convergence of method is guranteed.

Aknowledgements

The authors are thankful to the Islamic Azad University, central Tehran branch for their supports during this research and also the anonymous referee for careful reading and suggestion to improve the quality of this work.

References

[1] S. Abbasbandy, Homotopy analysis method for the Kawahara equation, Nonlinear Analysis: Real World Applications 11 (2010) 307-312.

[2] S. Abbasbandy, Solitary wave solutions to the modified form of CamassaHolm equation by means of the homotopy analysis method, Chaos, Solitons and Fractals 39 (2009) 428-435.

[3] J. Biazar, M. Eslami, A new homotopy perturbation method for solving system of partial differential equations, Computers and Mathematics with Applications 62 (2011) 225-234.

[4] J. Biazar, M. Eslami, H. Ghazvini, Homotopy perturbation method for system of partial differential equations, International Journal of Nonlinear Sciences and Numerical simulations 8 (3) (2007) 411-416.

[5] M.A. Fariborzi Araghi, A. Fallahzadeh, On the convergence of the Homotopy Analysis method for solving the Schrodinger Equation, Journal of Basic and Applied Scientific Research 2(6) (2012) 6076-6083.

[6] M.A. Fariborzi Araghi, A. Fallahzadeh, Explicit series solution of Boussinesq equation by homotopy analysis method, Journal of American Science, 8(11) (2012).

[7] T. Hayat, M. Khan, Homotopy solutions for a generalized second-grade fluid past a porous plate. Nonlinear Dyn 42 (2005) 395-405.

[8] S.J. Liao, Beyond pertubation: Introduction to the homotopy Analysis Method, Chapman and Hall/CRC Press, Boca Raton, (2003).

[9] S.J. Liao, Notes on the homotopy analysis method: some definitions and theorems, Communication in Nonlinear Science and Numnerical Simulation, 14 (2009) 983-997.

[10] P. Roul, P. Meyer, Numerical solution of system of nonlinear integro-differential equation by Homotopy perturbation method, Applied Mathematical Modelling 35 (2011) 4234-4242.

[11] A. Sami Bataineh, M.S.M. Noorani, I.Hashim, Approximation analytical solution of system of PDEs by homotopy analysis method, Computers and Mathematics with Applications 55 (2008) 2913-2923.

[12] F. Wang, Y. An, Nonnegative doubly periodic solution for nonlinear teleghraph system, J.math.Anal.Appl. 338 (2008) 91-100.

[13] A.M. Wazwaz, The variational iteration method for solving linear and nonlinear system of PDEs, Comput, Math, Appl 54 (2007) 895-902.

[14] W. Wu, Ch. Liou, Out put regulation of two-time-scale hyperbolic PDE systems, Journal of Process control 11 (2001) 637-647.

[15] W. Wu, S. Liao, Solving solitary waves with discontinuity by means of the homotopy analysis method. Chaos, Solitons & Fractals, 26 (2005) 177-185.

[16] E. Yusufoglu, An improvment to homotopy perturbation method for solving system of linear equations, Computers and Mathematic with Applications 58 (2009) 2231-2235.

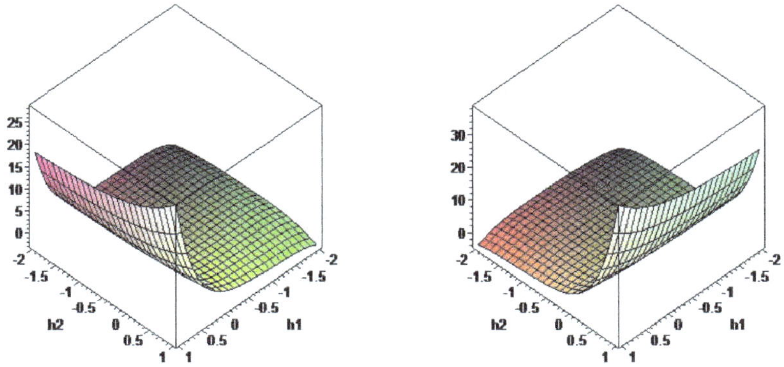

(a) H-surface of u (b) H-surface of v

Figure 1. H-Surfaces of u and v in example 4.1

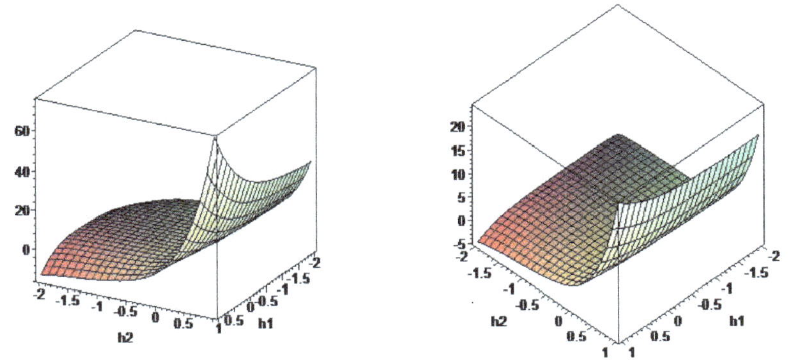

(a) H-surface v_t (b) H-surface v_x

Figure 2. H-Surfaces of v_t and v_x in example 4.1

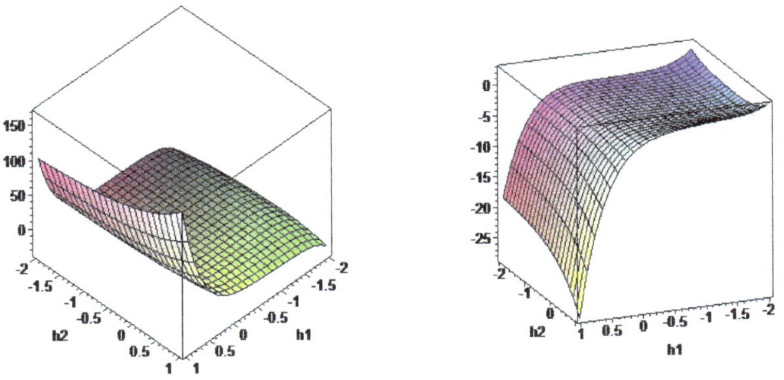

(a) H-surface u_{tt} (b) H-surface u_{xx}

Figure 3. H-Surfaces of u_{tt} and u_{xx} in example 4.1

5

On weakly eR-open functions

M. Özkoç[a*], B. S. Ayhan[a]

[a] *Department of Mathematics, Faculty of Science Muğla Sıtkı Koçman University, Menteşe-Muğla 48000 Turkey.*

Abstract. The main goal of this paper is to introduce and study a new class of function via the notions of e-θ-open sets and e-θ-closure operator which are defined by Özkoç and Aslım [10] called weakly eR-open functions and e-θ-open functions. Moreover, we investigate not only some of their basic properties but also their relationships with other types of already existing topological functions.

Keywords: e-Closed set, e-θ-open set, weakly eR-open function, e-θ-open function.

1. Introduction and Preliminaries

Throughout the present paper, X and Y always mean topological spaces on which no separation axioms are assumed unless explicitly stated. Let X be a topological space and A a subset of X. The closure and the interior of A are denoted by $cl(A)$ and $int(A)$, respectively. The family of all closed sets of X is denoted $C(X)$. A subset A is said to be regular open [12] (resp. regular closed [12]) if $A = int(cl(A))$ (resp. $A = cl(int(A))$). A point $x \in X$ is said to be δ-cluster point [13] of A if $int(cl(U)) \cap A \neq \emptyset$ for each open neigbourhood U of x. The set of all δ-cluster points of A is called the δ-closure [13] of A and is denoted by $cl_\delta(A)$. If $A = cl_\delta(A)$, then A is called δ-closed [13], and the complement of a δ-closed set is called δ-open [13]. A subset A is called semiopen [5] (resp. b-open [1], e-open [4], preopen [7], α-open [8]) if $A \subset cl(int(A))$ (resp. $A \subset cl(int(A)) \cup int(cl(A))$, $A \subset cl(int_\delta(A)) \cup int(cl_\delta(A))$, $A \subset int(cl(A))$,

*Corresponding author.
E-mail address: murad.ozkoc@mu.edu.tr (M. Özkoç).

$A \subset int(cl(int(A)))$. The complement of a semiopen (resp. b-open, e-open, preopen, α-open) set is called semiclosed [5](resp. b-closed [1], e-closed [4], preclosed [7], α-closed [8]). The intersection of all e-closed sets of X containing A is called the e-closure [4] of A and is denoted by $e\text{-}cl(A)$. The union of all e-open sets of X contained in A is called the e-interior [4] of A and is denoted by $e\text{-}int(A)$. A subset A is said to be e-regular [10] if it is e-open and e-closed.

A point x of X is called a b-θ-cluster [11] (e-θ-cluster [10], θ-cluster [13]) point of A if $bcl(U) \cap A \neq \emptyset$ ($e\text{-}cl(U) \cap A \neq \emptyset$, $cl(U) \cap A \neq \emptyset$) for every b-open (e-open, open) set U of X containing x, respectively. The set of all b-θ-cluster (e-θ-cluster, θ-cluster) points of A is called the b-θ-closure [11] (e-θ-closure [10], θ-closure [13]) of A and is denoted by $bcl_\theta(A)$ ($e\text{-}cl_\theta(A)$, $cl_\theta(A)$), respectively. A subset A is said to be b-θ-closed [11] (e-θ-closed [10], θ-closed [13]) if $A = bcl_\theta(A)$ ($A = e\text{-}cl_\theta(A)$, $A = cl_\theta(A)$), respectively. The complement of a b-θ-closed (e-θ-closed, θ-closed) set is called a b-θ-open [11] (e-θ-open [10], θ-open [13]) set. A point x of X said to be a b-θ-interior [11] (e-θ-interior [10], θ-interior [13]) point of a subset A, denoted by $bint_\theta(A)$ ($e\text{-}int_\theta(A)$, $int_\theta(A)$), if there exists a b-regular (e-regular, regular) set U of X containing x such that $U \subset A$, respectively. The family of all e-open (resp. e-closed, e-regular, b-θ-open, e-θ-open, b-θ-closed, e-θ-closed) subsets of X is denoted by $eO(X)$ (resp. $eC(X)$, $eR(X)$, $B\theta O(X)$, $e\theta O(X)$, $B\theta C(X)$, $e\theta C(X)$). The family of all e-open (e-closed, e-regular, b-θ-open, e-θ-open, b-θ-closed, e-θ-closed) sets of X containing a point x of X is denoted by $eO(X,x)$ (resp. $eC(X,x)$, $eR(X,x)$, $B\theta O(X,x)$, $e\theta O(X,x)$, $B\theta C(X)$, $e\theta C(X,x)$). Also it is noted in [10] that

$$e\text{-regular} \Rightarrow e\text{-}\theta\text{-open} \Rightarrow e\text{-open}.$$

We shall use the well-known accepted language almost in the whole of the article.

Definition 1.1 A function $f : (X,\tau) \to (Y,\sigma)$ is called:
(a) contra e-θ-open if $f(U)$ is e-θ-closed in Y for each open set U of X.
(b) contra e-θ-closed if $f(U)$ is e-θ-open in Y for each closed set U of X.
(c) strongly continuous [6] if for every subset A of X, $f(cl(A)) \subset f(A)$.
(d) weakly BR-open [2] if $f(U) \subset bint_\theta(f(cl(U)))$ for each open set U of X.

2. Weakly eR-open Functions

In this section, we define the concept of weakly eR-open and investigate some basic properties of them.

Definition 2.1 A function $f : X \to Y$ is said to be weakly eR-open if $f(U) \subset e\text{-}int_\theta(f(cl(U)))$ for each open set U of X.

Definition 2.2 A function $f : X \to Y$ is said to be e-θ-open if $f(U)$ is e-θ-open in Y for each open set U of X.

It is clear to see that every e-θ-open function is a weakly eR-open. However, a weakly eR-open function need not be e-θ-open as shown by the following example.

***Example* 2.3** Let $X = \{a,b,c,d\}$ and

$$\tau = \{\emptyset, X, \{a,d\}\} \quad \text{and} \quad \sigma = \{\emptyset, X, \{a\}, \{b\}, \{a,b\}, \{a,c\}, \{a,b,c\}, \{a,b,d\}\}.$$

The identity function $f : (X,\tau) \to (X,\sigma)$ is weakly eR-open, but it is not e-θ-open.

The notions of weakly eR-open function and weakly BR-open function are independent as shown by the following examples.

***Example* 2.4** Let $X = \{a, b, c, d, e\}$ and $\tau = \{\emptyset, X, \{a\}, \{c\}, \{a, c\}, \{c, d\}, \{a, c, d\}\}$. The identity function $f : (X, \tau) \to (X, \tau)$ is weakly eR-open, but it is not weakly BR-open.

***Example* 2.5** Let $X = \{a, b, c, d\}$ and $\tau = \{\emptyset, X, \{a\}, \{b\}, \{a, b\}, \{a, c\}, \{a, b, c\}, \{a, b, d\}\}$. $f = \{(a, d), (b, c), (c, b), (d, d)\}$ is weakly BR-open, but it is not weakly eR-open.

Lemma 2.6 [10] Let A be a subset of a space X. Then:
(1) $e\text{-}cl_\theta(A) = \cap\{V|(A \subset V)(V \in eR(X))\}$.
(2) $x \in e\text{-}cl_\theta(A)$ iff $A \cap U \neq \emptyset$ for each e-regular set U of X containing x.
(3) $e\text{-}cl_\theta(A)$ is e-θ-closed.
(4) Any intersections of e-θ-closed sets is e-θ-closed and any union of e-θ-open sets is e-θ-open.
(5) A is e-θ-open in X if and only if for each $x \in A$ there exists an e-regular set U containing x such that $x \in U \subset A$.

Theorem 2.7 Let $f : (X, \tau) \to (Y, \sigma)$ be a function. Then the following statements are equivalent:
(a) f is weakly eR-open,
(b) $f(int_\theta(A)) \subset e\text{-}int_\theta(f(A))$ for every subset A of X,
(c) $int_\theta(f^{-1}(B)) \subset f^{-1}(e\text{-}int_\theta(B))$ for every subset B of Y,
(d) $f^{-1}(e\text{-}cl_\theta(B)) \subset cl_\theta(f^{-1}(B))$ for every subset B of Y,
(e) $f(int(F)) \subset e\text{-}int_\theta(f(F))$ for each closed subset F of X,
(f) $f(int(cl(U))) \subset e\text{-}int_\theta(f(cl(U)))$ for each open subset U of X,
(g) $f(U) \subset e\text{-}int_\theta(f(cl(U)))$ for every regular open subset U of X,
(h) $f(U) \subset e\text{-}int_\theta(f(cl(U)))$ for every α-open subset U of X,
(i) For each $x \in X$ and each open set U of X containing x, there exists an e-θ-open set V of Y containing $f(x)$ such that $V \subset f(cl(U))$.

Proof. $(a) \Rightarrow (b)$: Let A be any subset of X and $x \in int_\theta(A)$.

$$\begin{aligned} x \in int_\theta(A) &\Rightarrow (\exists U \in \mathcal{U}(x))(x \in U \subset cl(U) \subset A) \\ &\left.\begin{aligned}\Rightarrow (\exists U \in \mathcal{U}(x))(f(x) \in f(U) \subset f(cl(U)) \subset f(A)) \\ f \text{ is weakly } eR\text{-open}\end{aligned}\right\} \Rightarrow \\ &\Rightarrow f(U) \subset e\text{-}int_\theta(f(cl(U))) \subset e\text{-}int_\theta(f(A)) \\ &\Rightarrow f(x) \in e\text{-}int_\theta(f(A)) \\ &\Rightarrow x \in f^{-1}(e\text{-}int_\theta(f(A))). \end{aligned}$$

$(b) \Rightarrow (c)$: Let B be any subset of Y.

$$\left.\begin{aligned} B \subset Y \Rightarrow f^{-1}(B) \subset X \\ (b) \end{aligned}\right\} \Rightarrow f\left(int_\theta(f^{-1}(B))\right) \subset e\text{-}int_\theta(f(f^{-1}(B))) \subset e\text{-}int_\theta(B)$$
$$\Rightarrow int_\theta(f^{-1}(B)) \subset f^{-1}(e\text{-}int_\theta(B)).$$

$(c) \Rightarrow (d)$: Let B be any subset of Y.

$$\left.\begin{array}{r} B \subset Y \Rightarrow Y \setminus B \subset Y \\ (c) \end{array}\right\} \begin{array}{l} \Rightarrow int_\theta(f^{-1}(Y \setminus B)) \subset f^{-1}(e\text{-}int_\theta(Y \setminus B)) \\ \Rightarrow int_\theta(X \setminus f^{-1}(B)) \subset f^{-1}(Y \setminus e\text{-}cl_\theta(B)) \\ \Rightarrow X \setminus cl_\theta(f^{-1}(B)) \subset X \setminus f^{-1}(e\text{-}cl_\theta(B)) \\ \Rightarrow f^{-1}(e\text{-}cl_\theta(B)) \subset cl_\theta(f^{-1}(B)). \end{array}$$

$(d) \Rightarrow (e)$: Let F be any closed set of X.

$$\left.\begin{array}{r} F \in C(X) \Rightarrow Y \setminus f(F) \subset Y \\ (d) \end{array}\right\} \Rightarrow$$
$$\Rightarrow f^{-1}(e\text{-}cl_\theta(Y \setminus f(F))) \subset cl_\theta(f^{-1}(Y \setminus f(F))) = cl_\theta(X \setminus f^{-1}(f(F))) \subset cl_\theta(X \setminus F)$$
$$\Rightarrow f^{-1}(Y \setminus e\text{-}int_\theta(f(F))) \subset cl_\theta(X \setminus F) = X \setminus int_\theta(F)$$
$$\left.\begin{array}{r} \Rightarrow X \setminus f^{-1}(e\text{-}int_\theta(f(F))) \subset X \setminus int_\theta(F) \\ F \in C(X) \Rightarrow int_\theta(F) = int(F) \end{array}\right\} \Rightarrow f\,(int(F)) \subset e\text{-}int_\theta(f(F)).$$

$(e) \Rightarrow (f)$, $(f) \Rightarrow (g)$: Obvious.
$(g) \Rightarrow (h)$: Let U be any α-open set of X.

$$\left.\begin{array}{r} U \in \alpha O(X) \Rightarrow (U \subset int(cl(int(U))))(int(cl(int(U))) \in RO(X)) \\ (g) \end{array}\right\} \Rightarrow$$
$$\begin{array}{l} \Rightarrow f(U) \subset f(int(cl(int(U)))) \subset e\text{-}int_\theta(f(cl(int(cl(int(U)))))) \\ \qquad = e\text{-}int_\theta(f(cl(int(U)))) \subset e\text{-}int_\theta(f(cl(U))). \end{array}$$

$(h) \Rightarrow (i)$: Straightforward.
$(i) \Rightarrow (a)$: Let U be an open set in X and $y \in f(U)$.

$$\left.\begin{array}{r} (U \in \tau)(y \in f(U)) \\ (i) \end{array}\right\} \Rightarrow \left.\begin{array}{r} (\exists V \in e\theta O(Y, y))(V \subset f(cl(U))) \\ y \in V \subset e\text{-}int_\theta(f(cl(U))) \end{array}\right\} \Rightarrow$$
$$\Rightarrow f(U) \subset e\text{-}int_\theta(f(cl(U))).$$ ■

Theorem 2.8 Let $f : (X, \tau) \to (Y, \sigma)$ be a bijective function. Then the following statements are equivalent:
(a) f is weakly eR-open,
(b) For each $x \in X$ and each open set U of X containing x, there exists an e-regular set V containing $f(x)$ such that $V \subset f(cl(U))$,
(c) $e\text{-}cl_\theta(f\,(int(cl(U)))) \subset f\,(cl(U))$ for each subset U of X,
(d) $e\text{-}cl_\theta(f\,(int(F))) \subset f\,(F)$ for each regular closed subset F of X,
(e) $e\text{-}cl_\theta(f\,(U)) \subset f\,(cl(U))$ for each open subset U of X,
(f) $e\text{-}cl_\theta(f\,(U)) \subset f\,(cl(U))$ for each preopen subset U of X,
(g) $f\,(U) \subset e\text{-}int_\theta(f\,(cl(U)))$ for each preopen subset U of X,
(h) $f^{-1}\,(e\text{-}cl_\theta(B)) \subset cl_\theta(f^{-1}\,(B))$ for each subset B of Y,
(i) $e\text{-}cl_\theta(f\,(U)) \subset f\,(cl_\theta(U))$ for each subset U of X,
(j) $e\text{-}cl_\theta(f\,(int(cl_\theta(U)))) \subset f\,(cl_\theta(U))$ for each subset U of X.

Proof. $(a) \Rightarrow (b)$: Let $x \in X$ and U be any open subset of X containing x.

$$\left.\begin{array}{r} x \in U \in \tau \\ (a) \end{array}\right\} \Rightarrow \left.\begin{array}{r} f(x) \in f\,(U) \subset e\text{-}int_\theta(f\,(cl(U))) \in e\theta O(Y, f(x)) \\ \text{Lemma 2.6(5)} \end{array}\right\} \Rightarrow$$
$$\Rightarrow (\exists V \in eR(Y, f(x)))(V \subset e\text{-}int_\theta(f\,(cl(U))) \subset f\,(cl(U))).$$

$(b) \Rightarrow (c)$: Let $x \in X$ and $U \subset X$.

$$\begin{aligned} f(x) \in Y \setminus f(cl(U)) = f(X \setminus cl(U)) &\Rightarrow x \in X \setminus cl(U) \\ &\Rightarrow (\exists G \in \mathcal{U}(x))(G \cap U = \emptyset) \\ &\left.\begin{aligned}&\Rightarrow (\exists G \in \mathcal{U}(x))(cl(G) \cap int(cl(U)) = \emptyset) \\ &(b)\end{aligned}\right\} \Rightarrow \\ &\Rightarrow (\exists V \in eR(Y, f(x)))(V \subset f(cl(G))) \\ &\Rightarrow (\exists V \in eR(Y, f(x)))(V \cap f(int(cl(U))) = \emptyset) \\ &\Rightarrow f(x) \notin e\text{-}cl_\theta(f(int(cl(U)))) \\ &\Rightarrow f(x) \in X \setminus e\text{-}cl_\theta(f(int(cl(U)))). \end{aligned}$$

$(c) \Rightarrow (d)$: Let F be any regular closed set of X.

$$\left.\begin{aligned} F \in RC(X) \Rightarrow e\text{-}cl_\theta(f(int(F))) = e\text{-}cl_\theta(f(int(cl(int(F))))) \\ (c) \end{aligned}\right\} \Rightarrow$$
$$\Rightarrow e\text{-}cl_\theta(f(int(F))) \subset f(cl(int(F))) = f(F).$$

$(d) \Rightarrow (e)$: Let U be any open subset of X.

$$\left.\begin{aligned} (U \in \tau)(cl(U) \in RC(X)) \\ (d) \end{aligned}\right\} \Rightarrow e\text{-}cl_\theta(f(U)) \subset e\text{-}cl_\theta(f(int(cl(U)))) \subset f(cl(U)).$$

$(e) \Rightarrow (f)$: Let U be any preopen subset of X.

$$\left.\begin{aligned} U \in PO(X) \Rightarrow (U \subset int(cl(U)))(int(cl(U)) \in \tau) \\ (e) \end{aligned}\right\} \Rightarrow$$
$$\Rightarrow e\text{-}cl_\theta(f(U)) \subset e\text{-}cl_\theta(f(int(cl(U)))) \subset f(cl(int(cl(U)))) \subset f(cl(U)).$$

$(f) \Rightarrow (g)$: Let U be any preopen subset of X.

$$\left.\begin{aligned} U \in PO(X) \Rightarrow X \setminus cl(U) \in \tau \\ (f) \end{aligned}\right\} \Rightarrow e\text{-}cl_\theta(f(X \setminus cl(U))) \subset f(cl(X \setminus cl(U)))$$
$$\begin{aligned} &\Rightarrow e\text{-}cl_\theta(Y \setminus f(cl(U))) \subset f(X \setminus int(cl(U))) = Y \setminus f(int(cl(U))) \\ &\Rightarrow Y \setminus e\text{-}int_\theta(f(cl(U))) \subset Y \setminus f(int(cl(U))) \\ &\Rightarrow f(U) \subset f(int(cl(U))) \subset e\text{-}int_\theta(f(cl(U))). \end{aligned}$$

$(g) \Rightarrow (h)$: Straightforward.
$(h) \Rightarrow (i)$: Let $U \subset X$.

$$\left.\begin{aligned} U \subset X \Rightarrow f(U) \subset Y \\ (h) \end{aligned}\right\} \Rightarrow f^{-1}(e\text{-}cl_\theta(f(U))) \subset cl_\theta(f^{-1}(f(U))) = cl_\theta(U)$$
$$\Rightarrow e\text{-}cl_\theta(f(U)) \subset f(cl_\theta(U)).$$

$(i) \Rightarrow (j)$: Let $U \subset X$.

$$\left.\begin{aligned} U \subset X \Rightarrow cl_\theta(U) \in C(X) \Rightarrow int(cl_\theta(U)) \subset X \\ (i) \end{aligned}\right\} \Rightarrow$$
$$\Rightarrow e\text{-}cl_\theta(f(int(cl_\theta(U)))) \subset f(cl_\theta(int(cl_\theta(U)))) = f(cl(int(cl_\theta(U)))) \subset f(cl_\theta(U)).$$

$(j) \Rightarrow (a)$: Straightforward. ■

Theorem 2.9 If X is a regular space and $f : (X, \tau) \to (Y, \sigma)$ is a bijective function, then the following statements are equivalent:
(a) f is weakly eR-open.
(b) For each θ-open set A in X, $f(A)$ is e-θ-open in Y.

(c) For any set B of Y and any θ-closed set A in X containing $f^{-1}(B)$, there exists an e-θ-closed set F in Y containing B such that $f^{-1}(F) \subset A$.

Proof. $(a) \Rightarrow (b)$: Let A be a θ-open set in X.

$$\left.\begin{array}{r} A \in \theta O(X) \Rightarrow Y \setminus f(A) \subset Y \\ (a)(\text{Theorem 2.7}(d)) \end{array}\right\} \Rightarrow f^{-1}(e\text{-}cl_\theta(Y \setminus f(A))) \subset cl_\theta(f^{-1}(Y \setminus f(A)))$$
$$\Rightarrow X \setminus f^{-1}(e\text{-}int_\theta(f(A))) \subset cl_\theta(X \setminus A) = X \setminus A$$
$$\Rightarrow A \subset f^{-1}(e\text{-}int_\theta(f(A)))$$
$$\Rightarrow f(A) \subset e\text{-}int_\theta(f(A)).$$

$(b) \Rightarrow (c)$: Let B be any set in Y and A be a θ-closed set in X such that $f^{-1}(B) \subset A$.

$$\left.\begin{array}{r} (B \subset Y)(A \in \theta C(X))(f^{-1}(B) \subset A) \Rightarrow (X \setminus A \in \theta O(X))(B \subset Y \setminus f(X \setminus A)) \\ (b) \end{array}\right\} \Rightarrow$$
$$\left.\begin{array}{r} \Rightarrow (f(X \setminus A) \in e\theta O(Y))(B \subset Y \setminus f(X \setminus A)) \\ F := Y \setminus f(X \setminus A) \end{array}\right\} \Rightarrow$$
$$\Rightarrow (F \in e\theta C(X))(B \subset Y)(f^{-1}(F) = f^{-1}(Y \setminus f(X \setminus A)) = f^{-1}(f(A)) = A.)$$

$(c) \Rightarrow (a)$: Let B be any set in Y.

$$\left.\begin{array}{r} (B \subset Y)(f^{-1}(B) \subset cl_\theta(f^{-1}(B))) \\ X \text{ is regular} \Rightarrow cl_\theta(f^{-1}(B)) \in \theta C(X) \\ (c) \end{array}\right\} \Rightarrow$$
$$\Rightarrow (\exists F \in e\theta C(Y))(B \subset F)(f^{-1}(F) \subset cl_\theta(f^{-1}(B)))$$
$$\Rightarrow (\exists F \in e\theta C(Y))(B \subset F)(f^{-1}(e\text{-}cl_\theta(B)) \subset f^{-1}(F) \subset cl_\theta(f^{-1}(B))).$$

Then from Theorem 2.8(h) f is weakly eR-open. ■

Theorem 2.10 If X is a regular space and $f : (X, \tau) \to (Y, \sigma)$ is a bijective function, then the following statements are equivalent:
(a) f is weakly eR-open.
(b) f is e-θ-open.
(c) For each $x \in X$ and each open set U of X containing x, there exists an e-open set V of Y containing $f(x)$ such that e-$cl(V) \subset f(U)$.

Proof. $(a) \Rightarrow (b)$: Let W be a nonempty open subset of X.

$$\left.\begin{array}{r} x \in W \in \tau \\ X \text{ is regular} \end{array}\right\} \Rightarrow (\exists U_x \in \mathcal{U}(x))(cl(U_x) \subset W)$$
$$\Rightarrow W = \cup\{U_x | x \in W\} = \cup\{cl(U_x) | x \in W\}$$
$$\left.\begin{array}{r} \Rightarrow f(W) = \cup\{f(U_x) | x \in W\} \\ f \text{ is weakly } eR\text{-open} \end{array}\right\} \Rightarrow f(W) = \cup\{f(U_x) | x \in W\}$$
$$\subset \cup\{e\text{-}int_\theta(f(cl(U_x))) | x \in W\}$$
$$\left.\begin{array}{r} \subset e\text{-}int_\theta(\cup\{f(cl(U_x)) | x \in W\}) \\ f \text{ is bijective} \end{array}\right\} \Rightarrow$$
$$\left.\begin{array}{r} \Rightarrow f(W) \subset e\text{-}int_\theta(f(\cup\{cl(U_x) | x \in W\})) = e\text{-}int_\theta(f(W)) \\ e\text{-}int_\theta(f(W)) \subset f(W) \end{array}\right\} \Rightarrow$$
$$\Rightarrow e\text{-}int_\theta(f(W)) = f(W)$$
$$\Rightarrow f(W) \in e\theta O(Y).$$

$(b) \Rightarrow (c)$ and $(c) \Rightarrow (a)$: Straightforward. ■

Theorem 2.11 If $f : (X, \tau) \to (Y, \sigma)$ is weakly eR-open and strongly continuous, then

f is e-θ-open.

Proof. Let U be any open subset of X.

$$\left.\begin{array}{r} U \in \tau \\ f \text{ is weakly } eR\text{-open} \end{array}\right\} \Rightarrow \left.\begin{array}{r} f(U) \subset e\text{-}int_\theta(f(cl(U))) \\ f \text{ is strongly continuous} \end{array}\right\} \Rightarrow$$
$$\Rightarrow f(U) \subset e\text{-}int_\theta(f(cl(U))) \subset e\text{-}int_\theta(f(U)).$$ ■

The following example shows that strong continuity is not decomposition of e-θ-openness. Namely, an e-θ-open function need not be strongly continuous.

***Example* 2.12** Let $X = \{a, b\}$ and τ be the indiscrete topology for X. Then the identity function $f : (X, \tau) \to (X, \tau)$ is an e-θ-open function but it is not strongly continuous.

Theorem 2.13 If $f : (X, \tau) \to (Y, \sigma)$ is contra e-θ-closed, then f is a weakly eR-open function.

Proof. Let U be any open subset of X.

$$\left.\begin{array}{r} U \in \tau \Rightarrow cl(U) \in C(X) \\ f \text{ is contra } e\text{-}\theta\text{-closed} \end{array}\right\} \Rightarrow f(cl(U)) \in e\theta O(Y)$$
$$\Rightarrow f(U) \subset f(cl(U)) = e\text{-}int_\theta(f(cl(U))).$$ ■

Theorem 2.14 If $f : (X, \tau) \to (Y, \sigma)$ is bijective contra e-θ-open, then f is a weakly eR-open function.

Proof. Let U be any open subset of X.
$$\left.\begin{array}{r} U \in \tau \\ f \text{ is contra } e\text{-}\theta\text{-open} \end{array}\right\} \Rightarrow f(U) \in e\theta C(Y) \Rightarrow e\text{-}cl_\theta(f(U)) = f(U) \subset f(cl(U)).$$
Then from Theorem 2.8(e) f is weakly eR-open. ■

Theorem 2.15 Let $f : (X, \tau) \to (Y, \sigma)$ be a bijective function. If $f(cl_\theta(U))$ is e-θ-closed in Y for every subset U of X, then f is weakly eR-open.

Proof. Let U be a subset of X.
$(U \subset X)(f(cl_\theta(U)) \in e\theta C(Y)) \Rightarrow e\text{-}cl_\theta(f(U)) \subset e\text{-}cl_\theta(f(cl_\theta(U))) = f(cl_\theta(U))$.
Then from Theorem 2.8(i) f is weakly eR-open. ■

Definition 2.16 A function $f : X \to Y$ is called complementary weakly eR-open (briefly c.w.eR-o) if for each open set U of X, $f(Fr(U))$ is e-θ-closed in Y, where $Fr(U)$ denotes the frontier of U.

Examples 2.17 and 2.18 show the independence of complementary weakly eR-openness and weakly eR-openness.

***Example* 2.17** Let $X = \{a, b, c, d\}$ and

$$\tau = \{\emptyset, X, \{a, d\}\} \quad \text{and} \quad \sigma = \{\emptyset, \{a\}, \{b\}, \{a, b\}, \{a, c\}, \{a, b, c\}, \{a, b, d\}, X\}.$$

The identity function $f : (X, \tau) \to (Y, \sigma)$ is weakly eR-open, but it is not c.w.eR-o.

***Example* 2.18** Let $X = \{a, b\}$, $\tau = \{\emptyset, X, \{a\}, \{b\}\}$ and $\sigma = \{\emptyset, X, \{b\}\}$. The identity function $f : (X, \tau) \to (X, \sigma)$ is c.w.eR-o., but it is not weakly eR-open.

Theorem 2.19 If $f : (X, \tau) \to (Y, \sigma)$ is bijective weakly eR-open and c.w.eR-o, then f is e-θ-open.

Proof. Let U be an open subset in X with $x \in U$. Since f is weakly eR-open, by Theorem 2.7(i) there exists an e-θ-open set V containing $f(x) = y$ such that $V \subset f(cl(U))$. Now $Fr(U) = cl(U) \setminus U$ and thus $x \notin Fr(U)$. Hence $y \notin f(Fr(U))$ and therefore $y \in V \setminus f(Fr(U))$. Put $V_y = V \setminus f(Fr(U))$. Now V_y is an e-θ-open set since f is c.w.eR-o. Since $y \in V_y$, then $y \in f(cl(U))$. But $y \notin f(Fr(U))$ and thus $y \notin f(Fr(U)) = f(cl(U)) \setminus f(U)$ which implies that $y \in f(U)$. Therefore $f(U) = \cup\{V_y | (V_y \in e\theta O(Y))(y \in f(U))\}$. Hence f is e-θ-open. ■

Recall that a space X is said to be e-connected [3] if X is not the union of two disjoint nonempty e-open sets.

Theorem 2.20 If $f : (X, \tau) \to (Y, \sigma)$ is a bijective weakly eR-open of a space X onto an e-connected space Y, then X is connected.

Proof. Let f be a bijective weakly eR-open of a space X onto an e-connected space Y and suppose that X is not connected.

$$\left.\begin{array}{r} X \text{ is not connected} \Rightarrow (\exists U_1, U_2 \in \tau \setminus \{\emptyset\})(U_1 \cap U_2 = \emptyset)(U_1 \cup U_2 = X) \\ f \text{ is bijective weakly } eR\text{-open} \end{array}\right\} \Rightarrow$$

$$\Rightarrow (f(U_i) \in \sigma \setminus \{\emptyset\})(\bigcap_i f(U_i) = \emptyset)(\bigcup_i f(U_i) = Y)(f(U_i) \subset e\text{-}int_\theta(f(cl(U_i)))$$
$$= e\text{-}int_\theta(f(U_i)))\ (i = 1, 2)$$
$$\Rightarrow (f(U_i) \in \sigma \setminus \{\emptyset\})(\bigcap_i f(U_i) = \emptyset)(\bigcup_i f(U_i) = Y)(f(U_i) = e\text{-}int_\theta(f(U_i)))\ (i = 1, 2)$$
$$\Rightarrow (f(U_i) \in e\theta O(Y) \setminus \{\emptyset\})(\bigcap_i f(U_i) = \emptyset)(\bigcup_i f(U_i) = Y)\ (i = 1, 2)$$

Then Y is not e-connected which is a contradiction. ■

Definition 2.21 A space X is said to be hyperconnected [9] if every nonempty open subset of X is dense in X.

Theorem 2.22 If X is a hyperconnected space, then a function $f : (X, \tau) \to (Y, \sigma)$ is weakly eR-open if and only if $f(X)$ is e-θ-open in Y.

Proof. *Sufficiency:* Obvious.
Necessity: Let U be a nonempty open subset of X.

$$\left.\begin{array}{r} (U \in \tau)(X \text{ is hyperconnected}) \Rightarrow cl(U) = X \Rightarrow e\text{-}int_\theta(f(cl(U))) = e\text{-}int_\theta(f(X)) \\ f \text{ is weakly } eR\text{-open} \end{array}\right\} \Rightarrow$$

$$\Rightarrow f(U) \subset f(X) = e\text{-}int_\theta(f(X)) = e\text{-}int_\theta(f(cl(U))).$$ ■

Theorem 2.23 Let $f : (X, \tau) \to (Y, \sigma)$ be a bijective weakly eR-open function. Then the following properties hold:
(a) If F is θ-closed in X, then $f(F)$ is e-θ-closed in Y.
(b) If F is θ-open in X, then $f(F)$ is e-θ-open in Y.

Proof. (a) Let $F \in \theta C(X)$.

$$\left.\begin{array}{r} F \in \theta C(X) \Rightarrow F = cl_\theta(F) \\ \text{Theorem 2.8}(i) \end{array}\right\} \Rightarrow \left.\begin{array}{r} e\text{-}cl_\theta(f(F)) \subset f(cl_\theta(F)) = f(F) \\ f(F) \subset e\text{-}cl_\theta(f(F)) \end{array}\right\} \Rightarrow$$

$$\Rightarrow f(F) = e\text{-}cl_\theta(f(F))$$
$$\Rightarrow f(F) \in e\theta C(Y)$$

(b) Similarly proved. ■

Acknowledgements

The authors are greatful to the referee for his/her careful reading and useful comments for the improvement of this paper. This study is dedicated to Professor Dr. Zekeriya Güney on the occasion of his 67th birthday.

References

[1] D.Andrijevic, On b-open sets. Mat. Vesnik., 48 (1996), 59-64.
[2] M. Caldas, E. Ekici, S. Jafari, R.M. Latif, On weakly BR-open functions and their characterizations in topological spaces. Demonstratio Math., 44 (1) (2011), 159-168.
[3] E. Ekici, New forms of contra continuity. Carpathian J. Math., 24 (1) (2008), 37-45.
[4] ———, On e-open sets, $\mathcal{DP}^*$-sets and $\mathcal{DPE}^*$-sets and decompositions of continuity. Arab. J. Sci. Eng. Sect. A Sci., 33 (2) (2008), 269-282.
[5] N. Levine, Semi-open sets and semi-continuity in topological spaces. Amer. Math. Monthly., 70 (1963), 36-41.
[6] ———, Strong continuity in topological spaces. Amer. Math. Monthly., 67 (1960), 269.
[7] A.S. Mashhour, M.E. Abd El-Monsef and S.N. El-Deeb, On precontinuous and weak precontinuous mappings. Proc. Math. Phys. Soc. Egypt., 53 (1982), 47-53.
[8] O. Njastad, On some classes of nearly open sets. Pacific J. Math., 15 (1965), 961-970.
[9] T. Noiri, A generalization of closed mappings. Atti. Accad. Naz. Lince Rend. Cl. Sci. Fis. Mat. Natur., 8 (1973), 210-214.
[10] M. Özkoç and G. Aslım, On strongly θ-e-continuous functions. Bull. Korean Math. Soc., 47 (5) (2010), 1025-1036.
[11] J.H. Park, Strongly θ-b-continuous functions. Acta Math. Hungar., 110 (4) (2006), 347-359.
[12] M. Stone, Application of the theory of Boolean ring to general topology. Trans. Amer. Math. Soc., 41 (1937), 374-481.
[13] N.V. Velicko, H-closed topological spaces. Amer. Math. Soc. Transl., 78 (1968), 103-118.

6

m-Projections involving Minkowski inverse and Range Symmetric property in Minkowski Space

M. Saleem Lone[a*] and D. Krishnaswamy[a]

[a]*Department of Mathematics, Annamalai University, Chidambaram, PO. Code 608002, Tamilnadu, India.*

Abstract. In this paper we study the impact of Minkowski metric matrix on a projection in the Minkowski Space $\mathcal{M}$ along with their basic algebraic and geometric properties.The relation between the m-projections and the Minkowski inverse of a matrix A in the minkowski space $\mathcal{M}$ is derived. In the remaining portion commutativity of Minkowski inverse in Minkowski Space $\mathcal{M}$ is analyzed in terms of m-projections as an analogous development and extension of the results on EP matrices.

Keywords: Minkowski inverse, m-projections, range symmetric, EP matrix.

1. Introduction and preliminaries

Let us denote by $M_{(m,n)}(\mathbb{C})$ the set of $m \times n$ matrices and when $m = n$ we write $M_n(\mathbb{C})$ for $M_{(n,n)}(\mathbb{C})$. The symbols $A^*, A^\sim, A^\oplus, \|A\|, A^\dagger$, $R(A)$, and $N(A)$ denote the conjugate transpose, Minkowski adjoint, Minkowski inverse, norm, Moore-Penrose inverse, range space and null space of a matrix A respectively. I_n denote the identity matrix of order $n \times n$ and R^+ denotes the set of positive real numbers. We use $\bar{P} = I_n - P$ (this notation has nothing to do with the complex conjugate of a matrix element). Also we use $\tilde{P}_A = I - AA^\oplus$ and $P_A = AA^\oplus$.

Indefinite inner product is a scalar product defined by

$$[u, v] = \langle u, Mv \rangle = u^* Mv,$$

*Corresponding author.
E-mail address: salemlone9@gmail.com (M. Saleem Lone).

where $\langle\ ,\rangle$ denotes the conventional Hilbert Space inner product and M is a Hermitian matrix. This hermitian matrix M is referred to as metric matrix. Minkowski space is an indefinite inner product space which is agreed upon to be the most suitable space for the study of Einstein's theory of special relativity and has been recently taken into consideration in a more generalized form by changing the dimension and the signature of the metric associated with its indefinite inner product. Moreover the matrix argument is taken into consideration. In Minkowski Space $\mathcal{M}$ the metric matrix is denoted by G and is defined as

$$G = \begin{bmatrix} 1 & 0 \\ 0 & -I_{n-1} \end{bmatrix} \text{ satisfying } G^2 = I_n \text{ and } G^* = G.$$

G is called the Minkowski metric matrix.
In case $u \in \mathbb{C}^n$, indexed as $u = (u_0, u_1, ..., u_{n-1})$, G is called the Minkowski metric tensor and is defined as $Gu = (u_0, -u_1, ..., -u_{n-1})$. For detailed study of indefinite linear algebra refer to [13]

An idempotent matrix is called a projection if it is a hermitian i,e. $A^2 = A = A^*$. Projections are widely used in the literature e.g. see [1, 7, 15, 16, 20, 23–27, 34]. There are many extensions and generalizations of the projections like generalized projections, k-generalized projections, hypergeneralized projections etc. see [5, 11, 14, 18]. Projections are also studied in indefinite inner product spaces see [10, and refrences therein]. In this paper we first take into account the impact of Minkowski metric matrix on a projection, giving rise to a new class of projections called m-projections.

Another part of this paper is devoted to study the commutativity of Minkowski inverse of a matrix A in terms of the m-projections. A matrix $A \in M_n(\mathbb{C})$ is said to be range hermitian or EP matrix if $N(A) = N(A^*)$ or equivalently $AA^\dagger = A^\dagger A$ [1, p.157]. For detailed study of EP matrices see [1, 6, 9, 12, 17, 28, 30–33]. The EP property was further extended to the elements of Banach Spaces, operators, Banach Algebras and in particular to C^* -Algebras see [8, and references therein]. Meenakshi in [2] introduced the concept of range symmetric matrices in the Minkowski space analogous to the EP matrices in the unitary space. $A \in M_n(\mathbb{C})$ is said to be range symmetric in Minkowski space $\mathcal{M}$ if and only if $N(A) = N(A^\sim)$ [2], where $A^\sim$ is the Minkowski adjoint of the matrix A. The Minkowski adjoint of a matrix $A \in M_{(m,n)}(\mathbb{C})$, denoted by $A^\sim$, is defined as $A^\sim = G_1 A^* G_2$, where $G_1 \in M_m(\mathbb{C})$ and $G_2 \in M_n(\mathbb{C})$ are the Minkowski metric matrices of respective size and A is said to be m-symmetric if $A^\sim = A$. We will show that the commutativity of a matrix A with its Minkowski inverse $A^\oplus$ is also related to the EP property in the Minkowski Space $\mathcal{M}$.

The Minkowski inverse of a matrix $A \in M_{(m,n)}(\mathbb{C})$ in the Minkowski Space $\mathcal{M}$ is defined as the unique matrix X satisfying the following four conditions

$$(1)\ AXA = A\ (2)\ XAX = X\ (3)\ (AX)^\sim = AX \text{ and } (4)\ (XA)^\sim = XA$$

In [22] the author has investigated certain properties of the Minkowski Inverse in the indefinite inner product space by generalizing the signature of the Minkowski metric matrix G. Unlike the Moore-Penrose inverse of matrices the Minkowski inverse of a matrix does not exist always. The following result of [3] gives the necessary and sufficient condition for the existence of the Minkowski inverse of a matrix A in the Minkowski Space $\mathcal{M}$

Lemma 1.1 For $A \in M_{(m,n)}(\mathbb{C})$, the Minkowski inverse $A^\oplus$ exists if and only if $rank(A) = rank(AA^\sim) = rank(A^\sim A)$. if $A^\oplus$ exists , then it is unique.

Definition 1.2 A matrix $A \in M_n(\mathbb{C})$ is said to be G-unitary if $AA^\sim = A^\sim A = I$.

Following result from [2] will also be used in the forth coming sections

Theorem 1.3 Let $A \in M_n(\mathbb{C})$, if $A^{\oplus}$ exists in $\mathcal{M}$ then $AA^{\oplus}$ and $A^{\oplus}A$ are projections on $R(A)$ and $R(A^{\sim})$ respectively.

Theorem 1.4 For $A \in M_n(\mathbb{C})$, the following statements are equivalent:

(i) A is range symmetric in $\mathcal{M}$
(ii) GA and AG are EP
(iii) $N(A^*) = N(AG)$.
(iv) $R(A) = R(A^{\sim})$

2. m-projections and their algebraic properties

The concept of m-projections roots from the Minkowski adjoint of a projection P satisfying the condition of being m-symmetric. We thus have a class of projections which satisfy the condition of being m-symmetric and we call such projections as m-projections (Minkowski projections). Hence we have the following definition

Definition 2.1 A Projection P is said to be m-projection if $P^2 = P = P^{\sim}$, where $P^{\sim} = GP^*G$ is the Minkowski adjoint of the projection P.

For the sake of completion we prove the following two results, concluding the basic algebraic perspective of the m-projection.

Lemma 2.2 If P is a m-projection then $I - P$ is also an m-projection

Proof. The proof is obvious from definition of the m - projection. ■

Theorem 2.3 Let S be the set of all m-projections in a Minkowski space M.Then for $P_1, P_2 \in S$ we have

(i) $P_1 + P_2 \in S$ if and only if $P_1P_2 = 0 = P_2P_1$
(ii) $P_1 - P_2 \in S$ if and only if $P_1P_2 = P_2P_1 = P_2$
(iii) $P_1P_2 \in S$ if and only if $P_1P_2 = P_2P_1$

Proof. (i) Clearly we have $(P_1 + P_2)^{\sim} = P_1 + P_2$ and

$$(P_1 + P_2)^2 = P_1^2 + P_1P_2 + P_2P_1 + P_2^2 = P_1 + P_2 \Leftrightarrow P_1P_2 + P_2P_1 = 0 \tag{1}$$

Premultiplying (1) by P_1 we get

$$P_1^2P_2 + P_1P_2P_1 = 0 \Rightarrow P_1^{\sim}P_2 + P_1P_2P_1 = 0 \tag{2}$$

Now postmultipling (1) by P_1 we get

$$P_1P_2P_1 + P_2P_1^{\sim} = 0 \tag{3}$$

Thus we have $P_1^{\sim}P_2 = P_2P_1^{\sim}$ which again on pre and post multiplying by $P_1^{\sim}$ gives $P_1P_2 = P_2P_1$ and the result follows. (ii) and (iii) are obvious. ■

Remark 1 *Since the Minkowski inverse of a matrix A exits if and only if rank($AA^{\sim}$) = rank($A^{\sim}A$) = rank(A) and this condition is trivially satisfied by a* m-*projection. For a* m-*projection P we have $PP^{\sim} = P^2 = P = P^2 = P^{\sim}P$ therefore* *we have rank*$(PP^{\sim}) = $ *rank*$(P^{\sim}P) = $ *rank* (P) *and hence* $P^{\oplus} = P$. *Given any matrix* $A \in \mathbb{C}^{m \times n}$, $AA^{\oplus}$ *is a* m*-projection*

The following result will be used as definition for the Range Symmetric matrix in $\mathcal{M}$

Theorem 2.4 Let $A \in M_{m,n}(\mathbb{C})$ and $A^{\oplus}$ exists in $\mathcal{M}$, then the following statements are equivalent:

(a) A is Range Symmetric (EP) in $\mathcal{M}$
(b) $AA^{\oplus} = A^{\oplus}A$

Proof. Assume that $A^{\oplus}$ exists and A is EP in $\mathcal{M}$. Using Theorem (1.3), we have $AA^{\oplus}$ is the projection on $R(A)$ and $A^{\oplus}A$ is the projection on $R(A^{\sim})$. Also from Theorem (1.4), statement (iv) we have $R(A) = R(A^{\sim})$, the result follows. ■

We recover the following corollary from [4]

Corollary 2.5 Let $A \in M_n(\mathbb{C})$ be an EP matrix of rank r, Then, there exists a unitary matrix P and a nonsingular matrix D such that $A = P(D \oplus 0)P^{\sim}$ in the Minkowski Space $\mathcal{M}$

Proof. Using Theorem (1.4) we have AG is range symmetric in $\mathcal{M}$. Therefore by [21, Theorem 1] AG has the following representation

$$AG = U \begin{bmatrix} A_1 & 0 \\ 0 & 0 \end{bmatrix} U^*, \tag{4}$$

where U is a unitary matrix. This gives

$$A = UG \begin{bmatrix} G_1A_1 & 0 \\ 0 & 0 \end{bmatrix} U^*G \tag{5}$$

Taking $G_1A_1 = D$ and $P = UG$ we get $P^{\sim} = U^*G$. Therefore we have

$$A = P \begin{bmatrix} D & 0 \\ 0 & 0 \end{bmatrix} P^{\sim}$$

■

The representation obtained in the corollary (2.5) eases to formulate the Minkowski inverse for the range symmetric matrices and in particular for the m-projections in the Minkowski space. From the definition of the G-unitary matrix it follows at once that a unitary matrix U is G-unitary if and only if $UG = GU$. We will use this assumption in formulating the Minkowski inverse of the m-projections in the forth coming results. Under this assumption the above corollary can be extended to the following equivalent statement:

Corollary 2.6 Let $A \in M_n(\mathbb{C})$ be a matrix of rank r. Then following conditions are equivalent:

(i) $AA^{\oplus} = A^{\oplus}A$
(ii) There exists a unitary matrix P and a nonsingular matrix D such that

$$A = P(D \oplus 0)P^{\sim}, \text{where } P \text{ is G-unitary.}$$

Proof. $(i) \Rightarrow (ii)$ is obvious from Theorem (2.4) and Corollary (2.5). Also $(ii) \Rightarrow (i)$ follows from direct verification by noting that the Minkowski inverse of the A is $A^{\oplus} = P(D^{-1} \oplus 0)P^{\sim}$, where P is G-unitary. ■

Let P be an m-projection in the Minkowski space $\mathcal{M}$. Taking into account Remark (1) and the observations made in the corollaries (2.5) and (2.6) of the Theorem (2.4), there exists a G-unitary matrix U such that

$$P = U \begin{bmatrix} I & 0 \\ 0 & 0 \end{bmatrix} U^{\sim} \tag{6}$$

The representation (6) can be used to determine the partitioning of any other m- projection say Q with the use of the same G-unitary matrix U such that

$$Q = U \begin{bmatrix} W & X \\ Y & Z \end{bmatrix} U^{\sim}. \tag{7}$$

Using the later part of the definition of a m-projection i.e. $Q = Q^{\sim}$, the representation (7) becomes

$$Q = U \begin{bmatrix} W & X \\ -G_1 X^{\sim} & Z \end{bmatrix} U^{\sim} \tag{8}$$

Where $W \in M_r(\mathbb{C})$ is m-symmetric, $Z \in M_{n-r}(\mathbb{C})$ is hermitian and G_1 is the Minkowski metric matrix of order $r \times r$. The following results give the relation between the submatrices W, X and Z of the matrix Q given in (8).

Lemma 2.7 Let Q be a m-projection as given in (8). Then:

(i) $W = W^2 - XG_1X^{\sim}$ or equivalently $W\bar{W} = -XG_1X^{\sim}$
(ii) $X = WX + XZ$ or equivalently $X^{\sim} = X^{\sim}W + G_1ZG_1X^{\sim}$
(iii) $G_1X^{\sim} = G_1X^{\sim}W + ZG_1X^{\sim}$ or equivalently $G_1X^{\sim}\bar{W} = ZG_1X^{\sim}$
(iv) $Z = Z^2 - G_1X^{\sim}X$ or equivalently $Z\bar{Z} = -G_1X^{\sim}X$

Proof. The proof of these relationships is a straightforward consequence of the condition $Q^2 = Q$ ■

Lemma 2.8 Let Q be a m-projection as given in (8). Then:

(i) $\bar{W} = \bar{W}^2 - XG_1X^{\sim}$
(ii) $WX = X\bar{Z}$
(iii) $XZ = \bar{W}X$
(iv) $G_1X^{\sim} = G_1X^{\sim}\bar{W} + \bar{Z}G_1X^{\sim}$
(v) $\bar{Z} = \bar{Z}^2 - G_1X^{\sim}X$

Proof. The proof follows by using Lemma (2.2) and the condition $\bar{Q}^2 = \bar{Q}$ ■

Theorem 2.9 Let Q be a m-projection as given in (8). Then:

(i) $WW^{\oplus}X = X$
(ii) $\bar{W}\bar{W}^{\oplus}X = X$
(iii) $ZZ^{\oplus}G_1X^{\sim} = G_1X^{\sim}$
(iv) $\bar{Z}\bar{Z}^{\oplus}G_1X^{\sim} = G_1X^{\sim}$
(v) $W^{\oplus}X = X\bar{Z}^{\oplus}$
(vi) $XZ^{\oplus} = \bar{W}^{\oplus}X$

Proof. The condition (i) follows on account of the condition (i) of the Lemma (2.7) and the fact that

$$R(W) = R(WW^* + XX^*) = R(WW^*) + R(XX^*) = R(W) + R(X) \tag{9}$$

Thus $R(X) \subseteq R(W)$, which can be expressed equivalently as $WW^{\oplus}X = X$. Analogously we can obtain the next three conditions. Also from the condition (ii) of the Lemma (2.7) we have $W^{\oplus}X = W^{\oplus}(WX + XZ)$. Using condition (i) of the theorem we get $W^{\oplus}X\bar{Z} = X$. Postmultiplying this equation by $\bar{Z}^{\oplus}$ and utilizing condition (iv), we obtain condition (v). The condition (vi) can be established similarly. ■

Theorem 2.10 Let Q be a m-projection as given in (8). Then:

(i) $P_w = W - X\bar{Z}^{\oplus}G_1X^{\sim}$ and $\tilde{P}_w = \bar{W} + X\bar{Z}^{\oplus}G_1X^{\sim}$
(ii) $P_{\bar{w}} = \bar{W} - XZ^{\oplus}G_1X^{\sim}$ and $\tilde{P}_{\bar{w}} = W + XZ^{\oplus}G_1X^{\sim}$
(iii) $P_z = Z - G_1X^{\sim}\bar{W}^{\oplus}X$ and $\tilde{P}_z = \bar{Z} + G_1X^{\sim}\bar{W}^{\oplus}X$
(iv) $P_{\bar{z}} = \bar{Z} - G_1X^{\sim}W^{\oplus}X$ and $\tilde{P}_{\bar{z}} = Z + G_1X^{\sim}W^{\oplus}X$

Proof. From the condition (i) of lemma (2.7) we have $W = W^2 - XG_1X^{\sim}$. Premultiplying on both sides by $\bar{W}^{\oplus}$ and using condition (vi) of Theorem (2.9) we get the later part of the point (i) and subtracting the obtained expression from I_r on both sides we get the remaining part. On the same lines the other points can be established. ■

3. m-projections onto certain subspaces

In this section we develop the representations of m-projectors onto certain subspace including their sum and intersection. The second lemma gives a powerful tool for constructing the m-projectors onto given spaces which is obtained as an analogous result from orthogonal projectors.

Lemma 3.1 Let Q be partitioned as in (6) . Then:

(i) $rk(\bar{W}) = r - rk(W) + rk(X)$
(ii) $rk(\bar{Z}) = n - r + rk(X) - rk(Z)$

Proof. Using (i) of Lemma (2.7) and the fact that $rk(W\bar{W}) = rk(W) + rk(\bar{W}) - r$, from (2.12) in [35] we get $rk(\bar{W}) = r + rk(X) - rk(W)$.
Condition (ii) follows analogously. ■

Lemma 3.2 Let $P,\ Q \in C_n^{mp}$. Then:

(i) $P + \bar{P}(\bar{P}Q)^{\oplus}$ is the m-projector onto $R(P) + R(Q)$
(ii) $P - P(P\bar{Q})^{\oplus}$ is the m-projector onto $R(P) \cap R(Q)$

Proof. The proof follows analogously from the equivalent conditions of Theorems (3) and (4) in [29]. ■

Using Lemma (3.2) we obtain the following representations of the m-projectors onto the sums and intersection of certain subspaces , including their dimensions.

Lemma 3.3 Let $P, Q \in C_n^{mp}$ and let $\mathbf{Q}$ be partitioned as in (8). Then:

(i) $P_{R(P)+R(Q)} = U \begin{bmatrix} I_r & 0 \\ 0 & P_z \end{bmatrix} U^{\sim}$, where $dim[R(P) + R(Q)] = r + rk(Z)$

(ii) $P_{R(P)+NQ)} = U \begin{bmatrix} I_r & 0 \\ 0 & P_{\bar{z}} \end{bmatrix} U^{\sim}$, where $dim[R(P) + N(Q)] = r + rk(X) + rk(Z)$

(iii) $P_{N(P)+R(Q)} = U \begin{bmatrix} P_w & 0 \\ 0 & I_{n-r} \end{bmatrix} U^{\sim}$, where $dim[N(P) + R(Q)] = n - r + rk(W)$

(iv) $P_{N(P)+N(Q)} = U \begin{bmatrix} P_{\bar{w}} & 0 \\ 0 & I_{n-r} \end{bmatrix} U^{\sim}$, where $dim[N(P) + N(Q)] = n - rk(W) + rk(X)$

Proof. For $P = U \begin{bmatrix} I_r & 0 \\ 0 & 0 \end{bmatrix} U^{\sim}$ and $Q = U \begin{bmatrix} W & X \\ -G_1X^{\sim} & Z \end{bmatrix} U^{\sim}$. We have $(\bar{P}Q) = U \begin{bmatrix} 0 & 0 \\ -G_1X^{\sim} & Z \end{bmatrix} U^{\sim}$. Now utilising the conditions (iv) of Lemma (2.7) and (iii) of Theorem (2.9), it can be easily verified that the minkowski inverse of $\bar{P}Q$ is

$$(\bar{P}Q)^{\oplus} = U \begin{bmatrix} 0 & XZ^{\oplus} \\ 0 & P_z \end{bmatrix} U^{\sim} \tag{10}$$

Hence, using the statement (i) of Lemma (3.2) and substituting (10), we obtain the m-projector as claimed in point (i). The remaining part is obvious from the representation of the projector. The remaining points can be obtained in a similar fashion. ∎

Lemma 3.4 Let $P,\ Q \in C_n^{mp}$ and let Q be partitioned as in (8). Then:

(i) $P_{R(P)\cap R(Q)} = U \begin{bmatrix} \tilde{P}_{\bar{w}} & 0 \\ 0 & 0 \end{bmatrix} U^{\sim}$, where $dim[R(P) \cap R(Q)] = rk(W) - rk(X)$

(ii) $P_{R(P)\cap N(Q)} = U \begin{bmatrix} \tilde{P}_w & 0 \\ 0 & 0 \end{bmatrix} U^{\sim}$, where $dim[R(P) \cap N(Q)] = r - rk(W)$

(iii) $P_{N(P)\cap R(Q)} = U \begin{bmatrix} 0 & 0 \\ 0 & \tilde{P}_{\bar{z}} \end{bmatrix} U^{\sim}$, where $dim[R(P) \cap R(Q] = rk(Z) - rk(X)$

(iv) $P_{N(P)\cap R(Q)} = U \begin{bmatrix} 0 & 0 \\ 0 & \tilde{P}_z \end{bmatrix} U^{\sim}$, where $dim[R(P) \cap R(Q)] = n - r - rk(Z)$

Proof. With given P and Q we have $P\bar{Q} = U \begin{bmatrix} \bar{W} & -X \\ 0 & 0 \end{bmatrix} U^{\sim}$. Using the conditions (i) and (ii) of Lemma (2.7) and Theorem (2.9) respectively direct verification shows that the Minkowski inverse $(P\bar{Q})^{\oplus}$ of $P\bar{Q}$ is

$$(P\bar{Q})^{\oplus} = U \begin{bmatrix} \bar{W}\bar{W}^{\oplus} & 0 \\ G_1X^{\sim}\bar{W}^{\oplus} & 0 \end{bmatrix} U^{\sim} \tag{11}$$

Now using the statement (ii) of the Lemma (3.2) and substituting (11) we obtain the representation claimed in point (i) of the lemma. Also $dim(P_{R(P)\cap R(Q)}) = rk(\tilde{P}_{\bar{w}}) = r - rk(\bar{W})$. Whereupon utilizing the rank equality obtained in statement (i) of the Lemma (3.1) we get the remaining part of the result. The remaining representations can be obtained similarly. ∎

4. Minkowski inverse and m-projections

In this section we characterize the relation between the m-projections and the Minkowski inverse of a matrix in the Minkowski Space $\mathcal{M}$. We denote by $\mathcal{M}^{-1}$ the set of all invertible elements in Minkowski space $\mathcal{M}$

Theorem 4.1 Let $A \in M_n(\mathbb{C})$ in the Minkowski Space $\mathcal{M}$, then the following conditions are equivalent:

(i) There exists a unique m-projection P such that $A + P \in \mathcal{M}^{-1}$ and $AP = PA = 0$
(ii) A is range symmetric in $\mathcal{M}$.

Proof. $(i) \Rightarrow (ii)$
Let P be an m-projection. Since $AP = 0$ and $P^2 = P$, we have $AP + P^2 = P$ which implies $(A + P)P = P$ Also $(A + P)^{-1}$ exists so we have $P = P(A + P)^{-1}$. Similarly we get $P = (A + P)^{-1}P$. we claim that

$$A\{(A + P)^{-1} - P\} = I - P \tag{12}$$

we have

$$\begin{aligned} A\{(A + P)^{-1} - P\} &= (A + P - P)\{(A + P)^{-1} - P\} \\ &= (A + P)\{(A + P)^{-1} - P\} - P\{(A + P)^{-1} - P\} \\ &= I - P. \end{aligned} \tag{13}$$

Similarly $\{(A + P)^{-1} - P\}A = I - P$. Now we prove that $A^{\oplus} = X = \{(A + P)^{-1} - P\}$ In order to prove this we show that X satisfies the conditions of the definition of the Minkowski inverse of A. Using (12) and the given condition that $AP = PA = 0$ we have

$$AXA = A\{(A + P)^{-1} - P\}A = (I - P)A = A - PA = A \tag{14}$$

Also

$$XAX = \{(A + P)^{-1} - P\}A\{(A + P)^{-1} - P\} = \{(A + P)^{-1} - P\} \tag{15}$$

We now prove the m-symmetric conditions i,e. $(AX)^{\sim} = AX$ and $(XA)^{\sim} = XA$
Again using (12) we have $(AX)^{\sim} = G[\{(A + P)^{-1} - P\}A]^*G = I - P = AX$
Analogously we can prove that $(XA)^{\sim} = XA$
Therefore $X = A^{\oplus}$. Also $AA^{\oplus} = I - P = A^{\oplus}A$ i,e. $AA^{\oplus} = A^{\oplus}A$. Therefore by Theorem (2.4) A is Range Symmetric i,e. A is EP in $\mathcal{M}$
$(ii) \Rightarrow (i)$ Let P be the m-projection defined by $P = I - AA^{\oplus}$.
Evidently we have $AP = 0$ and $PA = 0$. Also

$$(A + P)(A^{\oplus} + P) = AA^{\oplus} + AP + PA^{\oplus} + P = AA^{\oplus} + P = I \tag{16}$$

This shows that $(A + P)$ is invertible and $(A + P)^{-1} = (A^{\oplus} + P)$
Finally we show that the m-projection P is unique.
Assume that Q is another m-projection such that $AQ = 0 = QA$ and $A + Q \in \mathcal{M}^{-1}$ then,

$$A^{\oplus} = (A + P)^{-1} - P = (A + Q)^{-1} - Q \tag{17}$$

Premultiplying (10) by A we get

$$AA^{\oplus} = A(A + P)^{-1} - AP = A(A + Q)^{-1} - AQ \tag{18}$$

$$\Rightarrow A(A+P)^{-1} = A(A+Q)^{-1} \Rightarrow P = Q \tag{19}$$

This shows that the m-projection P is unique. ■

Remark 2 *For $A, B \in M_n(\mathbb{C})$ we have*

$$(A+B)^{\sim} = A^{\sim} + B^{\sim},\ (\lambda A)^{\sim} = \bar{\lambda} A^{\sim},\ (AB)^{\sim} = B^{\sim} A^{\sim} \text{and}\ (A^{\sim})^{\sim} = A. \tag{20}$$

It is interesting to observe from (20) that the mapping $A \mapsto A^{\sim}$ satisfies the conditions of being an involution. Under this involution the similar results will hold in the C^*-algebra.

Following the usual notation we also denote the unique m-projection $P = I - AA^{\oplus}$ by $\bar{P}_A$ and we have the following corollary

Corollary 4.2 Let $A \in M_{m,n}(\mathbb{C})$ and $A^{\oplus}$ exists in $\mathcal{M}$, then the following are equivalent

(i) $A^k = A^{\oplus}$
(ii) A is range symmetric and $A^{k+1} + \bar{P}_A = 1$

Proof. Clearly $A^k = A^{\oplus} \Rightarrow AA^{\oplus} = A^{\oplus}A$. Thus A is EP in $\mathcal{M}$. Now using the facts that $\bar{P}_A = I - AA^{\oplus}$ and $A^{\oplus} = (A+\bar{P}_A)^{-1} - \bar{P}_A$, we have

$$\begin{aligned} A^{\oplus} &= A^k \\ &\Leftrightarrow (A+\bar{P}_A)^{-1} - \bar{P}_A = A^k \\ &\Leftrightarrow (A+\bar{P}_A)^{-1} = A^k + \bar{P}_A \end{aligned} \tag{21}$$

This gives $(A+\bar{P}_A)(A^k+\bar{P}_A) = I_n \Leftrightarrow A^{k+1} + \bar{P}_A = I_n$. ■

Corollary 4.3 Let $A \in \mathcal{M}$ be EP, then

(i) $A^{\oplus} = (A+\bar{P}_A)^{-1}(I - \bar{P}_A)$
(ii) $\bar{P}_A$ is idempotent
(iii) $A\bar{P}_A = \bar{P}_A A = A^{\oplus}\bar{P}_A = \bar{P}_A A^{\oplus} = 0$
(iv) $\bar{P}_A = 0$ if and only if A is nonsingular
(v) $A = P(D \oplus 0)P^{\sim}$ then $\bar{P}_A = P(0 \oplus I_{n-r})P^{\sim}$
(vi) $rank(\bar{P}_A) = n - rank(A)$

Proof. Noting that $\bar{P}_A = I - AA^{\oplus}$, A is EP in $\mathcal{M}$ and using Theorem (2.4) we have

$$\begin{aligned} (A+\bar{P}_A)^{-1}(I-\bar{P}_A) &= (A+\bar{P}_A)^{-1} - (A+\bar{P}_A)^{-1}\bar{P}_A \\ &= (A+\bar{P}_A)^{-1} - (A+\bar{P}_A)^{-1}(I - AA^{\oplus}) \\ &= (A+\bar{P}_A)^{-1}(A+\bar{P}_A-\bar{P}_A)A^{\oplus} \\ &= (A+\bar{P}_A)^{-1}(A+\bar{P}_A)A^{\oplus} - (A+\bar{P}_A)^{-1}\bar{P}_A A^{\oplus} \\ &= A^{\oplus} - (A+\bar{P}_A)^{-1}(I-AA^{\oplus})A^{\oplus} \\ &= A^{\oplus} \end{aligned}$$

The remaining statements are obvious. ■

Remark 3 *If A is range symmetric in $\mathcal{M}$, then using Theorem (2.4), the definition of the Minkowski inverse X of a matrix A reduces to $AXA = A$, $XAX = X$, and $(AX)^{\sim} =$*

$(XA)^{\sim}$. which is analogous to the definition of the Group inverse in the Minkowski space $\mathcal{M}$. Also from [2, Theorem 2.9] we have if A is range symmetric in $\mathcal{M}$ and rank$(A^2) =$ rank(A) then $A^{\oplus}$ exists. But rank$(A^2) =$ rank(A) is the necessary and sufficient condition for the existence of the group inverse $A^{\sharp}$ of a matrix A [1, page156]. Thus in this case we have $A^{\oplus} = A^{\sharp}$

5. Commutativity with an range symmetric element

In this section some characterizations of the matrices which commute with their Minkowski inverse are obtained in terms of m-projections. Using the involution defined in the Remark (*2*) and the fact that every m-projection is also a projection we have an analogous results in the settings of a C^*-algebra. The following result of [19] is a very useful representation of the Weighted Minkowski inverse of a matrix $A \in M_{m,n}(\mathbb{C})$ and is used in this section.

Theorem 5.1 Let $A \in M_{m,n}(\mathbb{C}$ in $\mathcal{M}$ and $A = N^{-1}G_1A^*G_2M$ where $M \in M_m(\mathbb{C})$ and $N \in M_n(\mathbb{C})$ are positive definite matrices and G_1 and G_2 are Minkowski metric matrices of order $n \times n$ and $m \times m$ respectively such that $\sigma(A^{\alpha}) \subset R^+$. Then

$$A^{\oplus}_{M,N} = \lim_{t \to 0}(tI + A^{\sim}A)^{-1}A^{\sim}$$

In particular when $M = I_n$ and $N = I_n$, then $A^{\oplus}_{M,N}$ reduces to the Minkowski inverse $A^{\oplus}$ of A in $\mathcal{M}$

Lemma 5.2 Let P be an m-projection and A be Minkowski invertible i,e. $A^{\oplus}$ exists such that $PA = AP$ then

(i) $A^{\oplus}P = PA^{\oplus}$
(ii) PAP is Minkowski Invertible and $(PAP)^{\oplus} = PA^{\oplus}P$

Proof. (*i*) From $AP = PA$ and $P^{\sim} = P$ we have $A^{\sim}P = PA^{\sim}$. This gives

$$A^{\sim}PA = PA^{\sim}A \Rightarrow A^{\sim}AP = PA^{\sim}A.$$

Thus for some $t \in R^+$ we have

$$\Rightarrow P(A^{\sim}A + tI_n)^{-1} = (A^{\sim}A + tI_n)^{-1}P \tag{22}$$

Now using the Theorem (5.1) we deduce that $A^{\oplus}P = PA^{\oplus}$
(*ii*) It follows from the direct verification. ∎

Theorem 5.3 Let $A, B \in M_n(\mathbb{C})$ such that A is EP in $\mathcal{M}$ and $AB = BA = 0$. Then

(i) $\bar{P}_AB = B = B\bar{P}_A$
(ii) $A^{\oplus}B = BA^{\oplus} = 0$
(iii) $B^{\oplus}$ exists implies $AB^{\oplus} = B^{\oplus}A = 0$
(iv) $B^{\oplus}$ exists implies $(A + B)^{\oplus}$ exists and $(A + B)^{\oplus} = A^{\oplus} + B^{\oplus}$
(v) B is EP in $\mathcal{M}$ implies $A + B$ is EP in $\mathcal{M}$ and $\bar{P}_{A+B} = \bar{P}_A + \bar{P}_B - I$

Proof. (i) we have

$$\begin{aligned}(A+\bar{P}_A)(I-\bar{P}_A)B &= (A+\bar{P}_A-A\bar{P}_A-\bar{P}_A)B \\ &= AB+\bar{P}_AB-A\bar{P}_AB-\bar{P}_AB \\ &= 0.\end{aligned} \tag{23}$$

This gives $(A+\bar{P}_A)(I-\bar{P}_A)B=0$. But $(A+\bar{P}_A)\in\mathcal{M}^{-1}$. Therefore $(I-\bar{P}_A)B=0$ implies $\bar{P}_AB=B$. In a similar fashion, by using $B(I-\bar{P}_A)(A+\bar{P}_A)=0$ we get $B\bar{P}_A=B$. The equality (i) can also be obtained directly by using the give condition that A is EP in $\mathcal{M}$ and $\bar{P}_A=I-AA^{\oplus}$.

(ii) Recall that in Theorem (4.1) we have proved that $A^{\oplus}=(\bar{P}_A+A)^{-1}-\bar{P}_A$. Also $(\bar{P}_A+A)B=B$ implies $B=(\bar{P}_A+A)^{-1}B$. Therefore

$$\begin{aligned}A^{\oplus}B &= \{(\bar{P}_A+A)^{-1}-\bar{P}_A\}B \\ &= (\bar{P}_A+A)^{-1}B-\bar{P}_AB \\ &= 0\end{aligned} \tag{24}$$

This can also be proved by using Corollary (4.3).

(iii) From lemma (5.2), statement (ii) we have $(\bar{P}_AB\bar{P}_A)^{\oplus}=\bar{P}_AB^{\oplus}\bar{P}_A$. Also $\bar{P}_AB\bar{P}_A=B$. Therefore $B^{\oplus}=\bar{P}_AB^{\oplus}\bar{P}_A$ implies $AB^{\oplus}=A\bar{P}_AB^{\oplus}\bar{P}_A=0$. Similarly $B^{\oplus}A=0$.

(iv) This follows from direct verification.

(v) B is EP in $\mathcal{M}$ implies $BB^{\oplus}=B^{\oplus}B$. Also $(A+B)^{\oplus}=(A^{\oplus}+B^{\oplus})$. Using Theorem (2.4) we have $A+B$ is EP in $\mathcal{M}$. Also $\bar{P}_{A+B}=\bar{P}_A+\bar{P}_B-I$ follows by doing simple algebra. ∎

Let us define the norm of a matrix A as given in [34, page 49]. For $A\in M_{m,n}(\mathbb{C})$ lets define the norm of A as $\|A\|=\{tr(A^TA)\}^{1/2}=\sqrt{\sum_{i=1}^{m}\sum_{j=1}^{n}a_{ij}^2}$, then we have the following result used in next theorem from [34, page 49].

Lemma 5.4 $\|AB\|\leqslant\|A\|.\|B\|$ and $\|A+B\|\leqslant\|A\|+\|B\|$. Also if P is a projection then $\|AP\|\leqslant\|A\|$ with equality if $PA=A$.

Theorem 5.5 Let $A,B\in M_n(\mathbb{C})$ such that A is EP in $\mathcal{M}$, then $\|(I-\bar{P}_A)B\bar{P}_A\|\leqslant\|A^{\oplus}\|.\|AB-BA\|$. and $\|\bar{P}_AB(I-\bar{P}_A)\|\leqslant\|A^{\oplus}\|.\|AB-BA\|$

Proof. Since $\bar{P}_A$ and $(I-\bar{P}_A)$ are idempotent projections, we have $\|\bar{P}_A\|=\|(I-\bar{P}_A)\|=1$. Also $(I-\bar{P}_A)(AB-BA)\bar{P}_A=AB\bar{P}_A-\bar{P}_AAB\bar{P}_A+\bar{P}_ABA\bar{P}_A=AB\bar{P}_A$. This gives $\|(I-\bar{P}_A)(AB-BA)\bar{P}_A\|=\|AB\bar{P}_A\|$.

Thus we have

$$\begin{aligned}\|(I-\bar{P}_A)B\bar{P}_A\| &= \|\{I-(I-AA^{\oplus})\}B\bar{P}_A\| \\ &\leqslant \|A^{\oplus}\|.\|AB\bar{P}_A\| \\ &= \|A^{\oplus}\|.\|(I-\bar{P}_A)(AB-BA)\bar{P}_A\| \\ &= \|A^{\oplus}\|.\|AB-BA\|\end{aligned}$$

Analogously we can prove the second inequality. ∎

Corollary 5.6 Let $A, B \in M_n(\mathbb{C})$ such that A is EP in $\mathcal{M}$. If $AB = BA$ then $B\bar{P}_A = \bar{P}_A B$

Proof. The proof follows by using $AB = BA$ in the above proved inequalities. ■

Theorem 5.7 Let $A, B \in M_n(\mathbb{C})$ such that A and B are EP in $\mathcal{M}$ and $AB = BA$. Then

(i) $\bar{P}_A\bar{P}_B = \bar{P}_B\bar{P}_A$
(ii) $AB^{\oplus} = B^{\oplus}A$ and $A^{\oplus}B = BA^{\oplus}$
(iii) $A^{\oplus}B^{\oplus} = B^{\oplus}A^{\oplus} = (AB)^{\oplus}$

Proof. (i) Since A and B are both EP in $\mathcal{M}$. Therefore from Corollary (5.6) we have $\bar{P}_A B = B\bar{P}_A$ and $\bar{P}_B A = A\bar{P}_B$. Also from lemma (5.2), point (i) we have $B^{\oplus}\bar{P}_A = \bar{P}_A B^{\oplus}$. Now $\bar{P}_A\bar{P}_B = \bar{P}_A(I - BB^{\oplus}) = \bar{P}_A - \bar{P}_A BB^{\oplus} = \bar{P}_B\bar{P}_A$

(ii) Again using corollary (5.6) we have

$$AB + A\bar{P}_B = BA + \bar{P}_B A \tag{25}$$

which implies

$$A(B + \bar{P}_B) = (B + \bar{P}_B)A \tag{26}$$

Thus invertibility of $(B + \bar{P}_B)$ implies

$$(B + \bar{P}_B)^{-1}A = A(B + \bar{P}_B)^{-1} \tag{27}$$

Hence we have

$$\begin{aligned} AB^{\oplus} &= A\{(B + \bar{P}_B)^{-1} - \bar{P}_B\} \\ &= A(B + \bar{P}_B)^{-1} - A\bar{P}_B \\ &= B^{\oplus}A \end{aligned} \tag{28}$$

Similarly, $A^{\oplus}B = BA^{\oplus}$

(iii) we have

$$(B + \bar{P}_B)^{-1}A = A(B + \bar{P}_B)^{-1} \tag{29}$$

Similarly,

$$(B + \bar{P}_B)^{-1}\bar{P}_A = \bar{P}_A(B + \bar{P}_B)^{-1} \tag{30}$$

Adding equations (29) and (30), we get

$$(B + \bar{P}_B)^{-1}(A + \bar{P}_A)^{-1} = (A + \bar{P}_A)^{-1}(B + \bar{P}_B)^{-1} \tag{31}$$

Now

$$A^{\oplus}B^{\oplus} = \{(A + \bar{P}_A)^{-1} - \bar{P}_A\}\{(B + \bar{P}_B)^{-1} - \bar{P}_B\}$$

Using the equation (31) we get the result. ■

6. Conclusion

In this paper we have studied the algebraic and the geometric behavior of m-projections in Minkowski Space, and have established a relation between an m-projection and the Minkowski inverse associated with a matrix. Further we have characterized some Range Symmetric elements in Minkowski Space by using m-projections.The next step naturally will be directed towards establishing the full characterization of the Range Symmetric elements in the Minkowski Space.

Aknowledgement

Mohd Saleem Lone would like to express his sincere thanks to University Grants Commission for their financial support under BSR scheme through grant no. F25-1/2014-15(BSR)/7-254/2009(BSR).(20.01.2015).

References

[1] A. Ben-isreal, T. Greville, Generalized inverse: Theory and applications. New York, Springer Verlag, 2nd ed. 2003.
[2] A. R. Meenakshi, Range symmetric matrices in Minkowski space. Bulletin of Malaysian Math Sci Society, 23 (2000), 45-52.
[3] —, Generalized inverse of matrices in Minkowski space. Proc. Nat. Seminar Alg. Appln, 1 (2000), 1-14.
[4] A. R. Meenakshi and D. Krishnaswamy, Product of range symmetric block matrices in minkowski space. Bulletin of Malaysian Math Sci Society 29 (2006), 59-68.
[5] C. Schmoeger, Generalized projections in Banach algebra. Linear Algebra and its Applications, 430 (2009), 601-608.
[6] D. S. Djordevie, Product of EP operators on Hilbert space. Proc Amer Math Soc., 129 (2000), 1727-1731.
[7] D. S. Djordjevic, Y. Wei, Operators with equal projections related to their generalized inverse. Applied Mathematics and Computations, 155 (2004), 655-664.
[8] E. Boasso, On the moore penrose inverse, EP banach space operators and EP banach algebra elements. J Math. Anal. Appl., 339 (2008), 1003-1014.
[9] H. Schwerdtfeger, Introduction to linear algebra and the theory of matrices. Groningen, P. Noordhoff, 1950.
[10] H. Seppo and N. Kenneth, On projections in a space with an indefinite metric. Linear Algebra and its Applications. 208/209 (1994), 401-417.
[11] H. K. Du and Y. Li, The spectral characterization of generalized projections. Linear Algebra Appl. 400(2005), 313318.
[12] I. J. Katz, Wigmann type theorems for EP_r matrices. Duke Math J. 32 (1965), 423-428.
[13] I. Gohberg, P. Lancaster and L. Rodman, Indefinite linear algebra and applications. Basel, Boston, Berlin, Brikhauser verlag, 2005.
[14] J. Gro and K. Trenkler, Generalized and hypergeneralized projectors. Linear Algebra Appl. 364 (1997), 463-474.
[15] J. Gro, On the product of orthogonal projectors. Linear Algebra and its Applications, 289 (1999), 141-150.
[16] J. Rebaza, A first course in applied mathematics. New Jersey , Wiley 2012.
[17] J. J. Koliha, A simple proof of the product theorem for EP matrices. Linear Algebra and its Applications, 294 (1999), 213-215.
[18] J. K. Baksalary, O. M. Baksalary and X. Liu, Further properties of generalized and hypergeneralized projectors. Linear Algebra Appl. 389 (2004), 295-303.
[19] K. Adem, Z. A. Zhour, The representation and approximation for the weighted minkowski inverse in minkowski space. Mathematical and computer modeling, 47 (2008), 363-371.
[20] L. Lebtahi, N. Thome, A note on k-generalized projections. Linear Algebra Appl. 420 (2007), 572-575.
[21] M. Pearl, On normal and EP_r matrices. Michigan Math J. 6 (1959), 1-5.
[22] M. Z. Petrovic, S. Stanimirovic, Representation and computation of $\{2, 3^{\sim}\}$ and $\{2, 4^{\sim}\}$-inverses in indefinite inner product spaces. Applied Mathematics and Computation, 254 (2015), 157-171.
[23] O. Baksalary, G. Trenkler, Functions of orthogonal projectors involving the moore-penrose inverse. Comput. Math. Appl. 59 (2010), 764-778.
[24] O. M. Baksalary, G. Trenkler, Revisitation of the product of two orthogonal projectors. Linear Algebra and its Applications, 430 (2009), 2813-2833.
[25] P. R. Halmos, Finite-dimensional vector spaces. New York, Springer Verlag, 1974.
[26] R. Piziak, P. L. Odell and R. Hahn, Constructing projections on sums and intersections. Comput. Math. Appl. 37 (1999), 67-74.
[27] R. B. Bapat, J. S. Kirkland and K. M. Prasad, Combinatorial matrix theory and generalized inverses of matrices. New Delhi, Springer Verlag, 2013.

[28] R. E. Hartwig, I. J. Katz, On product of EP matrices. Linear Algebra and its Applications, 252 (1997), 339-345.
[29] R. Piziak et. al, Constructing projections on sums and intersections. Compt. Math. Appl. 37 (1999), 67-74.
[30] S. Cheng, Y. Tian, Two sets of new characterizations for normal and EP matrices. Linear Algebra and its Applications, 18 (1975), 327-333.
[31] S. L. Campbell, C. D. Meyer, EP operators and generalized inverse. Canad Math Bull., 18 (1975), 327-33.
[32] S. L. Campbell, C.D. Meyer, Generalized inverse of linear transformation. New York, Dover Publications, 1991.
[33] T. S. Baskett, I. J. Katz, Theorems on product of EP_r matrices. Linear Algebra and its Applications, 2 (1969), 87-103.
[34] Y. Haruo, et al. Projection matrices, generalized inverse matrices and singular value decomposition. New York , Springer Verlag, 2011.
[35] Y. Tian, G. P. H. Styan, Rank equalities for idempotent and involutory matrices. Linear Algebra Appl. 335 (2001), 101-117.

7

Normalized Laplacian spectrum of two new types of join graphs

M. Hakimi-Nezhaad[a], M. Ghorbani[a*]

[a] *Department of Mathematics, Faculty of Science, Shahid Rajaee Teacher Training University, Tehran, 16785-136, Iran.*

Abstract. Let G be a graph without an isolated vertex, the normalized Laplacian matrix $\tilde{\mathcal{L}}(G)$ is defined as $\tilde{\mathcal{L}}(G) = \mathcal{D}^{-\frac{1}{2}}\mathcal{L}(G)\mathcal{D}^{-\frac{1}{2}}$, where $\mathcal{D}$ is a diagonal matrix whose entries are degree of vertices of G. The eigenvalues of $\tilde{\mathcal{L}}(G)$ are called as the normalized Laplacian eigenvalues of G. In this paper, we obtain the normalized Laplacian spectrum of two new types of join graphs. In continuing, we determine the integrality of normalized Laplacian eigenvalues of graphs. Finally, the normalized Laplacian energy and degree Kirchhoff index of these new graph products are derived.

Keywords: Join of graphs, normalized Laplacian eigenvalue, integral eigenvalue.

1. Introduction

Let $G = (V, E)$ be a simple graph (namely a graph without loop and multiple edges) on n vertices and m edges and the degree of vertex $v \in V(G)$ is denoted by $deg(v)$. The eigenvalues of the adjacency matrix $\mathcal{A} = \mathcal{A}(G)$ of G are called the eigenvalues of G denoted by $\lambda_1(G) \geqslant \lambda_2(G) \geqslant \cdots \geqslant \lambda_n(G)$. Let $\lambda_1(G), \lambda_2(G), \ldots, \lambda_s(G)$ be the distinct eigenvalues of G with multiplicity $t_1, t_2, \ldots, t_s$, respectively. The multiset $\{[\lambda_1(G)]^{t_1}, [\lambda_2(G)]^{t_2}, \ldots, [\lambda_s(G)]^{t_s}\}$ of eigenvalues of $\mathcal{A}$ is called the spectrum of G.

*Corresponding author.
E-mail address: mghorbani@srttu.edu (M. Ghorbani).

The energy of graph G is a graph invariant introduced by Ivan Gutman as

$$\mathcal{E}(G) = \sum_{i=1}^{n} |\lambda_i(G)|,$$

see for more details [8, 10, 17]. Let $\mathcal{L}(G) = \mathcal{D}(G) - \mathcal{A}(G)$ be the Laplacian matrix of graph G, where $\mathcal{D}(G) = [d_{ij}]$ is the diagonal matrix whose entries are degree of vertices, namely, $d_{ii} = deg(v_i)$ and $d_{ij} = 0$ for $i \neq j$. The matrix $\mathcal{Q}(G) = \mathcal{D}(G) + \mathcal{A}(G)$ is called the sigenless Laplacian matrix of G. The sigenless Laplacian eigenvalues of G are denoted by $q_1(G), q_2(G), \ldots, q_n(G)$. If G is regular of degree r, then the eigenvalues of $\mathcal{Q}(G)$ are $2r, r + \lambda_2, \ldots, r + \lambda_n$, see [1, p.14]. In [15], Jooyandeh et al. introduced the concept of incidence energy of a graph as

$$\mathcal{RE}(G) = \sum_{i=1}^{n} \sqrt{q_i(G)}. \tag{1}$$

Let G be a graph without an isolated vertex, the normalized Laplacian matrix $\tilde{\mathcal{L}}(G)$ is defined as $\tilde{\mathcal{L}}(G) = \mathcal{D}^{-\frac{1}{2}}\mathcal{L}(G)\mathcal{D}^{-\frac{1}{2}}$ which implies that its $(i,j)-$entry is 1 if $i = j$, and it is $-1/\sqrt{deg(v_i)deg(v_j)}$, if $i \neq j$ and the vertices v_i, v_j are adjacent, and is zero otherwise. The eigenvalues of $\tilde{\mathcal{L}}(G)$ are the roots of $g(x) = det(\delta I - \tilde{\mathcal{L}}(G))$ and we call them as normalized Laplacian eigenvalues of G, denoted by $\delta_1(G), \delta_2(G), \ldots, \delta_n(G)$, where $\delta_n(G) = 0$ for all graphs, see for more details [2, 3, 6].

The complement of graph G is denoted by $\overline{G}$ and a complete graph on n vertices is denoted by K_n. Also a cycle graph with n vertices is denoted by C_n. Clearly, $\overline{K_n}$ is empty graph. The subdivision graph $S(G)$ of a graph G is obtained by inserting an additional vertex in the middle of each edge of G. Equivalently, each edge of G is replaced by a path of length 2, see Figure 1. It is a well-known fact that $P_{S(G)}(x) = x^{m-n}\mathcal{Q}_G(x^2)$, where $P_G(x)$ and $\mathcal{Q}_G(x)$ denote the characteristic polynomial and the signless Laplacian characteristic polynomial of G, respectively [7, p. 63]. From the definition, one can see that $\pm\sqrt{q_1(G)}, \pm\sqrt{q_2(G)}, \ldots, \pm\sqrt{q_n(G)}$ and $[0]^{m-n}$ are all eignvalues of $S(G)$.

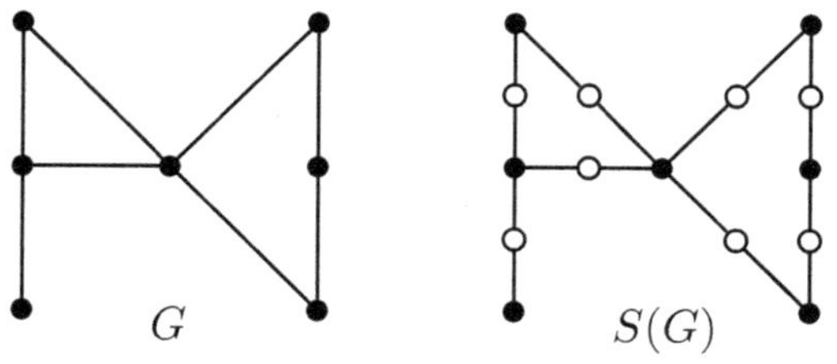

Figure 1. Two graphs G and subdivision $S(G)$.

Indulal in [13] defined two classes of join graphs as follows:

Definition 1.1 The S_{vertex} join of two graphs G_1 and G_2 denoted by $G_1\dot{\vee}G_2$ is obtained from $S(G_1)$ and G_2 by joining all vertices of G_1 with all vertices of G_2, see Figure 2.

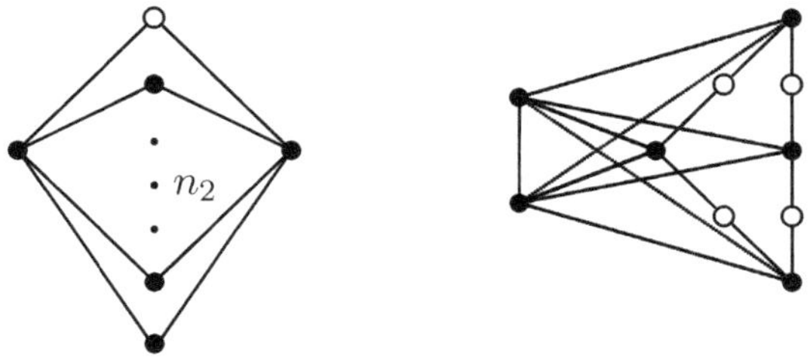

Figure 2. Graphs $K_2\dot{\vee}\overline{K}_{n_2}$ and $C_4\dot{\vee}K_2$.

Definition 1.2 The S_{edge} join of two graphs G_1 and G_2 denoted by $G_1 \underline{\vee} G_2$ is obtained from $S(G_1)$ and G_2 by joining all vertices of $S(G_1)$ corresponding to the edges of G_1 with all vertices of G_2, see Figure 3.

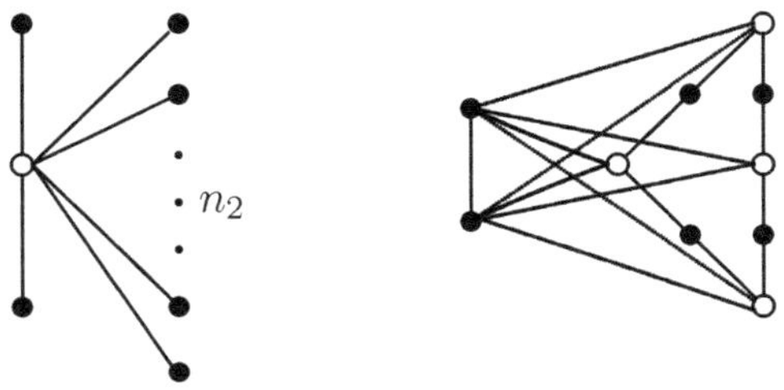

Figure 3. Graphs $K_2 \underline{\vee} \overline{K}_{n_2}$ and $C_4 \underline{\vee} K_2$.

The following results are crucial throughout this paper.

Lemma 1.3 [7] Let G be a $r-$regular $(n, m)-$graph with adjacency matrix $\mathcal{A}(G)$ and incidence matrix $\mathcal{R}$. Let $L(G)$ be its line graph. Then

$$\mathcal{R}\mathcal{R}^T = \mathcal{A}(G) + rI \quad \text{and} \quad \mathcal{R}^T\mathcal{R} = \mathcal{A}(\mathcal{L}(G)) + 2I.$$

Moreover, if G has an eigenvalue equals to $-r$ with an eigenvector $\mathcal{V}$, then G is bipartite and $\mathcal{R}^T\mathcal{V} = 0$.

Lemma 1.4 [7] Let G be an $r-$regular $(n, m)-$graph with eigenvalues $r, \lambda_2, \ldots, \lambda_n$. Then $\{[2r-2]^1, [\lambda_i + r - 2]^1, [-2]^{m-n}\}$, $(2 \leqslant i \leqslant n)$ consist the spectrum of line grah. Also $\mathcal{V}$ is an eigenvector belonging to the eigenvalue -2 if and only if $\mathcal{R}\mathcal{V} = 0$.

2. The normalized Laplacian spectra of two new join graphs

In this section, we determine the normalized Laplacian spectrum of $G_1\dot{\vee}G_2$ and $G_1 \underline{\vee} G_2$, where G_i is r_i-regular $(i = 1, 2)$. Consider two graphs G_1, G_2 with $V(G_1) = \{u_1, u_2, \ldots, u_{n_1}\}$, $E(G_1) = \{e_1, e_2, \ldots, e_{m_1}\}$ and $V(G_2) = \{v_1, v_2, \ldots, v_{n_2}\}$. Then $\mathbf{1}_k$ and $\mathbf{0}_k$ are two vectors of order k with all elements equal to 1 and 0, respectively. Moreover $\mathbf{0}_{i\times j}$ denotes an $i \times j$ matrix whose entries are zero, $\mathbf{J}_{i\times j}$ is one whose entries are 1 and $\mathbf{I}_n$ is the indentity matrix of order n.

Theorem 2.1 For $i = 1, 2$, let G_i be r_i-regular graph with n_i vertices with incidence matrix $\mathcal{R}$ and eigenvalues $r_i = \lambda_1(G_i) \geqslant \lambda_2(G_i) \geqslant \cdots \geqslant \lambda_n(G_i)$. Then the normalized

Laplacian spectrum of $G_1\dot{\vee}G_2$ is $\{[0]^1, [1 \pm \sqrt{\frac{\lambda_j(G_1)+r_1}{2(n_2+r_1)}}]^1, [1-\frac{\lambda_k(G_2)}{n_1+r_2}]^1, [1]^{m_1-n_1}\}$, $(2 \leqslant j \leqslant n_1)$, $(2 \leqslant k \leqslant n_2)$ together with the roots of

$$x^2 - \frac{3n_1+2r_2}{n_1+r_2}x + \frac{2n_1n_2+2r_1n_1+n_2r_2}{(n_1+r_2)(n_2+r_1)} = 0.$$

Proof. Let G_i be an r_i−regular graph with n_i vertices and an adjacency matrix $\mathcal{A}(G_i)$, $i = 1, 2$. Let $\mathcal{R}$ be the incidence matrix of G_1. Then by a proper labeling of vertices, the normalized Laplacian matrix of $G_1\dot{\vee}G_2$ can be written as

$$\begin{aligned}\tilde{\mathcal{L}}(G_1\dot{\vee}G_2) &= \mathcal{D}^{-\frac{1}{2}}\mathcal{L}(G_1\dot{\vee}G_2)\mathcal{D}^{-\frac{1}{2}} \\ &= \begin{bmatrix} \mathbf{I}_{n_1} & \frac{-\mathcal{R}_{n_1\times m_1}}{\sqrt{2(n_2+r_1)}} & \frac{-\mathbf{J}_{n_1\times n_2}}{\sqrt{(n_1+r_2)(n_2+r_1)}} \\ \frac{-\mathcal{R}^T_{m_1\times n_1}}{\sqrt{2(n_2+r_1)}} & \mathbf{I}_{m_1} & \mathbf{0}_{m_1\times n_2} \\ \frac{-\mathbf{J}_{n_2\times n_1}}{\sqrt{(n_1+r_2)(n_2+r_1)}} & \mathbf{0}_{n_2\times m_1} & \mathbf{I}_{n_2} - \frac{\mathcal{A}(G_2)}{n_1+r_2} \end{bmatrix},\end{aligned}$$

where

$$\mathcal{D} = \begin{bmatrix} (n_2+r_1)\mathbf{I}_{n_1} & \mathbf{0}_{n_1\times m_1} & \mathbf{0}_{n_1\times n_2} \\ \mathbf{0}_{m_1\times n_1} & 2\mathbf{I}_{m_1} & \mathbf{0}_{m_1\times n_2} \\ \mathbf{0}_{n_2\times n_1} & \mathbf{0}_{n_2\times m_1} & (n_1+r_2)\mathbf{I}_{n_2} \end{bmatrix},$$

and

$$\mathcal{L}(G_1\dot{\vee}G_2) = \begin{bmatrix} (n_2+r_1)\mathbf{I}_{n_1} & -\mathcal{R}_{n_1\times m_1} & -\mathbf{J}_{n_1\times n_2} \\ \mathcal{R}^T_{m_1\times n_1} & 2I_{m_1} & \mathbf{0}_{m_1\times n_2} \\ -\mathbf{J}_{n_2\times n_1} & \mathbf{0}_{n_2\times m_1} & (n_1+r_2)\mathbf{I}_{n_2} - \mathcal{A}(G) \end{bmatrix}.$$

$\mathbf{1}_{n_1}$ is an eigenvector corresponding to eigenvalue r_1 of G and the other eigenvectors are orthogonal to $\mathbf{1}_{n_1}$. Let X be an eigenvector of G_1 corresponding to an eigenvalue $\lambda_j(G_1) \neq r_1$, $2 \leqslant j \leqslant n_1$. Then $\Phi = [\alpha X \ \mathcal{R}^T X \ 0]^T$ is eigenvector of $\tilde{\mathcal{L}}(G_1\dot{\vee}G_2)$ corresponding to the normalized eigenvalue δ of graph $G_1\dot{\vee}G_2$. Then $\tilde{\mathcal{L}}.\Phi = \theta\Phi$. Thus

$$\begin{bmatrix} \mathbf{I}_{n_1} & \frac{-\mathcal{R}_{n_1\times m_1}}{\sqrt{2(n_2+r_1)}} & \frac{-\mathbf{J}_{n_1\times n_2}}{\sqrt{(n_1+r_2)(n_2+r_1)}} \\ \frac{-\mathcal{R}^T_{m_1\times n_1}}{\sqrt{2(n_2+r_1)}} & \mathbf{I}_{m_1} & \mathbf{0}_{m_1\times n_2} \\ \frac{-\mathbf{J}_{n_2\times n_1}}{\sqrt{(n_1+r_2)(n_2+r_1)}} & \mathbf{0}_{n_2\times m_1} & \mathbf{I}_{n_2} - \frac{\mathcal{A}(G_2)}{n_1+r_2} \end{bmatrix} \begin{bmatrix} \alpha X \\ \mathcal{R}^T X \\ 0 \end{bmatrix} = \theta \begin{bmatrix} \alpha X \\ \mathcal{R}^T X \\ 0 \end{bmatrix}.$$

This implies that

$$\begin{bmatrix} (\alpha - \frac{\mathcal{R}\mathcal{R}^T}{\sqrt{2(n_2+r_1)}})X \\ (1 - \frac{\alpha}{\sqrt{2(n_2+r_1)}})\mathcal{R}^T X \\ 0 \end{bmatrix} = \begin{bmatrix} \theta\alpha X \\ \theta\mathcal{R}^T X \\ 0 \end{bmatrix}.$$

The last equality follows from Lemma 1.3. By solving these equations, we get $\alpha =$

$\pm\sqrt{\lambda_j(G_1)+r_1}$, $2 \leqslant j \leqslant n_1$. Therefore, $\Phi_k = [\pm\sqrt{\lambda_j(G_1)+r_1}X\ \mathcal{R}^T X\ 0]^T$ is an eigenvector of $\tilde{\mathcal{L}}(G_1\dot{\vee}G_2)$ corresponding to the normalized eigenvalues θ_k of graph $G_1\dot{\vee}G_2$, where $\theta_k = 1 \pm \sqrt{\frac{\lambda_j(G_1)+r_1}{2(n_2+r_1)}}$ $(k = 1, 2)$.

Now consider the $m_1 - n_1$ linearly independent eigenvectors $\mathcal{Z}_l$ $(1 \leqslant l \leqslant m_1 - n_1)$ of $L(G_1)$ corresponding to the eigenvalue -2. Then by Lemma 1.4, $\mathcal{R}\mathcal{Z}_l = 0$. Consequently $\Omega_l = [0\ \mathcal{Z}_l\ 0]^T$ is an eigenvector of $\tilde{\mathcal{L}}(G_1\dot{\vee}G_2)$ with an eigenvalue 1, since

$$\tilde{\mathcal{L}}.\Omega = \begin{bmatrix} \mathbf{I}_{n_1} & \frac{-\mathcal{R}_{n_1\times m_1}}{\sqrt{2(n_2+r_1)}} & \frac{-\mathbf{J}_{n_1\times n_2}}{\sqrt{(n_1+r_2)(n_2+r_1)}} \\ \frac{-\mathcal{R}^T_{m_1\times n_2}}{\sqrt{2(n_2+r_1)}} & \mathbf{I}_{m_1} & \mathbf{0}_{m_1\times n_2} \\ \frac{-\mathbf{J}_{n_2\times n_1}}{\sqrt{(n_1+r_2)(n_2+r_1)}} & \mathbf{0}_{n_2\times m_1} & \mathbf{I}_{n_2} - \frac{\mathcal{A}(G_2)}{n_1+r_2} \end{bmatrix} \begin{bmatrix} 0 \\ \mathcal{Z}_l \\ 0 \end{bmatrix} = \mathbf{I}_{n_1\times m_1} \begin{bmatrix} 0 \\ \mathcal{Z}_l \\ 0 \end{bmatrix}.$$

On the other hand, 1_{n_2} is an eigenvector of G_2 corresponding to the eigenvalue r_2, in which all other eigenvectors are orthogonal to 1_{n_2}. Let Y be the eigenvector of G_2 corresponding to the eigenvalue $\lambda_j(G_2) \neq r_2$ $(2 \leqslant j \leqslant n_2)$. By a similar arguments we can deduce that $\Psi = [0\,0\,Y]^T$ is eigenvector of $\tilde{\mathcal{L}}(G_1\dot{\vee}G_2)$ corresponding to the eigenvalue $1 - \frac{\lambda_j(G_2)}{n_1+r_2}$ $(2 \leqslant j \leqslant n_2)$. Let $\Lambda = [\alpha\mathbf{J}_{n_1\times 1}\ \beta\mathbf{J}_{n_1\times 1}\ \gamma\mathbf{J}_{n_1\times 1}]^T$ for some vector $(\alpha, \beta, \gamma) \neq (0, 0, 0)$. The equation $\tilde{\mathcal{L}}(G_1\dot{\vee}G_2).\Lambda = x\Lambda$ yields that Λ is an eigenvector of $\mathcal{A}$ corresponding to eigenvalue x if and only if $\Lambda = [\alpha\ \beta\ \gamma]^T$ is an eigenvector of the matrix M, where

$$M = \begin{bmatrix} 1 & \frac{-r_1}{\sqrt{2(n_2+r_1)}} & \frac{-n_2}{\sqrt{(n_1+r_2)(n_2+r_1)}} \\ \frac{-2}{\sqrt{2(n_2+r_1)}} & 1 & 0 \\ \frac{-n_1}{\sqrt{(n_1+r_2)(n_2+r_1)}} & 0 & 1 - \frac{\mathcal{A}(G_2)}{n_1+r_2} \end{bmatrix}.$$

The characteristic polynomial of M is $x^3 - \frac{3n_1+2r_2}{n_1+r_2}x^2 + \frac{2n_1n_2+2r_1n_1+n_2r_2}{(n_1+r_2)(n_2+r_1)}x = 0$ and this completes the proof. ■

Corollary 2.2 If $G_2 \cong \overline{K}_{n_2}$ then the normalized Laplacian spectrum of $G_1\dot{\vee}G_2$ is $\{[0]^1$, $[1]^{m_1-n_1+n_2}$, $[1 \pm \sqrt{\frac{\lambda_j(G_1)+r_1}{2(n_2+r_1)}}]^1$, $[2]^1\}$, $(2 \leqslant j \leqslant n_1)$.

Theorem 2.3 For $i = 1, 2$, let G_i be an r_i-regular graph on n_i vertices with incidence matrix $\mathcal{R}$ and the eigenvalues $r_i = \lambda_1(G_i) \geqslant \lambda_2(G_i) \geqslant \cdots \geqslant \lambda_n(G_i)$. Then the normalized Laplacian spectrum of $G_1\veebar G_2$ is $\{[0]^1, 1\pm\sqrt{\frac{\lambda_j(G_1)+r_1}{r_1(n_2+2)}}, 1-\frac{\lambda_k(G_2)}{m_1+r_2}, [1]^{m_1-n_1}\}$, $(2 \leqslant j \leqslant n_1)$, $(2 \leqslant k \leqslant n_2)$, together with the roots of $x^2 - \frac{3m_1+2r_2}{m_1+r_2}x + \frac{2m_1n_2+4m_1+n_2r_2}{(m_1+r_2)(2+n_2)} = 0$.

Proof. It is not difficult to see that the normalized Laplacian matrix of $G_1 \veebar G_2$ can be written as follows:

$$\begin{aligned}\tilde{\mathcal{L}}(G_1 \veebar G_2) &= \mathcal{D}^{-\frac{1}{2}}\mathcal{L}(G_1 \veebar G_2)\mathcal{D}^{-\frac{1}{2}} \\ &= \begin{bmatrix} \mathbf{I}_{n_1} & \frac{-\mathcal{R}_{n_1\times m_1}}{\sqrt{r_1(n_2+2)}} & \mathbf{0}_{n_1\times n_2} \\ \frac{-\mathcal{R}^T_{m_1\times n_1}}{\sqrt{r_1(n_2+2)}} & \mathbf{I}_{m_1} & \frac{-\mathbf{J}_{m_1\times n_2}}{\sqrt{(m_1+r_2)(n_2+2)}} \\ \mathbf{0}_{n_2\times n_1} & \frac{-\mathbf{J}_{n_2\times m_1}}{\sqrt{(m_1+r_2)(n_2+2)}} & \mathbf{I}_{n_2} - \frac{\mathcal{A}(G_2)}{m_1+r_2} \end{bmatrix},\end{aligned}$$

where

$$\mathcal{D} = \begin{bmatrix} (n_2+2)\mathbf{I}_{n_1} & \mathbf{0}_{n_1\times m_1} & \mathbf{0}_{n_1\times n_2} \\ \mathbf{0}_{m_1\times n_1} & r_1\mathbf{I}_{m_1} & \mathbf{0}_{m_1\times n_2} \\ \mathbf{0}_{n_2\times n_1} & \mathbf{0}_{n_2\times m_1} & (m_1+r_2)\mathbf{I}_{n_2} \end{bmatrix},$$

and

$$\mathcal{L}(G_1 \veebar G_2) = \begin{bmatrix} (n_2+2)\mathbf{I}_{n_1} & -\mathcal{R}_{n_1\times m_1} & \mathbf{0}_{n_1\times n_2} \\ \mathcal{R}^T_{m_1\times n_1} & r_1 I_{m_1} & -\mathbf{J}_{m_1\times n_2} \\ \mathbf{0}_{n_2\times n_1} & -\mathbf{J}_{n_2\times m_1} & (m_1+r_2)\mathbf{I}_{n_2} - \mathcal{A}(G) \end{bmatrix}.$$

Continuing of the proof is similar to that of Theorem 2.1. ■

Corollary 2.4 If $G_2 \cong \overline{K}_{n_2}$, then the normalized Laplacian spectrum of $G_1 \veebar G_2$ is $\{[0]^1$, $[1]^{m_1-n_1+n_2}$, $[1 \pm \sqrt{\frac{\lambda_j(G_1)+r_1}{r_1(n_2+2)}}]^1$, $[2]^1\}$, $(2 \leqslant j \leqslant n_1)$.

3. Integrality of two new join graphs

The graph G is $\tilde{\mathcal{L}}-$integral if all its $\tilde{\mathcal{L}}-$eigenvalues are integral. Since all $\tilde{\mathcal{L}}-$eigenvalues of G are in interval $[0,2]$, we can deduced that G is $\tilde{\mathcal{L}}-$integral if and ony if its eigenvalues are 0, 1, or 2. The complete bipartite graphs are such graphs; in fact, no other connected graphs are $\tilde{\mathcal{L}}-$integral, see [14].

Proposition 3.1 [14] Let G be a connected graph. Then the following statements are equivalent.

1) G is bipartite with three $\tilde{\mathcal{L}}-$eigenvalues,
2) G is $\tilde{\mathcal{L}}-$integral,
3) G is complete bipartite.

The following propositions give a necessary and sufficient conditions in which S_{vertex} and S_{edge} joins of graphs are $\tilde{\mathcal{L}}-$integral.

Proposition 3.2 If G_i is r_i-regular connected graph $(i = 1, 2)$, then $G_1 \dot{\vee} G_2$ and $G_1 \veebar G_2$ are $\tilde{\mathcal{L}}-$integral if and only if $G_1 \cong K_2$ and $G_2 \cong \overline{K}_{n_2}$.

Proof.
Since $K_2 \dot{\vee} \overline{K}_{n_2} = K_{2,n_2}$ and $K_2 \veebar \overline{K}_{n_2} = K_{1,n_2+2}$ are two graphs of order n with the normalized Laplacian eigenvalues $\{[0]^1, [1]^{n-1}, [2]^1\}$, these graphs are $\tilde{\mathcal{L}}-$integral. Conversely, suppose that $G_1 \dot{\vee} G_2$ is $\tilde{\mathcal{L}}-$integral, $G_1 \ncong K_2$ and $G_2 \ncong \overline{K}_{n_2}$. By Theorem 2.1, the normalized Laplacian spectrum of $G_1 \dot{\vee} G_2$ is $\{[0]^1$, $[1 \pm \sqrt{\frac{\lambda_j(G_1)+r_1}{2(n_2+r_1)}}]^1$, $[1 - \frac{\lambda_k(G_2)}{n_1+r_2}]^1$, $[1]^{m_1-n_1}\}$, $(2 \leqslant j \leqslant n_1)$, $(2 \leqslant k \leqslant n_2)$ together with the roots of

$$x^2 - \frac{3n_1+2r_2}{n_1+r_2}x + \frac{2n_1n_2+2r_1n_1+n_2r_2}{(n_1+r_2)(n_2+r_1)} = 0.$$

Since G_1 is connected graph, $G_1 \dot{\vee} G_2$ is connected. Hence, $[0]^1$ is an eigenvalue of $G_1 \dot{\vee} G_2$. Thus, the other eigenvalues must be equal to $[1]^{n-1}$, $[2]^1$. Since $1 - \frac{\lambda_j(G_2)}{n_1+r_2} < 2$, $2 \leqslant j \leqslant n_2$, we have $\frac{\lambda_j(G_2)}{n_1+r_2} = 0$. So, $\lambda_j(G_2) = 0$, $2 \leqslant j \leqslant n_2$ and it implies that $G_2 \ncong \overline{K}_{n_2}$, a contradiction. Since $1-\sqrt{\frac{\lambda_j(G_1)+r_1}{2(n_2+r_1)}}<2$, $2\leqslant j\leqslant n_1$, we acheive $\sqrt{\frac{\lambda_j(G_1)+r_1}{2(n_2+r_1)}}=0$ and thus $\lambda_j(G_1)=-r_1$, $2\leqslant j\leqslant n_1$, a contradiction. On the other hand, $1<1+\sqrt{\frac{\lambda_j(G_1)+r_1}{2(n_2+r_1)}}\leqslant 2$, $2\leqslant j\leqslant n_1$, yields that $\sqrt{\frac{\lambda_j(G_1)+r_1}{2(n_2+r_1)}}=1$. So, $\lambda_j(G_1)=-r_1+2n_2>0$, $2\leqslant j\leqslant n_1$, a contradiction. As well as, for two eigenvalues that are the roots of

$$x^2-\frac{3n_1+2r_2}{n_1+r_2}x+\frac{2n_1n_2+2r_1n_1+n_2r_2}{(n_1+r_2)(n_2+r_1)}=0. \tag{2}$$

Since, G is connected and the normalized Laplacian eigenvalues are integral, one can deduce that Eq.(2) is equal to $(x-1)^2=x^2-2x+1$ or $(x-1)(x-2)=x^2-3x+2$. In the first case, one can conclude easily that $\frac{2n_1n_2+2r_1n_1+n_2r_2}{(n_1+r_2)(n_2+r_1)}=1$ which yields that $n_1n_2=r_2-n_1$, for $r_1\geqslant 1$ which comes to a contradiction, because for example, there is no an 8-regular graph of order 5. Thus, the second case holds, namely

$$x^2-\frac{3n_1+2r_2}{n_1+r_2}x+\frac{2n_1n_2+2r_1n_1+n_2r_2}{(n_1+r_2)(n_2+r_1)}=x^2-3x+2$$

and the proof is completed. ■

Proposition 3.3 Let G_1 be an empety graph of order n_1 and G_2 be $r-$regular graph of order n_2, then $G_1\dot{\vee}G_2$ and $G_1\veebar G_2$ are $\tilde{\mathcal{L}}-$integral if and only if $G_2\cong\overline{K}_{n_2}$ or G_2 is complete bipartite graph.

Proof. By using Proposition 3.1, Theorems 2.1 and 2.3, the proof is stright forward. ■

The normalized Laplacian energy of G is defined as $\tilde{\mathcal{LE}}(G)=\sum_{i=1}^{n}|\delta_i(G)-1|$, see [4, 12] for details.

Proposition 3.4 Let G_1 is an r_1-regular of graph of order n_1 and $G_2\cong\overline{K}_{n_2}$. Then

1) $\tilde{\mathcal{LE}}(G_1\dot{\vee}G_2)=2\Big(1+\frac{\mathcal{RE}(G_1)-\sqrt{2r_1}}{\sqrt{2(n_2+r_1)}}\Big)$,

2) $\tilde{\mathcal{LE}}(G_1\veebar G_2)=2\Big(1+\frac{\mathcal{RE}(G_1)-\sqrt{2r_1}}{\sqrt{r_1(n_2+2)}}\Big)$.

Proof. By Corollary 2.2, Corollary 2.4, respectively, we have

$$\begin{aligned}\tilde{\mathcal{LE}}(G_1\dot{\vee}G_2)&=\sum_{i=1}^{n}|\delta_i(G_1\dot{\vee}G_2)-1|\\&=1+\sum_{j=2}^{n_1}|\pm\sqrt{\frac{\lambda_j(G_1)+r_1}{2(n_2+r_1)}}|+1\\&=2+2\sum_{j=2}^{n_1}\sqrt{\frac{\lambda_j(G_1)+r_1}{2(n_2+r_1)}}\\&=2\Big(1+\frac{\mathcal{RE}(G_1)-\sqrt{2r_1}}{\sqrt{r_1(n_2+2)}}\Big),\end{aligned}$$

$$\begin{aligned}\tilde{\mathcal{LE}}(G_1 \veebar G_2) &= \sum_{i=1}^{n} |\delta_i(G_1 \veebar G_2) - 1| \\ &= 1 + \sum_{j=2}^{n_1} |\pm\sqrt{\frac{\lambda_j(G_1) + r_1}{r_1(n_2+2)}}| + 1 \\ &= 2 + 2\sum_{j=2}^{n_1}\sqrt{\frac{\lambda_j(G_1) + r_1}{r_1(n_2+2)}} \\ &= 2\Big(1 + \frac{\mathcal{RE}(G_1) - \sqrt{2r_1}}{\sqrt{r_1(n_2+2)}}\Big),\end{aligned}$$

where the last equality follows from $\mathcal{RE}(G_1) - \sqrt{2r_1} = \sum_{j=2}^{n_1}\sqrt{\lambda_j(G_1) + r_1}$. ■

The degree Kirchhoff index of graph G is defined by Chen et al. in [5] as

$$Kf^*(G) = \sum_{i<j} d_i r_{ij} d_j,$$

where r_{ij} denotes the resistance-distance between vertices v_i and v_j in a graph G. They proved that $Kf^*(G) = 2m\sum_{i=2}^{n}\frac{1}{\delta_i(G)}$, where $\delta_i(G)$ is normalized Laplacian eigenvalues of G of order n $(2 \leqslant i \leqslant n)$, see [11, 18] for details.

Proposition 3.5 Let G_1 is an r_1-regular graph of order n_1 with m_1 edges and $G_2 \cong \overline{K}_{n_2}$. Then

$$Kf^*(G_1\dot{\vee}G_2) = 2(2m_1 + n_1n_2)\Big(\frac{1}{2} + m_1 - n_1 + n_2 + \frac{(n_1-1)\sqrt{2(n_2+r_1)}}{(n_1-1)\sqrt{2(n_2+r_1)} \pm (\mathcal{RE}(G_1) - \sqrt{2r_1})}\Big),$$

$$Kf^*(G_1 \veebar G_2) = 2m_1(2+n_2)\Big(\frac{1}{2} + m_1 - n_1 + n_2 + \frac{(n_1-1)\sqrt{r_1(n_2+2)}}{(n_1-1)\sqrt{r_1(n_2+2)} \pm (\mathcal{RE}(G_1) - \sqrt{2r_1})}\Big).$$

Proof. It is not difficult to see that the number of edges of $G_1\dot{\vee}G_2$ and $G_1 \veebar G_2$ are $E(G_1\dot{\vee}G_2) = 2m_1 + n_1n_2$ and $E(G_1 \veebar G_2) = m_1(2+n_2)$, respectively. On the other hand, since G_1 is r_1-regular by using Eq.(1),

$$\mathcal{RE}(G_1) = \sum_{j=1}^{n_1}\sqrt{\lambda_j(G_1) + r_1}$$

and so

$$\mathcal{RE}(G_1) - \sqrt{2r_1} = \sum_{j=2}^{n_1}\sqrt{\lambda_j(G_1) + r_1}.$$

Now, Corollaries 2.2, 2.4 complete the proof. ■

References

[1] A. E. Brouwer, W. H. Haemers, Spectra of Graphs, Universitext, Springer, New York, 2012.
[2] S. Butler, Eigenvalues and Structures of Graphs, Ph.D. dissertation, University of California, San Diego, 2008.
[3] M. Cavers, The normalized Laplacian matrix and general Randić index of graphs, Ph.D. University of Regina, 2010.
[4] M. Cavers, S. Fallat, S. Kirkland, On the normalized Laplacian energy and general Randic index R_{-1} of graphs, Linear Algebra Appl. 433 (2010), 172-190.
[5] H. Chen, F. Zhang, Resistance distance and the normalized Laplacian spectrum, Discr. Appl. Math., 155 (2007), 654-661.
[6] F. R. K. Chung, Spectral Graph Theory, American Math. Soc. Providence, 1997.
[7] D. Cvetković, M. Doob, H. Sachs, Spectra of Graphs: Theory and Applications, Academic Press, New York, 1980.
[8] I. Gutman, The energy of a graph, Steiermrkisches Mathematisches Symposium (Stift Rein, Graz, 1978), Ber. Math. Statist. Sekt. Forsch. Graz 103 (1978) 1-22.
[9] I. Gutman, B. Mohar, The Quasi-Wiener and the Kirchhoff indicescoincide, J. Chem. Inf. Comput. Sci. 36 (1996), 982-985.
[10] I. Gutman, The energy of a graph: old and new results, in: A. Betten, A. Kohner, R. Laue, A. Wassermann (Eds.), Algebraic Combinatorics and Applications, Springer, Berlin, 2001, 196-211.
[11] M. Hakimi-Nezhaad, A. R. Ashrafi, I. Gutman, Note on degree Kirchhoff index of graphs, Trans. Comb. 2 (3) (2013), 43-52.
[12] M. Hakimi-Nezhaad, A. R. Ashrafi, A note on normalized Laplacian energy of graphs, J. Contemp. Math. Anal. 49 (5) (2014), 207-211.
[13] G. Indulal, Spectrum of two new joins of graphs and infinite families of integral graphs, Kragujevac J. Math. 36 (1) (2012), 133-139.
[14] E. R. Van Dam, G. R. Omidi, Graphs whose normalized Laplacian has three eigenvalues, Linear Algebra Appl. 435 (10) (2011), 2560-2569.
[15] M. R. Jooyandeh, D. Kiani, M. Mirzakhah, Incidence energy of a graph, MATCH Commun. Math. Comput. Chem. 62 (2009), 561-572.
[16] D. J. Klein, M. Randić, Resistance distance, J. Math. Chem. 12 (1993), 81-95.
[17] X. Li, Y. Shi, I. Gutman, Graph energy, Springer, New York, 2012.
[18] B. Zhou, N. Trinajstić, On resistance-distance and Kirchhoff index, J. Math. Chem. 46 (2009), 283-289.

8

On dual shearlet frames

M. Amin khah[a*], A. Askari Hemmat[b] and
R. Raisi Tousi[c]

[a] *Department of Application Mathematics, Kerman Graduate University of High Technology, PO. Code 76315-115, Iran.*;

[b] *Department of Mathematics, Shahid Bahonar University of Kerman, PO. Code 76175-133, Iran.*;

[c] *Department of Mathematics, Ferdowsi University of Mashhad, PO. Code 1159-91775, Iran.*

Abstract. In This paper, we give a necessary condition for function in L^2 with its dual to generate a dual shearlet tight frame with respect to admissibility.

Keywords: Dual shearlet frame, Bessel sequence, admissible shearlet.

1. Introduction

We begin by recalling some notations and denitions[1, 2, 4]. For $j, k \in \mathbb{Z}$, let

$$A_{a_0^j} = \begin{bmatrix} {a_0}^j & 0 \\ 0 & {a_0}^{\frac{j}{2}} \end{bmatrix} \quad , \qquad S_k = \begin{bmatrix} 1 & k \\ 0 & 1 \end{bmatrix}.$$

where $A_{a_0^j}$ and S_k are called *parabolic scaling matrices* and *shearing matrix*, respectively.
For $\psi \in L^2(\mathbb{R}^2)$, a *discrete shearlet system* associated with ψ is defined by

$$\{\psi_{j,k,m} = a_0^{-\frac{3}{4}j}\psi(S_k A_{a_0^{-j}} \cdot - m) : \ j, k \in \mathbb{Z}, m \in \mathbb{Z}^2\}, \tag{1}$$

*Corresponding author.
E-mail address: m.aminkhah@student.kgut.ac.ir (M. Amin khah).

with $a_0 > 0$.

The *discrete shearlet transform* of $f \in L^2(\mathbb{R}^2)$ is the mapping defined by

$$f \mapsto \mathcal{SH}_\psi f(j,k,m),$$

where

$$\mathcal{SH}_\psi f(j,k,m) = \langle f, \psi_{j,k,m}\rangle, \quad (j,k,m) \in \mathbb{Z} \times \mathbb{Z} \times \mathbb{Z}^2.$$

If $\psi \in L^2(\mathbb{R}^2)$ satisfies

$$c_\psi := \int_{\mathbb{R}^2} \frac{|\widehat{\psi}(\xi)|^2}{|\xi_1|^2} d\xi < \infty, \tag{2}$$

it is called an admissible shearlet.

Throughout this paper, we assume that H is a measurable subset of $\mathbb{R}^2$ such that

$$\chi_H(x) = \chi_{S_{-1}^T A_{2^{-1}} H}(x) \quad a.e. \quad \text{and} \quad |\mathrm{H} \setminus \mathrm{H}^\circ| = 0,$$

where H° denotes the interior of H, $H \setminus H^\circ := \{x \in \mathbb{R}^2 : x \in H \text{ and } \mathrm{x} \notin \mathrm{H}^\circ\}$, and $|H \setminus H^\circ|$ denotes the Lebesgue measure of $H \setminus H^\circ$. We consider the subspace $L^2(H)^\vee$ of $L^2(\mathbb{R}^2)$ defined as

$$L^2(H)^\vee = \{f : f \in L^2(\mathbb{R}^2) : \mathrm{supp}\widehat{\mathrm{f}} \subseteq \mathrm{H}\}.$$

Also, we will use the notation of the cube

$$\Theta_a(v) := \{w \in \mathbb{R}^2 : |w_i - v_i| \leqslant a, i = 1,2\}, \tag{3}$$

with radius a and center at $v = (v_1, v_2)$, where $w = (w_1, w_2)$.

To define a dual shearlet tight frame (DSTF) in $L^2(H)^\vee$, we need to recall a shearlet frame in $L^2(H)^\vee$.

A discrete shearlet system $\{\psi_{j,k,m}\}_{j,k,m}$ as defined in (1) is called a shearlet frame for $L^2(H)^\vee$, if there exist constants $0 < A \leqslant B < \infty$ such that for all $f \in L^2(H)^\vee$,

$$A\|f\|^2 \leqslant \sum_{j,k\in\mathbb{Z}} \sum_{m\in\mathbb{Z}^2} |\langle f, \psi_{j,k,m}\rangle|^2 \leqslant B\|f\|^2, \qquad f \in L^2(H)^\vee \tag{4}$$

A discrete shearlet system $\{\psi_{j,k,m}\}_{j,k,m}$ forms a Bessel sequence for $L^2(H)^\vee$, if only the right hand side inequality in (4) holds.

We say that ψ with $\tilde{\psi}$ generates a DSTF in $L^2(H)^\vee$ if ψ and $\tilde{\psi}$ are a Bessel sequences and for some non-zero constant B,

$$B\langle f, g\rangle = \sum_{j,k\in\mathbb{Z}} \sum_{m\in\mathbb{Z}^2} \langle f, \psi_{j,k,m}\rangle \langle \tilde{\psi}_{j,k,m}, g\rangle, \qquad f, g \in L^2(H)^\vee. \tag{5}$$

2. Main results

In this section, we discuss a necessary condition for ψ with $\tilde{\psi}$ in $L^2(H)^\vee$ to generate a DSTF via admissibility.

Proposition 2.1 If $\{\psi_{j,k,m}\}_{j,k,m}$ forms a Bessel sequence with Bessel bound B, then

$$\sum_{j,k\in\mathbb{Z}} |\widehat{\psi}(S^T_{-k}A_{2^{-j}}\xi|^2 \leqslant B \tag{6}$$

and ψ is admissible shearlet.

Proof. First, we observe, using (4), that

$$\sum_{j,k\in\mathbb{Z}}\sum_{m\in\mathbb{Z}^2} |\langle \hat{f}, \hat{\psi}_{j,k,m}\rangle|^2 \leqslant B\|\hat{f}\|^2, \tag{7}$$

for all $f \in L^2(H)^\vee$ and for any $j, k \in \mathbb{Z}$, we have

$$\sum_{m\in\mathbb{Z}^2} |\langle \hat{f}, \hat{\psi}_{j,k,m}\rangle|^2 = 2^{\frac{3}{2}j} \sum_{m\in\mathbb{Z}^2} |\int_{[0,2\pi]^2} \sum_{l\in\mathbb{Z}^2} \hat{f}(A_{2^j}S^T_k(w+2\pi l))\overline{\hat{\psi}}(w+2\pi l)e^{2\pi i m^T.w}dw|^2 \tag{8}$$

$$= 2^{\frac{3}{2}j} \int_{\mathbb{R}^2} |\sum_{l\in\mathbb{Z}^2} \hat{f}(A_{2^j}S^T_k(w+2\pi l))\overline{\hat{\psi}}(w+2\pi l)|^2 dw,$$

where the last equality in (8) is obtained by the Parseval equality.

Then by (7) and (8), we have

$$\sum_{j,k\in\mathbb{Z}} 2^{\frac{3}{2}j} \int_{\mathbb{R}^2} |\sum_{l\in\mathbb{Z}^2} \hat{f}(A_{2^j}S^T_k(w+2\pi l))\overline{\hat{\psi}}(w+2\pi l)|^2 dw \leqslant B\|\hat{f}\|^2, \tag{9}$$

for all $f \in L^2(H)^\vee$, consider $v \in \mathbb{R}^2$ and the function

$$\hat{f}(\xi) = \frac{1}{2\varepsilon}\chi_{\Theta_\varepsilon(v)}(\xi), \tag{10}$$

where $\varepsilon > 0$, χ_Θ denotes the characteristic function of a set Θ and $\Theta_\varepsilon(v)$ is defined by (3).

For any positive integer N and all sufficiently small $\varepsilon > 0$, in (9) we obtain

$$\sum_{k\in\mathbb{Z}}\sum_{|j|\leqslant N} 2^{\frac{3}{2}j} \int_{\Theta_{2^{-\frac{3}{2}j}\varepsilon}(S^T_{-k}A_{2^{-j}}v)} |\hat{\psi}(w)|^2 dw \leqslant B.$$

Hence, by taking $\varepsilon \to 0$ and $N \to \infty$, (6) follows. ■

By using proposition 2.1, we obtain the following result which gives a necessary condition for ψ with $\tilde{\psi}$ to generate a DSTF.

Theorem 2.2 Let ψ with $\tilde{\psi}$ in $L^2(H)^\vee$ generate a DSTF in $L^2(H)^\vee$ with bound B, then we have

$$\sum_{j,k\in\mathbb{Z}} \overline{\hat{\psi}}(S_{-k}^T A_{2^{-j}}\xi)\hat{\tilde{\psi}}(S_{-k}^T A_{2^{-j}}\xi) = B\chi_H(\xi) \quad a.e.. \tag{11}$$

In particular, ψ is admissible.

Proof. Let $H_0 := H^\circ \setminus \{0\}$. From the assumption $|H \setminus H^\circ| = 0$, to prove (11) it suffices to prove that

$$\sum_{j,k\in\mathbb{Z}} \overline{\hat{\psi}}(S_{-k}^T A_{2^{-j}}\xi)\hat{\tilde{\psi}}(S_{-k}^T A_{2^{-j}}\xi) = B \qquad a.e.\ \xi \in H_0. \tag{12}$$

By the Parseval equality and the polarization identity, setting $T := [0, 2\pi)^2$, we have the equality

$$\sum_{j,k\in\mathbb{Z}} \sum_{m\in\mathbb{Z}^2} \langle f, \psi_{j,k,m}\rangle\langle\tilde{\psi}_{j,k,m}, g\rangle$$

$$= \sum_{j,k\in\mathbb{Z}} 2^{-\frac{3}{2}j} \int_T [\hat{f}(A_{2^j} S_k^T \cdot), \hat{\psi}](\eta)[\hat{\tilde{\psi}}, \hat{g}(A_{2^j} S_k^T \cdot)](\eta)d\eta, \quad f, g \in L^2(\mathbb{R}^2), \tag{13}$$

where the bracket product is defined as

$$[f, g](\eta) = \sum_{m\in\mathbb{Z}^2} f(\eta + 2\pi m)\overline{g(\eta + 2\pi m)}.$$

by definition, ψ with $\tilde{\psi}$ satisfies Equation (5). By (13), we can rewrite (5) as

$$B\langle\hat{f}, \hat{g}\rangle = \sum_{j,k\in\mathbb{Z}} 2^{-\frac{3}{2}j} \int_T [\hat{f}(A_{2^j} S_k^T \cdot), \hat{\psi}](\eta)[\hat{\tilde{\psi}}, \hat{g}(A_{2^j} S_k^T \cdot)](\eta)d\eta, \quad f, g \in L^2(H)^\vee. \tag{14}$$

For any fixed $k \in \mathbb{Z}$, we consider

$$M^j := A_{2^j} = \begin{bmatrix} 2^j & 0 \\ 0 & 2^{\frac{j}{2}} \end{bmatrix}.$$

Now, let $\hat{f}(\zeta) = \hat{g}(\zeta) = \frac{1}{\sqrt{|D_l(\xi,\gamma_l)|}}\chi_{D_l(\xi,\gamma_l)}(\zeta)$, where for $l \in \mathbb{Z}$ and $\gamma_l \in \mathbb{Z}^2$, we define

$$D_l(\xi, \gamma_l) := \{M^l[S_k^T(x + 2\pi\gamma_l)] \ : \ x \in T\}, \quad \xi \in H_0.$$

Since $\xi \neq 0$ and $\xi \in H^\circ$, we can choose $l_\xi < 0$ such that

$$M^j D_l(\xi, \gamma_l) \cap D_l(\xi, \gamma_l) = \emptyset, \quad \forall j < 0,\ l \leqslant l_\xi,\ j, l \in \mathbb{Z}. \tag{15}$$

For a detailed proof of (15), the reader is referred to [3].

It is obvious that $f, g \in L^2(H)^\vee$. Hence (14) yields

$$B = B\langle \hat{f}, \hat{g}\rangle$$

$$= \sum_{k\in\mathbb{Z}} \left[\sum_{j\geqslant l-l_N} 2^{-\frac{3}{2}j} \int_T [\hat{f}(A_{2^j} S_k^T \cdot), \hat{\psi}](\eta)[\hat{\hat{\psi}}, \hat{g}(A_{2^j} S_k^T \cdot)](\eta) d\eta \right. \tag{16}$$

$$\left. + \sum_{j<l-l_N} 2^{-\frac{3}{2}j} \int_T [\hat{f}(A_{2^j} S_k^T \cdot), \hat{\psi}](\eta)[\hat{\hat{\psi}}, \hat{g}(A_{2^j} S_k^T \cdot)](\eta) d\eta \right],$$

with the integer $l_N < 0$ depending only on N. Since $\hat{f}(\zeta) = \hat{g}(\zeta) = \frac{1}{\sqrt{|D_l(\xi,\gamma_l)|}} \chi_{D_l(\xi,\gamma_l)}(\zeta)$, then for any $j \geqslant l - l_N$, we obtain

$$[\hat{f}(A_{2^j} S_k^T \cdot), \hat{\psi}](\eta)[\hat{\hat{\psi}}, \hat{g}(A_{2^j} S_k^T \cdot)](\eta) = \frac{1}{|D_l(\xi, \gamma_l)|}[\overline{\hat{\psi}}\hat{\hat{\psi}}, \chi_{D_l(\xi,\gamma_l)}(A_{2^j} S_k^T \cdot)](\eta). \tag{17}$$

Hence, By (15) and (17), in (16) we have

$$B = \lim_{l\to\infty} \frac{1}{|D_l(\xi,\gamma_l)|} \int_{D_l(\xi,\gamma_l)} \sum_{k\in\mathbb{Z}} \sum_{j\leqslant l_N - l} \overline{\hat{\psi}}(A_{2^j} S_k^T \eta)\hat{\hat{\psi}}(A_{2^j} S_k^T \eta) d\eta$$

$$= \sum_{k\in\mathbb{Z}} \sum_{j\in\mathbb{Z}} \overline{\hat{\psi}}(A_{2^j} S_k^T \eta)\hat{\hat{\psi}}(A_{2^j} S_k^T \eta),$$

then the result follows. ■

References

[1] C. K. Chui, X. Shi, On a LittlewoodPaley identity and characterization of wavelets, Math. Anal. Appl. 177 (1993) 608–626.
[2] I. Daubechies, B. Han, Pairs of dual wavelet frames from any two renable functions, Constr. Appr.,to appear.
[3] B. Han, On dual wavelet tight frames, Appl. Comput. Harmon. Anal. 4 (1997) 380–413.
[4] G. Kutyniok, D. Labate, Shearlets: Multiscale Analysis for Multivariate Data, Birkhauser, Basel, 2012.

9

Application of triangular functions for solving Vasicek model

Z. Sadati[a*], Kh. Maleknejad[a]

[a]*Department of Mathematics, Khomein Branch, Islamic Azad University, Khomein, Iran.*

Abstract. This paper introduces a numerical method for solving the vasicek model by using a stochastic operational matrix based on the triangular functions (TFs) in combination with the collocation method. The method is stated by using conversion the the vasicek model to a stochastic nonlinear system of $2m + 2$ equations and $2m + 2$ unknowns. Finally, the error analysis and some numerical examples are provided to demonstrate applicability and accuracy of this method.

Keywords: Triangular functions, Stochastic operational matrix, Vasicek model, collocation method.

1. Introduction

The vasicek model is a mathematical model describing the evolution of interest rates where play important role in finance. This model can be used for interest rate derivative valuation and also adapted for credit market. It is based on the ornstein-uhlenbeck process (is the first account of a bond pricing model), that incorporates stochastic interest rate and can be also seen as a stochastic investment model. The short-term interest rate process $(X(t))_{t\in\mathbf{R}^+}$ solves the equation

$$dX(t) = \alpha(\beta - X(t))dt + \sigma dB(t),$$

*Corresponding author.
E-mail address: z.sadati@khomein.ac.ir (Z. Sadati).

where $dX(t)$ be the change in the short-term interest rate, α be the speed of mean reversion, β be the average interest rate and σ be the volatility of the short rate.
The main disadvantage is that, under vasicek model, it is theoretically possible for the interest rate to become negative. This shortcoming was fixed in the Cox-Ingersoll-Ross (CIR) model. The CIR process is a markov process with continuous paths defined by the following SDE:

$$dX(t) = \alpha(\beta - X(t))dt + \sigma\sqrt{X(t)}dB(t),$$

or

$$X(t) = X_0 + \int_0^t \alpha(\beta - X(s))ds + \int_0^t \sigma\sqrt{X(s)}dB(s), \tag{1}$$

where $B(s)$ is a standard Brownian motion (SBM) defined on a complete probability space $(\Omega, \mathcal{F}, \{\mathcal{F}_t\}_{t\geqslant 0}, P)$ with natural filtration $\{\mathcal{F}_t\}_{t\geqslant 0}$ and $X(t)$ is unknown stochastic processes defined on same probability space.

In the last years, many methods are proposed and applied for numerical solutions of stochastic differential equations [5, 8, 9, 10], because these kinds of equations can not be solved analytically. Hence, it is importance to provide their numerical solutions.
In this work, we reduce the Eq. (1) to the stochastic nonlinear system of $2m+2$ equations and $2m+2$ unknowns without integration by using operational matrices based on the TFs in combination with the collocation technique, with several advantages in reducing computations and making convergence faster than the other methods.

The results of the paper are organized as follows: In Section 2, we state some essential preliminaries which play fundamental role in our method. In Section 3, we solve Eq. (1) by using the stochastic operational matrix based on the TFs in combination with the collocation method. In Sections 4 and 5, we provide the error analysis and some numerical examples to demonstrate the applicability and accuracy of presented method. Finally, in Section 6, we give a brief conclusion.

2. Preliminaries

The first, we state the basic properties of the SBM that play important role in solving Eq. (1). For more details see [2, 12].

Let $h(t, X),\ g(t, X) : (0, T) \times R \to R$ be measurable functions and continuous with the main properties as follows:
$\mathbf{A}_1$. There are constants $k_1,\ k_2 > 0$, such that:

$$\begin{cases} |g(t, X) - g(t, Y)| \leqslant k_1|X - Y|, & (lipschitz\ continuity), \\ |g(t, X)| < k_2(1 + |X|), & (lineargrowth). \end{cases}$$

$\mathbf{A}_2$. There are constants $k_3,\ k_4 > 0$, such that:

$$\begin{cases} |h(t, X) - h(t, Y)| < k_3|X - Y|, & (lipschitz\ continuity), \\ |h(t, X)| < k_4(1 + |X|), & (lineargrowth). \end{cases}$$

For $X,\ Y \in R$ and $t \in (0, T)$.

Theorem 2.1 [2, 12] Let $g(t, X(t))$ and $h(t, X(t))$ hold in conditions $\mathbf{A}_1$, $\mathbf{A}_2$ and $E \mid X_0 \mid^2 < \infty$, then, there exists a unique solution for Eq. (1).

Definition 2.2 The SBM $\{B(t),\ t \geqslant 0\}$ is the stochastic process with main properties as follows:
1. The process has independent increments for $0 \leqslant t_0 \leqslant t_1 \leqslant ... \leqslant t_n \leqslant T$.
2. $B(t+h) - B(t)$ is normally distributed with mean 0 and variance h, for all $t \geqslant 0$, $h > 0$.
3. $B(t)$ is a continuous function.

Definition 2.3 Let $\nu = \nu(S, T)$ be the class of functions $\alpha(t, \omega) : [0, \infty) \times \Omega \longrightarrow R$ such that:
1. The function $\alpha(t, \omega)$ be $\beta \times F$ measurable.
2. The function $\alpha(t, \omega)$ is $\mathcal{F}_t$-adapted.
3. $E\big[\int_S^T \alpha^2(t, \omega)dt\big] < \infty$.

Theorem 2.4 [2, 12] Let $f \in \nu(S, T)$, then

$$E\big[\big(\int_S^T f(t,\omega)dB(t)(\omega)\big)^2\big] = E\big[\int_S^T f^2(t,\omega)dt\big].$$

Finally, we introduce some essential properties of the TFs that are needful for this paper. For more details see [1, 3, 4, 6, 7, 11].
1. The 1D-TF vector are defined as follows:

$$T(t) = \begin{pmatrix} T1(t) \\ T2(t) \end{pmatrix},$$

where

$$\begin{cases} T1(t) = [T_0^1(t), \dots, T_i^1(t), \dots, T_{m-1}^1(t)]^T, \\ T2(t) = [T_0^2(t), \dots, T_i^2(t), \dots, T_{m-1}^2(t)]^T, \end{cases}$$

with

$$T_i^1(t) = \begin{cases} 1 - \frac{t-ih}{h} & ih \leqslant t < (i+1)h, \\ 0 & otherwise, \end{cases}$$

and

$$T_i^2(t) = \begin{cases} \frac{t-ih}{h} & ih \leqslant t < (i+1)h, \\ 0 & otherwise, \end{cases}$$

where $h = \frac{T}{m}$.
2. Let the function $f(t) \in L^2\big((0, T)\big)$, then

$$f(t) \approx F^T.T(t),$$

where $F=[f_1,f_2]^T$, $f_1=(f(ih))_{m\times 1}$, $f_2=(f((i+1)h))_{m\times 1}$ and $i=0,1,..,m-1$.
3.

$$\int_0^t T(s)ds \approx P_T.T(t),$$

where

$$P_T=\begin{pmatrix} P1 & P2 \\ P1 & P2 \end{pmatrix},$$

with

$$P1=\begin{pmatrix} 0 & \frac{h}{2} & \frac{h}{2} & \dots & \frac{h}{2} \\ 0 & 0 & \frac{h}{2} & \dots & \frac{h}{2} \\ 0 & 0 & 0 & \dots & \frac{h}{2} \\ \vdots & \vdots & \vdots & \ddots & \vdots \\ 0 & 0 & 0 & \dots & 0 \end{pmatrix}_{m\times m},$$

and

$$P2=\begin{pmatrix} \frac{h}{2} & \frac{h}{2} & \frac{h}{2} & \dots & \frac{h}{2} \\ 0 & \frac{h}{2} & \frac{h}{2} & \dots & \frac{h}{2} \\ 0 & 0 & \frac{h}{2} & \dots & \frac{h}{2} \\ \vdots & \vdots & \vdots & \ddots & \vdots \\ 0 & 0 & 0 & \dots & \frac{h}{2} \end{pmatrix}_{m\times m}.$$

4.

$$\int_0^t T(s)dB(s) \approx P_S.T(t),$$

where

$$P_S=\begin{pmatrix} P1_S & P1_S \\ P2_S & P2_S \end{pmatrix},$$

with

$$P1_S=\begin{pmatrix} \alpha(0) & \beta(0) & \beta(0) & \dots & \beta(0) \\ 0 & \alpha(1) & \beta(1) & \dots & \beta(1) \\ 0 & 0 & \alpha(2) & \dots & \beta(2) \\ \vdots & \vdots & \vdots & \ddots & \vdots \\ 0 & 0 & 0 & \dots & \beta(m-2) \\ 0 & 0 & 0 & \dots & \alpha(m-1) \end{pmatrix}_{m\times m},$$

$$P2_S = \begin{pmatrix} \gamma(0) & \rho(0) & \rho(0) & \dots & \rho(0) \\ 0 & \gamma(1) & \rho(1) & \dots & \rho(1) \\ 0 & 0 & \gamma(2) & \dots & \rho(2) \\ \vdots & \vdots & \vdots & \ddots & \vdots \\ 0 & 0 & 0 & \dots & \rho(m-2) \\ 0 & 0 & 0 & \dots & \gamma(m-1) \end{pmatrix}_{m\times m},$$

and

$$\begin{cases} \alpha(i) = (i+1)[B((i+0.5)h) - B(ih)] - \int_{ih}^{(i+0.5)h} \frac{s}{h} dB(s), \\ \beta(i) = (i+1)[B((i+1)h) - B(ih)] - \int_{ih}^{(i+1)h} \frac{s}{h} dB(s), \\ \gamma(i) = -i[B((i+0.5)h) - B(ih)] + \int_{ih}^{(i+0.5)h} \frac{s}{h} dB(s), \\ \rho(i) = -i[B((i+1)h) - B(ih)] + \int_{ih}^{(i+1)h} \frac{s}{h} dB(s). \end{cases}$$

3. Solving NSDE by using the TFs

To find a solution for the Eq. (1) we can write it as follows:

$$X(t) = X_0 + \int_0^t p(s)ds + \int_0^t q(s)dB(s). \tag{2}$$

with

$$\begin{cases} \alpha(\beta - X(s)) = g(s, X(s)) = p(s), \\ \sigma\sqrt{X(s)} = h(s, X(s)) = q(s), \end{cases} \tag{3}$$

with substituting (3) in Eq. (2), we get

$$X(t) = X_0 + \int_0^t p(s)ds + \int_0^t q(s)dB(s). \tag{4}$$

By using properties of the TFs, we can write

$$\begin{cases} p(s) \approx P^T.T(s), \\ q(s) \approx Q^T.T(s), \end{cases} \tag{5}$$

where

$$P = (p_i)_{2m\times 1} = \big(p(0), p(h), \dots, p((m-1)h), p(h), p(2h), \dots, p(mh)\big)_{2m\times 1},$$

and

$$Q = (q_i)_{2m\times 1} = \big(q(0), q(h), \dots, q((m-1)h), q(h), q(2h), \dots, q(mh)\big)_{2m\times 1}.$$

With substituting (5) in (4), we get

$$X(t) \approx X_0 + \int_0^t P^T.T(s)ds + \int_0^t Q^T.T(s)dB(s), \tag{6}$$

or

$$X(t) \approx X_0 + P^T P_T T(t) + Q^T P_S T(t). \tag{7}$$

Also, by substituting (7) in (3), we obtain

$$\begin{cases} p(t) \approx g(t, X_0 + P^T P_T T(t) + Q^T P_S T(t)), \\ q(t) \approx h(t, X_0 + P^T P_T T(t) + Q^T P_S T(t)). \end{cases} \tag{8}$$

Now, with replacing $\approx$ by $=$, the relation (8) is approximated via the collocation method in $m+1$ nodes $t_j = \frac{j}{\frac{1}{T}m+1}$ $(j = 0, 1, \dots, m)$, as follows:

$$\begin{cases} p(t_j) = g(t_j, X_0 + P^T P_T T(t_j) + Q^T P_S T(t_j)), \\ q(t_j) = h(t_j, X_0 + P^T P_T T(t_j) + Q^T P_S T(t_j)), \end{cases} \tag{9}$$

or

$$\begin{cases} P^T T(t_j) = g(t_j, X_0 + P^T P_T T(t_j) + Q^T P_S T(t_j)), \\ Q^T T(t_j) = h(t_j, X_0 + P^T P_T T(t_j) + Q^T P_S T(t_j)), \end{cases} \tag{10}$$

where Eq. (10) is the nonlinear system of $2m+2$ equations and $2m+2$ unknowns. From solving Eq. (10), we can conclude

$$X(t) = X_m(t) = X_0 + P^T P_T T(t) + Q^T P_S T(t). \tag{11}$$

4. Error analysis

Theorem 4.1 Let $f(t)$ be an arbitrary real bounded function on $(0,1)$, $|f'(t)| \leqslant M$ and $e(t) = f(t) - \hat{f}(t)$ that $\hat{f}(t)$ denotes the TFs of $f(t)$. Then,

$$||e(t)||^2 \leqslant O(h^2),$$

where $||e(t)||^2 = \int_0^1 |e(t)|^2 dt$.

Proof. By using properties of the TFs, we can write

$$|e(t)| = |f(t) - \hat{f}(t)| = |f(t) - \sum_{i=0}^{m-1} f(ih)(1 - \frac{t-ih}{h}) + f((i+1)h)(\frac{t-ih}{h})|.$$

Let $t \in (ih,\ (i+1)h)$, so, we get

$$|e(t)| = |f(t) - \hat{f}(t)| = |f(t) - f(ih)(1 - \frac{t-ih}{h}) - f((i+1)h)(\frac{t-ih}{h})| =$$
$$|f(t) - f(ih) + \big(f(ih) - f((i+1)h)\big)(\frac{t-ih}{h})| \leqslant |f(t) - f(ih)| + |f(ih) -$$
$$f((i+1)h)||\frac{t-ih}{h}| \leqslant |f(t) - f(ih)| + |f(ih) - f((i+1)h)|,$$

by using the mean value theorem, we get

$$|e(t)| \leqslant |f'(\alpha)|(t-ih) + |f'(t)h| \leqslant Mh,$$

consequently

$$||e(t)||^2 = \int_0^1 |e(t)|^2 dt \leqslant M^2h^2 \leqslant O(h^2). \qquad \blacksquare$$

Let

$$\begin{cases} p^m(t) = g(t, X_m(t)), \\ q^m(t) = h(t, X_m(t)), \end{cases} \tag{12}$$

and

$$\begin{cases} \hat{p}(t) = \hat{g}(t, X_m(t)), \\ \hat{q}(t) = \hat{h}(t, X_m(t)), \end{cases} \tag{13}$$

where $\hat{p}(t)$ and $\hat{q}(t)$ are defined by properties of the TFs. Also, let $X_m(t)$ be numerical solution of Eq. (1) defined in Eq. (11).

Theorem 4.2 Let $X_m(t)$ be the numerical solution of Eq. (1) defined in Eq. (11) and let conditions $\mathbf{A}_1$, $\mathbf{A}_2$ and $E \mid X_0 \mid^2 < \infty$ hold. Then,

$$\| X(t) - X_m(t) \|^2 \leqslant O(h^2), \qquad t \in (0,1), \tag{14}$$

where $\| X \|^2 = E[X^2]$.

Proof

$$X(t) - X_m(t) = \int_0^t (p(s) - \hat{p}(s))ds + \int_0^t (q(s) - \hat{q}(s))dB(s), \tag{15}$$

via $(x+y)^2 \leqslant 2(x^2+y^2)$ and the property of the Itô isometry for the SBM, we can write

$$\| X(t) - X_m(t) \|^2 \leqslant 2(\| \int_0^t (p(s) - \hat{p}(s))ds \|^2 + \| \int_0^t (q(s) - \hat{q}(s))dB(s) \|^2) \leqslant 2$$
$$(\int_0^t \| p(s) - \hat{p}(s) \|^2 ds + \| \int_0^t (q(s) - \hat{q}(s))ds \|^2) \leqslant 2(\int_0^t \| p(s) - \hat{p}(s) \|^2 ds +$$
$$\int_0^t \| q(s) - \hat{q}(s) \|^2 ds) \leqslant 2(2 \int_0^t \| p(s) - p^m(s) \|^2 ds + 2 \int_0^t \| p^m(s) - \hat{p}(s) \|^2 ds$$
$$+2 \int_0^t \| q(s) - q^m(s) \|^2 ds + 2 \int_0^t \| q^m(s) - \hat{q}(s) \|^2 ds) \leqslant 4(\int_0^t \| p(s) - p^m(s) \|^2$$
$$ds + \int_0^t \| p^m(s) - \hat{p}(s) \|^2 ds + \int_0^t \| q(s) - q^m(s) \|^2 ds + \int_0^t \| q^m(s) -$$
$$\hat{q}(s) \|^2 ds). \quad (16)$$

By using Theorem (4.1), we have

$$\begin{cases} \| p^m(s) - \hat{p}(s) \|^2 \leqslant L_1 h^2, & L_1 > 0, \\ \| q^m(s) - \hat{q}(s) \|^2 \leqslant L_2 h^2, & L_2 > 0. \end{cases} \quad (17)$$

Also, by using conditions $\mathbf{A}_1$ and $\mathbf{A}_2$, we have

$$\begin{cases} \int_0^t \| p(s) - p^m(s) \|^2 ds \leqslant L_3 \int_0^t \| X(s) - X_m(s) \|^2 ds, \\ \int_0^t \| q(s) - q^m(s) \|^2 ds \leqslant L_4 \int_0^t \| X(s) - X_m(s) \|^2 ds. \end{cases} \quad (18)$$

With substituting (17) and (18) in (16), we obtain

$$\| X(t) - X_m(t) \|^2 \leqslant 4(L_1 h^2 + L_3 \int_0^t \| X(s) - X_m(s) \|^2 ds + L_2 h^2 +$$
$$L_4 \int_0^t \| X(s) - X_m(s) \|^2 ds), \quad (19)$$

or

$$\eta(t) \leqslant m + n \int_0^t \eta(s) ds,$$

where $m = 4(L_1 h^2 + L_2 h^2)$, $n = 4(L_3 + L_4)$ and $\eta(s) = \| X(s) - X_m(s) \|^2$. Furthermore, from Gronwall inequality, we get

$$\eta(t) \leqslant m(1 + n \int_0^t \exp\big(n(t-s)\big)ds), \quad t \in (0,1),$$

so

$$\| X(t) - X_m(t) \|^2 \leqslant O(h^2).$$ ■

5. Numerical examples

In this section, we give some numerical results to illustrate our main results. The numerical results have been shown in Figures (1-4) via a comparison between numerical solution of deterministic model and numerical solution of stochastic model. Also, the numerical solution of deterministic model has been approximated using properties of the TFs. In addition, we assume $X_0 = 0.03$, $\alpha = 0.05$, $\beta = 0.3$ and $\sigma = 0.002$ in Figures (1-2) and $X_0 = 0.5$, $\alpha = 0.2$, $\beta = 0.005$ and $\sigma = 0.002$ in Figures (3-4).

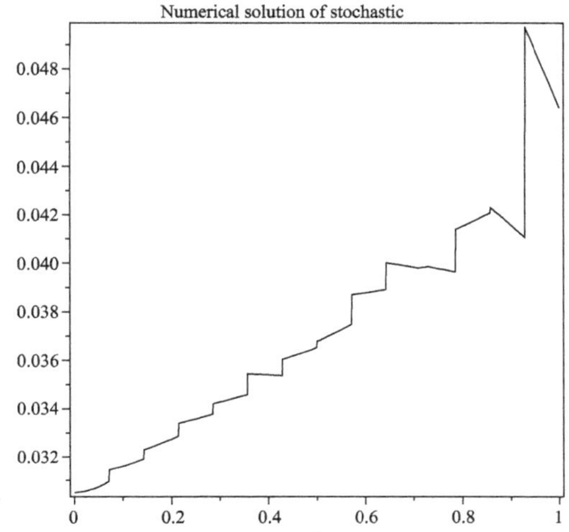

Fig.1

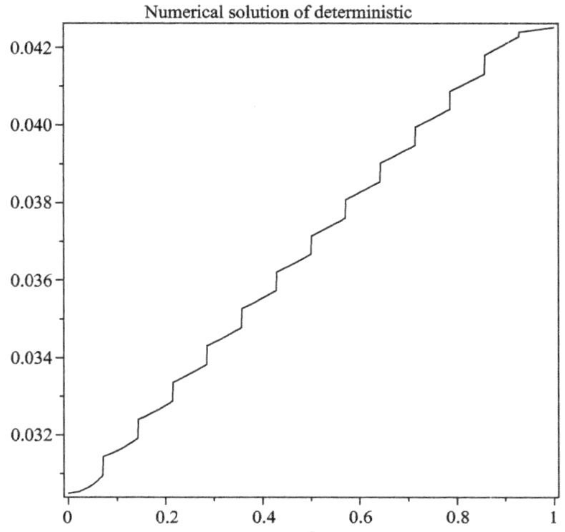

Fig.2

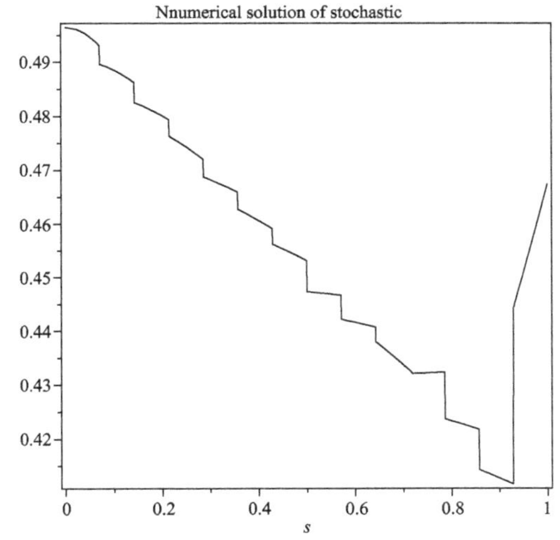

Fig.3

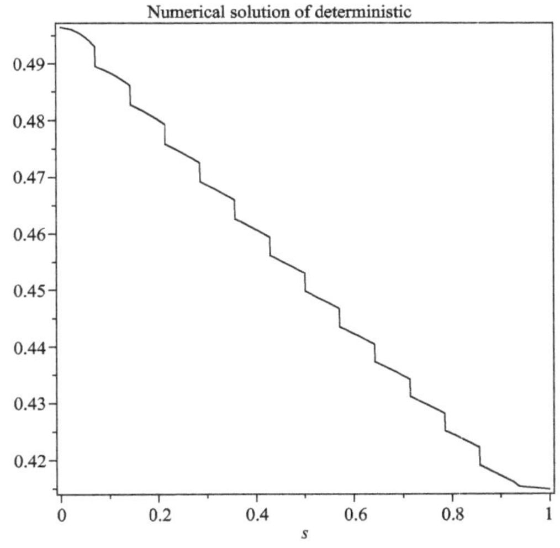

Fig.4

6. Conclusion

In this paper, we introduce the numerical method based on the TFs for solving the vasicek model. With using this method, we reduce Eq. (1) to the stochastic nonlinear system. Also, numerical simulations are provided to accuracy of presented method.

Acknowledgements

The authors are extending their heartfelt thanks to the reviewers for their valuable suggestions for the improvement of the article. Also, the authors thank Islamic Azad University for supporting this work.

References

[1] A. Deb, A. Dasgupta and G. Sarkar, A new set of orthogonal functions and its application to the analysis of dynamic systems. *J. Franklin Inst,* vol. 343, (2006) 1-26.
[2] B. Oksendal, *Stochastic Differential Equations,* An Introduction with Applications, Fifth Edition, Springer-Verlag, New York, 1998.
[3] E. Babolian, H. R. Marzban, and M. Salmani, Using triangular orthogonal functions for solving fredholm integral equations of the second kind, *Appl. Math. Comput,* vol. 201, (2008) 452-456.
[4] E. Babolian, Z. Masouri and S. Hatamzadeh-Varmazya, A direct method for numerically solving integral equations system using orthogonal triangular functions, *Int. J. Industrial Mathematics,* vol. 1, no. 2, (2009) 135-145, 2009.
[5] E. Pardoux and P. Protter, Stochastic volterra equations with anticipating coefficients, *Ann. Probab,* vol. 18, (1990) 1635-1655.
[6] K. Maleknejad, H. Almasieh and M. Roodaki, Triangular functions (TF) method for the solution of volterra-fredholm integral equations, *Communications in Nonlinear Science and Numerical Simulation,* vol. 15, no. 11, (2009) 3293-3298.
[7] K. Maleknejad, and Z. Jafari Behbahani, Applications of two-dimensional triangular functions for solving nonlinear class of mixed volterra-fredholm integral equations, *Mathematical and Computer Modelling.*
[8] K. Maleknejad, M. Khodabin, and M. Rostami, A numerical method for solving m-dimensional stochastic Itvolterra integral equations by stochastic operational matrix, *Computers and Mathematics with Applications,* vol. 63, (2012) 133-143.
[9] K. Maleknejad, M. Khodabin and M. Rostami, Numerical solution of stochastic volterra integral equations by stochastic operational matrix based on block pulse functionsx, *Mathematical and Computer Modelling,* vol. 55, (2011) 791-800.
[10] M. A. Berger and V.J. Mizel, Volterra equations with Ito integrals I, *J. Integral Equations* vol. 2, no. 3, (1980) 187-245.
[11] M. Khodabin, K. Maleknejad and F. Hosseini, Application of triangular functions to numerical solution of stochastic volterra integral equations, *IAENG International Journal of Applied Mathematics,* 2013, IJAM-43-1-01.
[12] P. E. Kloeden and E. Platen, Numerical solution of stochastic differential equations, *Applications of Mathematics, Springer-Verlag, Berlin,* 1999.

10

Lie higher derivations on $B(X)$

S. Ebrahimi*

Department of Mathematics, Payame Noor University, P.O. Box 19395-3697, Tehran, Iran.

Abstract. Let X be a Banach space of $dimX > 2$ and $B(X)$ be the space of bounded linear operators on X. If $L : B(X) \to B(X)$ be a Lie higher derivation on $B(X)$, then there exists an additive higher derivation D and a linear map $\tau : B(X) \to \mathbb{F}I$ vanishing at commutators $[A, B]$ for all $A, B \in B(X)$ such that $L = D + \tau$.

Keywords: Lie higher derivations, higher derivations.

2010 AMS Subject Classification: Primary: 46M10 Secondary: 46H25, 46M18.

1. Introduction

Let X be a Banach space over $\mathbb{F}$, where $\mathbb{F}$ is the real number field $\mathbb{R}$ or the complex field $\mathbb{C}$. Recall that an additive map $L : B(X) \to B(X)$ is called a derivation if

$$L(AB) = L(A)B + AL(B) \qquad \forall A, B \in B(X).$$

More generally, an additive map $L : B(X) \to B(X)$ is called a Lie derivation if $L([A, B]) = [L(A), B] + [A, L(B)]$ for all $A, B \in B(X)$. In recent years, Lie derivations has attracted the attentions of many researchers(see [1, 2, 8], and references therein). On the other hand, higher derivations were introduced and studied mainly in commutative rings and later, also in non-commutative rings and some operator algebras (see, for example, [3, 11] and the references therein). We first recall the concepts of higher derivations and Lie higher derivations.

*Corresponding author.
E-mail address: seebrahimi2272@gmail.com (S. Ebrahimi).

Definition 1.1 Let $L = (L_i)_{i\in\mathbb{N}}$ ($\mathbb{N}$ denotes the set of natural numbers including 0) be a sequence of additive maps of a ring $\mathcal{A}$ such that $L_0 = \mathrm{id}_{\mathcal{A}}$. L is said to be a higher derivation if for every $n \in \mathbb{N}$ we have $L_n(AB) = \sum_{i+j=n} L_i(A)L_j(B)$ for all $A, B \in \mathcal{A}$; a Lie higher derivation if for every $n \in \mathbb{N}$ we have $L_n([A, B]) = \sum_{i+j=n}[L_i(A), L_j(B)]$ for all $A, B \in \mathcal{A}$.

It is clear that all higher derivations are Lie higher derivations. However, the converse is not true in general. Assume that $D = (D_n)_{n\in\mathbb{N}}$ is a higher derivation on a ring $\mathcal{A}$. For any $n \in \mathbb{N}$, let $L_n = D_n + h_n$, where h_n is an additive map from $\mathcal{A}$ into its center vanishing on every commutator. It is easily seen that $(L_n)_{n\in\mathbb{N}}$ is a Lie higher derivation, but not a higher derivation if $h_n \neq 0$ for some n. Then a natural question is to ask whether or not every Lie higher derivations have the above form?

In [10], Fei and Chen discussed the properties of Lie higher derivations on nest algebras. Generalized higher derivations, jordan higher derivations and higher derivations were studied by many authors (see [6, 7, 9, 11, 12]). The purpose of this paper is to show that every Lie higher derivations on $B(X)$ is proper.

2. Lie higher derivations on $B(X)$

In this section, we give a characterization of Lie higher derivations of $B(X)$. The following is our main result which generalizes the main result in [4].

Theorem 2.1 Let X be a Banach space of $dimX > 2$ and $L : B(X) \to B(X)$ be a Lie higher derivation on $B(X)$. Then there exists an additive higher derivation D and a linear map $\tau : B(X) \to \mathbb{F}I$ vanishing at commutators $[A, B]$ for all $A, B \in B(X)$ such that $L = D + \tau$.

Proof. In the following, we always assume that $L = (L_n)_{n\in\mathbb{N}}$ is an Lie higher derivation of $B(X)$. We proceed by induction on $n \in \mathbb{N}$.

If $n = 1$, then by the definition of Lie higher derivations, L_1 is a Lie derivation of $B(X)$. So by ([4, Theorem 1.1]), there exists an additive derivation D and a linear map $\tau : B(X) \to \mathbb{F}I$ vanishing at commutators $[A, B]$ for all $A, B \in B(X)$ such that $L = D+\tau$.

For the convenience, in the sequel, take $x_0 \in X, f_0 \in X^*$ satisfying $f_0(x_0) = 1$. Let $P = x_0\otimes f_0$ and $Q = I-P$ be idempotent of $B(X)$, it is obvious that $PQ = QP = 0$. Then $B(X) = B_{11}+B_{12}+B_{21}+B_{22}$, where $B_{11} = PB(X)P$, $B_{12} = PB(X)Q$, $B_{21} = QB(X)P$, $B_{22} = QB(X)Q$.

Thus by ([4, Lemmas 2.2-2.12]) we have;

$$\mathbf{P_1}: \begin{cases} PL_1(P)P + QL_1(P)Q \in \mathbb{F}I; \\ PL_1(Q)P + QL_1(Q)Q \in \mathbb{F}I; \\ \Delta_1(PAQ + QAP) = P\Delta_1(A)Q + Q\Delta_1(A)P \quad \text{where} \\ \Delta_1(A) = L_1(A) - (AT - TA) \quad \text{and} \quad T = PL_1(P)Q - QL_1(P)P; \\ \Delta_1(P) \in \mathbb{F}I, \quad \Delta_1(Q) \in \mathbb{F}I; \\ \Delta_1(\mathrm{B_{ij}}) \subseteq \mathrm{B_{ij}} \text{ for } 1 \leqslant i \neq j \leqslant 2; \\ \Delta_1(X_{ii}) \in B_{ii} + \mathbb{F}I. \end{cases}$$

Assume that $L = (L_n)$ is a Lie higher derivation of $B(X)$. We proceed by induction on $n \in N$. When $n = 1$, the conclusion is true by discussion. We now assume that $L_m(x) = D_m(x) + \tau_m(x)$ holds for all $x \in B(X)$ and for all $m < n \in N$, where τ_m :

$B(X) \to Z(B(X))$ is such that $\tau_m([x,y]) = 0$ for all $x, y \in B(X)$ and $D_m(xy) = \sum_{i+j=m} D_i(x)D_j(y)$ for all $x, y \in B(X)$.
Moreover, we have the following properties;

$$\mathbf{P_m}: \begin{cases} PL_m(P)P + QL_m(P)Q \in \mathbb{F}I; \\ PL_m(Q)P + QL_m(Q)Q \in \mathbb{F}I; \\ \Delta_m(PAQ + QAP) = P\Delta_m(A)Q + Q\Delta_m(A)P \quad where \\ \Delta_m(A) = L_m(A) - (AT - TA) \quad and \quad T = PL_m(P)Q - QL_m(P)P; \\ \Delta_m(P) \in \mathbb{F}I, \quad \Delta_m(Q) \in \mathbb{F}I; \\ .\Delta_m(\mathrm{B_{ij}}) \subseteq \mathrm{B_{ij}} for \ 1 \leqslant i \neq j \leqslant 2; \\ \Delta_m(X_{ii}) \in B_{ii} + \mathbb{F}I. \end{cases}$$

Our aim is to show that L_n also satisfies the similar properties and that $L_n(x) = D_n(x) + \tau_n(x)$ holds for all $x \in B(X)$ and τ_n is linear map from $B(X)$ into its center satisfying $\tau_n([x,y]) = 0$ for all $x, y \in B(X)$. Therefore, by induction, the theorem is true.
We will prove it by several claims.

Claim 1. $PL_n(P)P + QL_n(P)Q \in \mathbb{F}I$ and $PL_n(Q)P + QL_n(Q)Q \in \mathbb{F}I$.

proof. Let $x \in X$, $f \in X^*$. Then

$$\begin{aligned} L_n(Px \otimes Q^*f) &= L_n([P, Px \otimes Q^*f]) \\ &= \sum_{i+j=n} [L_i(P), L_j(Px \otimes Q^*f)] \\ &= [L_n(P), Px \otimes Q^*f] + [P, L_n(Px \otimes Q^*f)] + \sum_{i+j=n; i \neq 0,n} [L_i(P), L_j(Px \otimes Q^*f)] \\ &= L_n(P)Px \otimes Q^*f - Px \otimes Q^*fL_n(P) + PL_n(Px \otimes Q^*f) \\ &\quad - L_n(Px \otimes Q^*f)P + \sum_{i+j=n; i \neq 0,n} [L_i(P), L_j(Px \otimes Q^*f)] \end{aligned}$$

Multiplying this equation by P from the left and by Q from the right, we get, for all $x \in X$ and $f \in X^*$, that

$$PL_n(P)P(x \otimes f)Q = P(x \otimes f)QL_n(P)Q - P \sum_{i+j=n; i \neq 0,n} [L_i(P), L_j(Px \otimes Q^*f)]Q.$$

By $\mathbf{P_m}$, $PL_i(P)P + QL_i(P)Q \in \mathbb{F}I$ and $PL_j(P)P + QL_j(P)Q \in \mathbb{FI}$.
So $P\sum_{i+j=n \ \ i\neq 0,n}[L_i(P), L_j(Px \otimes Q^*f)]Q = 0$.
Thus $PL_n(P)P = \mu P$ for some $\mu \in \mathbb{F}$. Hence $QL_n(P)Q = \mu Q$, which implies that $PL_n(P)P + QL_n(P)Q = \mu I$. Similarly we can prove $PL_n(Q)P + QL_n(Q)Q \in \mathbb{F}I$.
Now we put $T = PL_n(P)Q - QL_n(P)P$. For $A \in B(X)$, define $\Delta_n(A) = L_n(A) - (AT - TA)$. Also we can easily checked that $\Delta_n[A, B] = [\Delta_n(A), B] + [A, \Delta_n(B)]$ for all $A, B \in B(X)$.

Claim 2.$\Delta_n(P) \in \mathbb{F}I$.

proof. Using Claim 1, we have

$$\Delta_n(P) = L_n(P) - (PT - TP) = PL_n(P)P + QL_n(P)Q \in \mathbb{F}I.$$

Claim 3. $\Delta_n(PAQ + QAP) = P\Delta_n(A)Q + Q\Delta_n(A)P$ for all $A \in B(X)$.

proof. Let $A \in B(X)$, then

$$\begin{aligned}
\Delta_n(PAQ + QAP) &= \Delta_n([P, [P, A]]) \\
&= \sum_{i+j=n} [\Delta_i(P), \Delta_j([P, A]]) \\
&= [\Delta_n(P), [P, A]] + [p, \Delta_n([P, A])] + \sum_{i+j=n \quad i \neq 0,n} [\Delta_i(P), \Delta_j([P, A])] \\
&= [P, [P, \Delta_n(A)]] + \sum_{i+j=n; i \neq 0,n} [\Delta_i(P), \Delta_j([P, A])]
\end{aligned}$$

Since by Claim 2, $\Delta_i(P) \in \mathbb{F}I$, for $i < n$, so $\sum_{i+j=n;i\neq 0,n}[\Delta_i(P), \Delta_j([P, A])] = 0$
So

$$\Delta_n(PAQ + QAP) = [P, [P, \Delta_n(A)]] = P\Delta_n(A)Q + Q\Delta_n(A)P.$$

Claim 4. $\Delta_n(Q) \in \mathbb{F}I$.

proof. Applying Claim 1, we have

$$P\Delta_n(Q)Q + Q\Delta_n(Q)P = \Delta_n(PQQ + QQP) = 0.$$

Thus $\Delta_n(Q) = PL_n(Q)Q + QL_n(Q)Q \in \mathbb{F}$.

Claim 5. $\Delta_n(B_{ij}) \subseteq \mathrm{B_{ij}}$ for $1 \leqslant i \neq j \leqslant 2$.

proof. For any $X \in B_{12}$, we have

$$\begin{aligned}
\Delta_n(X) &= \Delta_n([P, X]) \\
&= \sum_{i+j=n} [\Delta_i(P), \Delta_j(X)] \\
&= [P, \Delta_n(X)] + \sum_{i+j=n; i \neq 0,n} [\Delta_i(P), \Delta_j(X)] \\
&= [P, \Delta_n(X)] + \sum_{j; j \neq 0,n} [P, \Delta_j(X)]
\end{aligned}$$

Since $\Delta_i(P) \in \mathbb{F}I$ is already shown, $\sum_{i+j=n;i\neq 0,n}[\Delta_i(P), \Delta_j(X)] = 0$ immediately follows.

So

$$\Delta_n(X) = P\Delta_n(x)Q - Q\Delta_n(X)P$$

Multiplying above equation by P from the right, we get

$$P\Delta_n(X)P = Q\Delta_n(X)P = Q\Delta_n(X)Q = 0$$

Thus

$$\Delta_n(X) = P\Delta_n(X)P \in \mathrm{B}_{12}$$

Similarly we can prove that $\Delta_n(y) \subseteq \mathrm{B}_{21}$ for any $y \in \mathrm{B}_{21}$.

Claim 6. There is a functional $f_{ni} : \mathrm{B_{ii}} \to \mathbb{F}$ such that $\Delta_n(X_{ii}) - f_{ni}(X_{ii})I \in \mathrm{B_{ii}}$ for all $X_{ii} \in \mathrm{B_{ii}}$, $1 \leqslant i \leqslant 2$.

proof. Let $X_{ii} \in \mathrm{B_{ii}}$, Applying Claim 3, we obtain

$$0 = \Delta_n(0) = \Delta_n(PX_{ii}Q + QX_{ii}P) = P\Delta_n(X_{ii})Q + Q\Delta_n(X_{ii})P$$

Then we can assume that $\Delta_n(X_{11}) = a_{11}+a_{22}$ and $\Delta_n(X_{22}) = b_{11}+b_{22}$, where $a_{11}, b_{11} \in \mathrm{B}_{11}$, $a_{22}, b_{22} \in \mathrm{B}_{22}$. Also

$$\begin{aligned} 0 &= \Delta_n([X_{11}, X_{22}]) \\ &= \sum_{i+j=n} [\Delta_i(X_{11}), \Delta_j(X_{22})] \\ &= [a_{22}, X_{22}] + [X_{11}, a_{11}] + \sum_{i+j=n; i\neq 0,n} [\Delta_i(X_{11}), \Delta_j(X_{22})]. \end{aligned}$$

By the property $\mathbf{P_m}$, $\Delta_i(X_{11}) \in B_{11} + \mathbb{F}I$ and $\Delta_j(X_{22}) \in \mathrm{B}_{22} + \mathbb{F}I$.
one gets $\sum_{i+j=n;i\neq 0,n}[\Delta_i(X_{11}), \Delta_j(X_{22})] = 0$ which implies that $[a_{22}, X_{22}] = 0$ for all $X_{22} \in \mathrm{B}_{22}$ and $[X_{11}, a_{11}] = 0$ for all $X_{11} \in \mathrm{B}_{11}$.
Therefore, there exist scalars $f_{n1}(X_{11})$ and $f_{n2}(X_{22})$ such that $a_{22} = f_{n1}(X_{11})Q$ and $b_{11} = f_{n2}(X_{22})P$. So $\Delta_n(X_{11}) - f_{n1}(X_{11})I \in B_{11}$ and $\Delta_n(X_{22}) - f_{n2}(X_{22})I \in \mathrm{B}_{22}$.

Claim 7.Δ_n is additive on B_{12} and B_{21}.

proof. Let $X_{22} \in \mathrm{B}_{22}$,$X_{12} \in \mathrm{B}_{12}$ and $Y_{21} \in \mathrm{B}_{21}$, by applying Claim 5, 6 we have

$$\begin{aligned}\Delta_n[X_{22}+X_{21},Y_{21}] &= \sum_{i+j=n}[\Delta_i(X_{22}+X_{21}),\Delta_j(Y_{21})]\\ &= [\Delta_n(X_{22}+X_{21}),Y_{21}]+[X_{22}+X_{21},\Delta_n(Y_{21})]\\ &\quad+\sum_{i+j=n;i\neq 0,n}[\Delta_i(X_{22}+X_{21}),\Delta_j(Y_{21})]\\ &= [\Delta_n(X_{22}+X_{21}),Y_{21}]+[X_{22},\Delta_n(Y_{21})]\\ &\quad+\sum_{i+j=n;i\neq 0,n}[\Delta_i(X_{22}+X_{21}),\Delta_j(Y_{21})]\\ \Delta_n[X_{22}+X_{21},Y_{21}] &= \Delta_n[X_{22},Y_{21}]\\ &= \sum_{i+j=n}[\Delta_i(X_{22}),\Delta_j(Y_{21})]\\ &= [\Delta_n(X_{22})+\Delta_n(X_{21}),Y_{21}]+[X_{22},\Delta_n(Y_{21})]\\ &\quad+\sum_{i+j=n;i\neq 0,n}[\Delta_i(X_{22})+\Delta_i(X_{21}),\Delta_j(Y_{21})]\end{aligned}$$

which implies that

$$[\Delta_n(X_{22}+X_{21})-\Delta_n(X_{22})-\Delta_n(X_{21}),Y_{21}]+\sum_{i+j=n;i\neq 0,n}[\Delta_i(X_{22}+X_{21}),\Delta_j(Y_{21})]$$
$$-\sum_{i+j=n;i\neq 0,n}[\Delta_i(X_{22}),\Delta_j(Y_{21})]=0$$

But by $\mathbf{P_m}$, we have

$$\sum_{i+j=n;i\neq 0,n}[\Delta_i(X_{22}+X_{21}),\Delta_j(Y_{21})]=\sum_{i+j=n;i\neq 0,n}[\Delta_i(X_{22})+\Delta_i(X_{22}),\Delta_j(Y_{21})].$$

Thus we conclude that $[\Delta_n(X_{22}+X_{21})-\Delta_n(X_{22})-\Delta_n(X_{21}),Y_{21}]=0$. By a similar way to we can prove that Δ_n is additive on B_{21} and similarly additivity of Δ_n on B_{12} can be deduced easily.

Now we define $D_n(A)=\Delta_n(PAQ)+\Delta_n(QAP)+\Delta_n(PAP)+\Delta_n(QAQ)-(f_1(PAP)-f_2(QAQ))I$. Then by Claim 5, 6 we have

Claim 8. For $B_{ij}\in\mathrm{B_{ij}}$, $1\leqslant i,j\leqslant 2$ we have

(1) $D_n(B_{ij})\in\mathrm{B_{ij}}$, $1\leqslant i,j\leqslant 2$;
(2) $D_n(B_{12})=\Delta_n(\mathrm{B}_{12})$ and $D_n(B_{21})=\Delta_n(\mathrm{B}_{21})$;
(3) $D_n(B_{11}+B_{12}+B_{21}+B_{22})=D_n(B_{11})+D_n(B_{12})+D_n(B_{21})+D_n(B_{22})$.

The following claims immediately follows from Claim 7 and Claim 8.

Claim 9. D_n is additive on B_{12} and B_{21}.

Claim 10. D_n is additive on B_{11} and B_{22}.
It follows similarly to proof of Lemma 2.12 in [4].

Claim 11. D_n is additive.
It follows similarly to proof of Lemma 2.13 in [4].

Claim 12. D_n has the following properties:

(1) $D_n(A_{ii}B_{ij}) = \sum_{t+k=n} D_t(A_{ii})D_k(B_{ij}) \quad 1 \leqslant i \neq j \leqslant 2$
(2) $D_n(B_{ij}A_{jj}) = \sum_{t+k=n} D_t(B_{ij})D_k(A_{jj}) \quad 1 \leqslant i \neq j \leqslant 2$
(3) $D_n(A_{ii}B_{ii}) = \sum_{t+k=n} D_t(A_{ii})D_k(B_{ii})$

proof. Let $A_{ii} \in \mathrm{B_{ii}}$ and $B_{ij} \in \mathrm{B_{ij}}$, then

$$\begin{aligned} D_n(A_{ii}B_{ij}) &= \Delta_n(A_{ii}B_{ij}) \\ &= \sum_{t+k=n} [D_t(A_{ii}), D_k(B_{ij})] \\ &= \sum_{t+k=n} [D_t(A_{ii}) + \tau_t(A_{ii}), D_k(B_{ij}) + \tau_t(B_{ij})] \\ &= \sum_{t+k=n} [D_t(A_{ii}), D_k(B_{ij})] \end{aligned}$$

Since $D_t(A_{ii}) \in \mathrm{B_{ii}} + \mathbb{F}I$ and $D_k(B_{ij}) \in \mathrm{B_{ij}}$, then

$$D_n(A_{ii}B_{ij}) = \sum_{t+k=n} D_t(A_{ii})D_k(B_{ij}).$$

Also

$$\begin{aligned} D_n(B_{ij}A_{jj}) &= \Delta_n(B_{ij}A_{jj}) \\ &= \sum_{t+k=n} [D_t(B_{ij}), D_k(A_{jj})] \\ &= \sum_{t+k=n} [D_t(B_{ij}) + \tau_t(A_{jj}), D_k(B_{ij}) + \tau_t(A_{jj})] \\ &= \sum_{t+k=n} [D_t(B_{ij}), D_k(A_{jj})] \end{aligned}$$

Since $D_k(A_{jj}) \in \mathrm{B_{jj}} + \mathbb{F}I$ and $D_t(B_{ij}) \in \mathrm{B_{ij}}$, then

$$D_n(B_{ij}A_{jj}) = \sum_{t+k=n} D_t(B_{ij})D_k(A_{jj}).$$

Furthermore

$$D_n(A_{ii}B_{ii}C_{ij}) = \sum_{t+k=n} D_t(A_{ii}B_{ii})D_k(C_{ij})$$
$$= \sum_{t+k=n,k\neq 0} D_t(A_{ii}B_{ii})D_k(C_{ij}) + D_n(A_{ii}B_{ii})C_{ij}$$

But

$$D_n(A_{ii}B_{ii}C_{ij}) = \sum_{t+k=n} D_t(A_{ii})D_k(B_{ii}C_{ij})$$
$$= \sum_{t+k+l=n} D_t(A_{ii})D_k(B_{ij})D_l(C_{ij})$$
$$= \sum_{t+k+l=n,L\neq 0} D_t(A_{ii})D_k(B_{ij})D_l(C_{ij}) + \sum_{t+k=n,l\neq 0} D_t(A_{ii})D_k(B_{ij})C_{ij}$$

Thus

$$D_n(A_{ii}B_{ii})C_{ij} = \sum_{t+k=n} D_t(A_{ii})D_k(B_{ij})C_{ij}$$

Which implies that

$$D_n(A_{ii}B_{ii}) = \sum_{t+k=n} D_t(A_{ii})D_k(B_{ij}).$$

Claim 14. $\Delta_n(B_{11}+C_{22}) - \Delta_n(B_{11}) - \Delta_n(C_{22}) \in \mathbb{F}I$.

Claim 15. $\Delta_n(A_{ij}B_{ji}) = \sum_{i+j=n}[\Delta_i(A_{ij}, \Delta_j(B_{ji})] \quad 1 \leqslant i \neq j \leqslant 2$

proof. Note that $[A_{12}, B_{21}] = [[A_{12},, P_2], B_{21}]$ holds true for all $A_{12} \in \mathrm{B}_{12}$ and $B_{21} \in \mathrm{B}_{21}$. So we have

$$\Delta_n([B_{21}, C_{12}]) - D_n([B_{21}, C_{12}]) = \sum_{i+j=n} [\Delta_i(B_{21}, \Delta_j(C_{12})] - D_n(B_{21})C_{12} - D_n(C_{12}B_{21})$$
$$= [\Delta_n(B_{21}), C_{12}] + [B_{21}, \Delta_n(C_{12})]$$
$$+ \sum_{i+j=n;i\neq 0,n} [\Delta_i(B_{21}), \Delta_j(D_{12})] - D_n(B_{21})C_{12} - D_n(C_{12}B_{21})$$
$$= [D_n(B_{21}), C_{12}] + [B_{21}, D_n(C_{12})]$$
$$+ D_n(B_{21}C_{12}) - D_n(B_{21}C_{12})$$
$$= D_n(A_{12})B_{21} - P_2D_n(A_{12})B_{21} - B_{21}D_n(A_{12})P_2 + B_{21}D_n(A_{12})$$
$$+ A_{12}D_n(B_{21}) - D_n(B_{21})A_{12} - D_n(A_{12}B_{21}) + D_n(B_{21}A_{12})$$

Which implies that

$(D_n(A_{12})B_{21}+A_{12}D_n(B_{21})-D_n(A_{12}B_{21}))+(D_n(B_{21}A_{12})-D_n(B_{21})A_{12}-B_{21}D_n(A_{12}))$
$= K \in \mathbb{F}I.$

Using Claim 12 and multiplying B_{22} from the left in the above equality, We obtain

$$B_{22}D_n(B_{21}A_{12}) - B_{22}D_n(B_{21})A_{12} - B_{22}B_{21}D_n(A_{12}) = B_{22}K.$$

So

$K = D_n(A_{12}B_{21}+A_{12}D_n(B_{21})-D_n(A_{12}B_{21}))+(D_n(B_{21}A_{12})-D_n(B_{21})A_{12}-B_{21}D_n(A_{12})$
$= 0.$

Therefore

$$D_n(A_{12})B_{21}+A_{12}D_n(B_{21})-D_n(A_{12}B_{21}) = -(D_n(B_{21}A_{12})-D_n(B_{21})A_{12}-B_{21}D_n(A_{12})) \in \mathrm{B}_{11}\bigcap \mathrm{B}_{22}$$

So we have
$D_n(A_{12})B_{21} + A_{12}D_n(B_{21}) - D_n(A_{12}B_{21}) = 0$ and $D_n(B_{21}A_{12}) - D_n(B_{21})A_{12} - B_{21}D_n(A_{12}) = 0$
So D_n is a derivation.

Claim 16. For all $A \in B(X)$, $\Delta_n(A) - \Delta_n(PAQ) - \Delta_n(QAP) \in \mathbb{F}I$.
For any $A \in B(X)$ and Claim it is easy to checked that $\Delta_n(A)-\Delta_n(PAQ)-\Delta_n(QAP) \in \mathbb{F}I$.

Claim 17. $\tau_n([X,Y]) = 0$ for all $X, Y \in B(X)$.

proof. For any $X, Y \in B(X)$, we have

$$\begin{aligned}\tau_n([X,Y]) &= \Delta_n([X,Y]) - D_n([X,Y])\\ &= \sum_{i+j=n} [\Delta_i(X), \Delta_j(Y)] - D_n(XY) - D_n(YX)\\ &= \sum_{i+j=n} [D_i(X)+\tau_i(X), D_j(Y)+\tau_j(Y)] - \sum_{i+j=n} D_i(X)D_j(Y) - D_j(Y)D_j(X)\\ &= \sum_{i+j=n} [D_i(X), D_j(Y) - \sum_{i+j=n} D_i(X)D_j(Y) - D_j(Y)D_j(X)\\ &= 0\end{aligned}$$

The proof of theorem is completed. ■

References

[1] W. Cheung, *Lie derivations on triangular matrices*, Linear Multilinear Algebra **55** (2007), 619.626.
[2] Y. Du, Y. Wang, *Lie derivations of generalized matrix algebras*, Linear Algebra Appl. **433** (2012), 2719-2726.
[3] M. Ferrero, C. Haetinger, *Higher derivations of semiprime rings*, Comm. Algebra. **30** (2002), 2321-2333.
[4] F. Lu ., B. Liu, *Lie derivable maps on $B(X)$* , J. Math. Anals. Appl. **372**(2010), 369-376. 2321-2333

[5] Y. B. Li, Z. K. Xiao, *Additivity of maps on generalized matrix algebras*, Electron. J. Linear Algebra **22**, (2011) 743757.
[6] A. Nakajima, it On generalized higher derivations, Turk. J. Math. **24**(2000), 295-311.
[7] X. F. Qi,*chacterization of Lie higher derivation on triangular algebras*, Acta Mathematica Sinica. **26** (2013), 1007-1018.
[8] G. A. Swain, P.S. Blau,*Lie derivations in prime rings with involution* , Canad. Math. Bull. **42**, (1999) 401-411.
[9] Z.-K Xiao, F. Wei,*Jordan higher derivations on triangular algebras*, Linear Algebra and its Applications. **432** (2010), 26152622.
[10] Q. X.- Fei, H. J. Chen, *Lie higher derivations on nest algebras*, Comm. Math. Research.(2010) 131-143.
[11] Z.-K Xiao, F. Wei , *Higher derivations of triangular algebras and its generalizations*, Linear Algebra Appl. **435** (2011),1034-1054.
[12] W. Y. Yu, J. H. Zhang, *Chacterization of Lie higher and Lie triple derivation on triangular algebras*, J. Korean Math. Soc. **49** (2012), 419-433.

11

Subcategories of topological algebras

Vijaya L. Gompa*

Department of Mathematics, Troy University, Dothan, AL 36304.

Abstract. In addition to exploring constructions and properties of limits and colimits in categories of topological algebras, we study special subcategories of topological algebras and their properties. In particular, under certain conditions, reflective subcategories when paired with topological structures give rise to reflective subcategories and epireflective subcategories give rise to epireflective subcategories.

Keywords: Monotopolocial category, topological category, topological functors, universal algebra, topological algebra, reflective subcategory, coreflective subcategory, epireflective subcategory.

Essentially a topological algebra is a universal algebra endowed with a topological structure so that algebraic operations are continuous in all variables together. Wyler has generalized the construction of categories of topological algebras (see [15]), by obtaining from what he calls a "top" category (which is equivalent to the concept topological category) $\boldsymbol{C}^s$ and an operational category $\boldsymbol{A}$ over a category $\boldsymbol{C}$ a new category $\boldsymbol{A}^\tau$ which is "top" over $\boldsymbol{A}$ and operational over $\boldsymbol{C}^s$, with a pullback property. Fay further generalized the categories of topological algebras using a concept called topologically algebraic situation (see [4]). Later Nel ([11]) and Koslowski ([9]) have given descriptions that are adopted in our work. First, let us describe some concepts used in this work.

*Corresponding author.
E-mail address: vgompa@troy.edu (Vijaya L. Gompa).

1. Preliminaries

There are several definitions for the term "algebraic functor" in the literature, all of which are equivalent in some special categories, but not in general. We choose to adopt the following popular definition [8, page 243]. A functor $U : \boldsymbol{X} \to \boldsymbol{Y}$ is called **algebraic** iff U has a left adjoint and preserves and reflects regular epimorphisms.

An algebraic functor is faithful ([8], 32.17). If $U : \boldsymbol{X} \longrightarrow \boldsymbol{Y}$ is an algebraic functor and the category $\boldsymbol{X}$ has coequalizers, then $(\boldsymbol{X}, U)$ is called an **algebraic category over $\boldsymbol{Y}$**. Algebraic category over $\boldsymbol{Set}$ is equivalent to regularly algebraic category over $\boldsymbol{Set}$ in the sense of [1, 20.35, 23.38, 23.39]. The functor from the category $\boldsymbol{Grp}$ of groups into $\boldsymbol{Set}$ is algebraic but not topological. It is known (see [7]) that every algebraic functor into $\boldsymbol{Set}$ is topologically algebraic.

A functor $U : \boldsymbol{A} \to \boldsymbol{X}$ **creates isomorphisms** iff to each $\boldsymbol{X}$-isomorphism $f : X \longrightarrow UA$ corresponds a unique $\boldsymbol{A}$-morphism $g : B \longrightarrow A$ such that $Ug = f$ and g is an $\boldsymbol{A}$-isomorphism. A functor $U : \boldsymbol{A} \to \boldsymbol{X}$ is called **(generating, monosource) - factorizable** if for every source $(X, f_i : X \to UA_i)_{i \in I}$ there exists a generating map $e : X \to UA$ (i.e., e is an $\boldsymbol{X}$-morphism such that for any two $\boldsymbol{X}$-morphisms $r : UA \longrightarrow Y$ and $s : UA \longrightarrow Y$, the equality $r \circ e = s \circ e$ implies $r = s$) and a monosource $(A, m_i : A \to A_i)_{i \in I}$ (i.e., for any two $\boldsymbol{A}$-morphisms $u : B \longrightarrow A$ and $v : B \longrightarrow A$, the equalities $m_i \circ u = m_i \circ v$ for all $i \in I$ implies $u = v$) such that the diagram

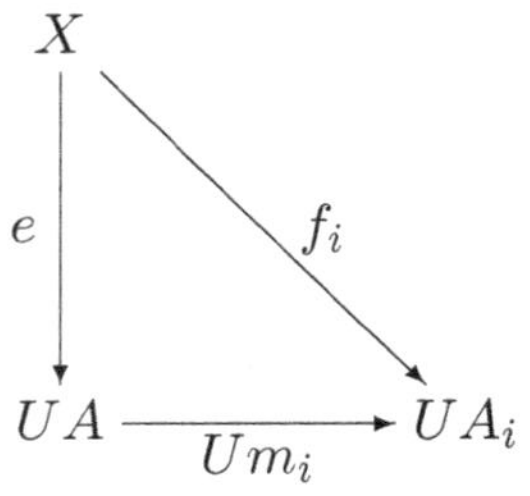

commutes. A functor $U : \boldsymbol{X} \longrightarrow \boldsymbol{Y}$ is said to be **essentially algebraic** [6] provided that it creates isomorphisms and is (generating, monosource) - factorizable. If $U : \boldsymbol{X} \longrightarrow \boldsymbol{Y}$ is an essentially algebraic functor and faithfull, then $(\boldsymbol{X}, U)$ is called an **essentially algebraic category over $\boldsymbol{Y}$**.

A functor $U : \boldsymbol{A} \longrightarrow \boldsymbol{X}$ is called **uniquely transportable** iff any $\boldsymbol{X}$-isomorphism $f : X \longrightarrow UA$ can be lifted via U to a unique $\boldsymbol{A}$-isomorphism $g : A' \longrightarrow A$. For a later use, we will formulate a result in the following Theorem, whose proof can be found in [1, 23.2].

Theorem 1.1

The following conditions are equivalent for a uniquely transportable (generating, monosource) - factorizable functor $U : \boldsymbol{X} \longrightarrow \boldsymbol{Y}$.

(a) U is essentially algebraic.
(b) U reflects isomorphisms.
(c) U reflects limits.
(d) U reflects equalizers.
(e) U reflects extremal epimorphisms and is faithful.
(f) Every monosource in $\boldsymbol{X}$ is U-initial. ∎

A family $\Omega = (n_j)_{j \in J}$ of natural numbers indexed by some set J is called a **type**. The index set J is called the **order** of Ω. In the following, we let a type $\Omega = (n_j)_{j \in J}$ be fixed. A pair $(|A|, (\omega_j)_{j \in J})$ of a set $|A|$ and a family $\omega_j : |A|^{n_j} \to |A|$ $(j \in J)$ of mappings is called an Ω-**algebra** (see, for example, [3]). For the sake of simplicity, we write A instead of the pair $(|A|, (\omega_j)_{j\in J})$ and $\omega_{j,A}$ for the $\mathbf{n}_j$**-ary operation** ω_j on A. If the Ω-algebra A is clear from the context, we drop the suffix A in denoting its n_j-ary ($j \in J$) operation. If A and B are Ω-algebras, then a mapping $f : |A| \to |B|$ is said to be an Ω-**morphism** $f : A \to B$ iff for each $j \in J$, $f \circ \omega_{j,A} = \omega_{j,B} \circ f^n$ where $n = n_j$ and $f^n : |A|^n \to |B|^n$ is the mapping with the obvious definition $(a_1, \ldots, a_n) \to (fa_1, \ldots, fa_n)$.

The symbol $\boldsymbol{Alg(\Omega)}$ denotes the category whose objects are Ω-algebras and whose morphisms are Ω-morphisms. $\boldsymbol{Alg(\Omega)}$ is algebraic over $\boldsymbol{Set}$ (see [1, 7.72 (3), 23.6 (1), 23E (a)]).

A subcategory of an algebraic category, in general, may not be algebraic. However, this is guaranteed by several equivalent conditions for a special type of subcategory. To state this result we need the definition of an isomorphism closed subcategory. A subcategory $\boldsymbol{A}$ of $\boldsymbol{B}$ is called **isomorphism closed** iff every isomorphism in $\boldsymbol{B}$ whose domain or codomain belongs to $\boldsymbol{A}$ is a morphism in $\boldsymbol{A}$. We state two results in this direction whose proofs can be found in the stated references.

Theorem 1.2 If $(\boldsymbol{B}, U)$ is an algebraic category and $\boldsymbol{A}$ is a full isomorphism closed subcategory of $\boldsymbol{B}$ with embedding $E : \boldsymbol{A} \hookrightarrow \boldsymbol{B}$ such that $\boldsymbol{A}$ is closed under the formation of subobjects in $\boldsymbol{B}$, then the following are equivalent [8, 38.2]:
(a) $(\boldsymbol{A}, U \circ E)$ is algebraic.
(b) $\boldsymbol{A}$ is reflective in $\boldsymbol{B}$.
(c) $\boldsymbol{A}$ is a complete subcategory of $\boldsymbol{B}$.
(d) $\boldsymbol{A}$ is closed under the formation of products in $\boldsymbol{B}$. ∎

Theorem 1.3 An isomorphism closed full subcategory $\boldsymbol{A}$ of $\boldsymbol{Alg(\Omega)}$ is epireflective in $\boldsymbol{Alg(\Omega)}$ iff $\boldsymbol{A}$ is closed under the formation of products and subalgebras ([7], [1, 20.18, 23.6(1), 23.12(1), 16.9]). ∎

Now we can conclude, as a consequence of these two results, that a full isomorphism closed epireflective subcategory $\boldsymbol{A}$ of $\boldsymbol{Alg(\Omega)}$ is algebraic over $\boldsymbol{Set}$ and hence admits free Ω-algebras and has regular factorizations because algebraic category over $\boldsymbol{Set}$ means regularly algebraic category over $\boldsymbol{Set}$ in the sense of [1, 23.35,23.38, 23.39]. A full isomorphism closed epireflective subcategory of $\boldsymbol{Alg(\Omega)}$ is usually referred to as an **SP-class** or as a **quasiprimitive category** of algebras. A full subcategory $\boldsymbol{A}$ of the category $\boldsymbol{Alg(\Omega)}$ is a variety (in the sense of [3]) iff $\boldsymbol{A}$ is closed under subalgebras, homomorphic images and direct products. A variety is also called an **HSP-class** or a **primitive category** of algebras.

A variety is an epireflective subcategory of $\boldsymbol{Alg(\Omega)}$ (by Theorem 3) and is algebraic over $\boldsymbol{Set}$ (by Theorem 2). Thus every nontrivial variety has free algebras. Since every algebraic construct is topologically algebraic, both SP-classes and HSP-classes are topologically algebraic over $\boldsymbol{Set}$.

The following theorem, whose proof can be found in [1, 23.8, 23.13], sheds some light on essentially algebraic subcategories of $\boldsymbol{Alg(\Omega)}$.

Theorem 1.4

A concrete category $(\boldsymbol{A}, U)$ is essentially algebraic iff U creates isomorphisms, U is adjoint, and $\boldsymbol{A}$ is (epi, monosource) - factorizable.

Every essentially algebraic construct is complete, cocomplete, and wellpowered. ∎

Thus products, equalizers, coequalizers, intersections and free objects exist in any essentially algebraic subcategory of $\boldsymbol{Alg(\Omega)}$. Moreover, in such categories any monosource is point separating (because essentially algebraic functors preserve monosources, see [1,

23A]) and products are concrete (since any essentially algebraic functor preserves products). Since any category that has (epi, monosource) - factorizations is an (extremal epi, monosource) - category (see [1, 15.10]), an essentially algebraic construct is an (extremal epi, monosource) - category.

2. Paired Categories

Let $\boldsymbol{X}$ be a construct with finite concrete powers and $\boldsymbol{A}$ be a subcategory of $\boldsymbol{Alg(\Omega)}$. By a **paired object** (from $\boldsymbol{X}$ and $\boldsymbol{A}$) is meant an ordered pair (X, A) where X and A are objects in $\boldsymbol{X}$ and $\boldsymbol{A}$ respectively with the same underlying set such that, for each $j \in J$, the $n(=n_j)$-ary operation $\omega_{j,A} : |A|^n \to |A|$ on A is an $\boldsymbol{X}$-morphism $\omega_{j,A} : X^n \to X$. In this case, we write $\omega_{j,X}$ for the $\boldsymbol{X}$-morphism from X^n to X whose underlying function is $\omega_{j,A}$. If (X, A) and (X', A') are two paired objects (from $\boldsymbol{X}$ and $\boldsymbol{A}$), then an $\boldsymbol{X}$-morphism $f : X \to X'$ that is also an $\boldsymbol{A}$-morphism $f : A \to A'$ is called a **paired morphism** (from $\boldsymbol{X}$ and $\boldsymbol{A}$) and is denoted by $f : (X, A) \to (X', A')$. The category of all paired objects (from $\boldsymbol{X}$ and $\boldsymbol{A}$) together with paired morphisms (from $\boldsymbol{X}$ and $\boldsymbol{A}$) is called the **paired category** (from $\boldsymbol{X}$ and $\boldsymbol{A}$). We denote this category by $\boldsymbol{X \diamond A}$.

In this work, we assume that all subcategories are full isomorphism closed. The fact that the most of the natural subcategories fall into this class justifies our assumption. Unless otherwise stated, $\boldsymbol{X}$ and $\boldsymbol{Y}$ denote arbitrary constructs with finite concrete powers, and $\boldsymbol{A}$ represents any subcategory of $\boldsymbol{Alg(\Omega)}$. For the sake of simplicity, we will denote an object (X, A) in the paired category $\boldsymbol{X \diamond A}$ (from $\boldsymbol{X}$ and $\boldsymbol{A}$) either by X or by A. We will use a similar identification for morphisms in the paired category.

To see some examples of paired categories, notice that the category of topological groups with continuous homomorphisms is the paired category $\boldsymbol{Top \diamond Grp}$ from $\boldsymbol{Top}$ and $\boldsymbol{Grp}$.

***Example* 2.1** The category $\boldsymbol{Ab}$ of abelian groups and group homomorphisms can be viewed as the paired category from $\boldsymbol{Grp}$ and $\boldsymbol{Grp}$.

Indeed, Suppose G is a set, and $(G, \cdot, ^{-1})$ and $(G, \odot, *)$ are groups, the first operation is group multiplication and the second operation is group inversion, such that

$$\odot : (G, \cdot, ^{-1}) \times (G, \cdot, ^{-1}) \to (G, \cdot, ^{-1})$$

and

$$* : (G, \cdot, ^{-1}) \to (G, \cdot, ^{-1})$$

are group homomorphisms. If e and E are the identity elements in $(G, \cdot, ^{-1})$ and $(G, \odot, *)$ respectively, then $e \odot e = e$ and $e* = e$, because (e, e) is the identity element in $(G, \cdot, ^{-1}) \times (G, \cdot, ^{-1})$ and any group homomorphism maps the identity element in the domain group to the identity element of the codomain group. Combining these two equalities, we have

$$E = e \odot e* = e \odot e = e.$$

For any x and x' in G, since

$$\odot((x, e) \cdot (e, x')) = (x \odot e) \cdot (e \odot x'),$$

we have

$$x \odot x' = x \cdot x'.$$

This shows that $\odot = \cdot$ and $* = ^{-1}$. Consequently, the group inversion being a group homomorphism, group $(G, \cdot, ^{-1})$ has to be abelian.

Example* 2.2** Similarly the paired category ***Grp◇Rng is nothing but ***Ab***. (Note that, in a ring, if $(a+b)\cdot(c+d) = a\cdot c + b\cdot d$, then, taking $b = c = 0$, $a \cdot d = 0$ for all a and d.)

For the sake of simplicity, we assume that all the subcategories are isomorphism closed. The fact that most of the natural subcategories fall into this class justifies our assumption. We also use the convention that using the same symbol for morphisms in different constructs indicates that their underlying functions are the same [thus underlying sets for the domain objects (respectively, for the codomain objects) are the same]. However, for the sake of clarity, in some instances we may use different symbols for morphisms in different categories with the same underlying functions.

Unless otherwise stated, $\boldsymbol{X}$ and $\boldsymbol{Y}$ are assumed to be constructs admitting concrete finite powers while $\boldsymbol{A}$ and $\boldsymbol{B}$ are subcategories of $\boldsymbol{Alg}(\boldsymbol{\Omega})$.

3. Subcategories

Let us first discuss the construction of subcategories of $\boldsymbol{X}\Diamond\boldsymbol{A}$ from subcategories of $\boldsymbol{X}$ and of $\boldsymbol{A}$. It is clear that if $\boldsymbol{B}$ is a subcategory of $\boldsymbol{A}$, then $\boldsymbol{X}\Diamond\boldsymbol{B}$ is a subcategory of $\boldsymbol{X}\Diamond\boldsymbol{A}$. On the other hand, if $\boldsymbol{Y}$ is a subcategory of $\boldsymbol{X}$ then the category $\boldsymbol{Y}\Diamond\boldsymbol{A}$ need not be a subcategory of $\boldsymbol{X}\Diamond\boldsymbol{A}$ because concrete powers in $\boldsymbol{Y}$ need not agree with the concrete powers in $\boldsymbol{X}$. Here is an example:

***Example* 3.1** The additive group $\mathbb{R}$ of real numbers with its usual topology, i.e., with the nearness structure

$$\xi := \{\mathcal{A} \in \boldsymbol{P}^2(\mathbb{R}) : \cap\{\bar{A} : A \in \mathcal{A}\} \neq \phi\}$$

constitutes a counterexample since $\mathbb{R}$ is a topological group but not a nearness group: The addition $+ : \mathbb{R} \times \mathbb{R} \longrightarrow \mathbb{R}$ is not uniformly continuous with respect to the ***Near*** product structure on $\mathbb{R} \times \mathbb{R}$ (for a detailed proof, see [2]). ❙

However, we have the following result.

Theorem 3.2 If $\boldsymbol{Y}$ is a subcategory of $\boldsymbol{X}$ such that concrete powers in $\boldsymbol{Y}$ agree with concrete powers in $\boldsymbol{X}$ then $\boldsymbol{Y}\Diamond\boldsymbol{A}$ is a subcategory of $\boldsymbol{X}\Diamond\boldsymbol{A}$.

In particular, if $\boldsymbol{Y}$ is an epireflective subcategory of $\boldsymbol{X}$, then $\boldsymbol{Y}\Diamond\boldsymbol{A}$ is a subcategory of $\boldsymbol{X}\Diamond\boldsymbol{A}$.

Proof. Let (Y, A) be any object in $\boldsymbol{Y}\Diamond\boldsymbol{A}$. For each $j \in J$, the n_j-th product Y^{n_j} of Y in the category $\boldsymbol{Y}$ is the same as the n_j-th product of Y in the category $\boldsymbol{X}$ and the n_j-ary operation $\omega_{j,Y} : Y^{n_j} \to Y$, being a morphism in $\boldsymbol{Y}$, must be a morphism in $\boldsymbol{X}$. Thus (Y, A) is also an $\boldsymbol{X}\Diamond\boldsymbol{A}$-object. Obviously $\boldsymbol{Y}\Diamond\boldsymbol{A}$-morphisms are also morphisms in $\boldsymbol{X}\Diamond\boldsymbol{A}$.

If $\boldsymbol{Y}$ is an epireflective subcategory of $\boldsymbol{X}$, then the products in $\boldsymbol{Y}$ do agree with those in $\boldsymbol{X}$ so that $\boldsymbol{Y}\Diamond\boldsymbol{A}$ is a subcategory of $\boldsymbol{X}\Diamond\boldsymbol{A}$ by what was proved above. ∎

If $\boldsymbol{Y}$ is a coreflective subcategory of $\boldsymbol{X}$, then any object in $\boldsymbol{X}\diamond\boldsymbol{A}$ is also in $\boldsymbol{Y}\diamond\boldsymbol{A}$ if its first part is already in $\boldsymbol{Y}$. In other words:

Theorem 3.3 If $\boldsymbol{Y}$ is a coreflective subcategory of $\boldsymbol{X}$ and the pair (X, A) is an object in $\boldsymbol{X}\diamond\boldsymbol{A}$ such that X is an object in $\boldsymbol{Y}$ then (X, A) is also an object in $\boldsymbol{Y}\diamond\boldsymbol{A}$.

Proof. We need to prove that each algebraic operation on A is a $\boldsymbol{Y}$-morphism. Let $j \in J$ and $n = n_j$. To avoid ambiguity, let us use the symbol Y to indicate X regarded as a $\boldsymbol{Y}$-object and write Y^n and X^n for the products of X to itself n times in the categories $\boldsymbol{Y}$ and $\boldsymbol{X}$ respectively. Since $\boldsymbol{Y}$ is coreflective in $\boldsymbol{X}$, the $\boldsymbol{Y}$-product Y^n is the $\boldsymbol{Y}$-coreflection of the $\boldsymbol{X}$-product X^n. Therefore, any $\boldsymbol{X}$-morphism $X^n \longrightarrow X$ is also a $\boldsymbol{Y}$-morphism $Y^n \longrightarrow$Y. In particular, the n_j-ary operation on A being an $\boldsymbol{X}$-morphism $\omega_{j,X} : X^n \longrightarrow X$ is indeed a $\boldsymbol{Y}$-morphism $\omega_{j,Y} : Y^n \longrightarrow$Y. ■

Fay [4] proved, among other things, an (E, M)-topological version of Wyler's taut lift theorem (see [15]). Tholen (see [14]) generalized and discussed some applications of Wyler's theorem. In this section we will show another application, namely, an essentially algebraic functor between two subcategories of $\boldsymbol{Alg}(\Omega')$ and $\boldsymbol{Alg}(\boldsymbol{\Omega})$ can be extended to the associated paired categories. We use (E, M)-topological version of Wyler's taut lift theorem due to Fay ([4, Theorem (5.3)]) restated here as a lemma in a slightly different form using the hypothesis that is necessary so that the proof given by him works for the lemma. The lemma as stated here is also a consequence of the proof of Theorem (4.1) in [4]. Although it appears closer to Tholen's Theorem (4.1) in [4], we chose to use Fay's result because of its notational advantage. First we will explain a concept.

If $T : \boldsymbol{L} \longrightarrow \boldsymbol{K}$ is a functor and M is a class of sources in $\boldsymbol{K}$, then M_T denotes the class of all T-initial sources $(L, f_i : L \longrightarrow L_i)_{i\in I}$ in $\boldsymbol{L}$ with $(TL, Tf_i : TL \longrightarrow TL_i)_{i\in I}$ a source in M. A pair (g, L) of an $\boldsymbol{L}$-object L and $\boldsymbol{K}$-morphism $g : K \longrightarrow TL$ is said to be **T-universal map for $\boldsymbol{K}$** ([8, 26.1]) iff for each $\boldsymbol{L}$-object L' and each $\boldsymbol{K}$-morphism $f : K \longrightarrow TL'$, there exists a unique $\boldsymbol{L}$-morphism $\bar{f} : L \longrightarrow L'$ such that the triangle

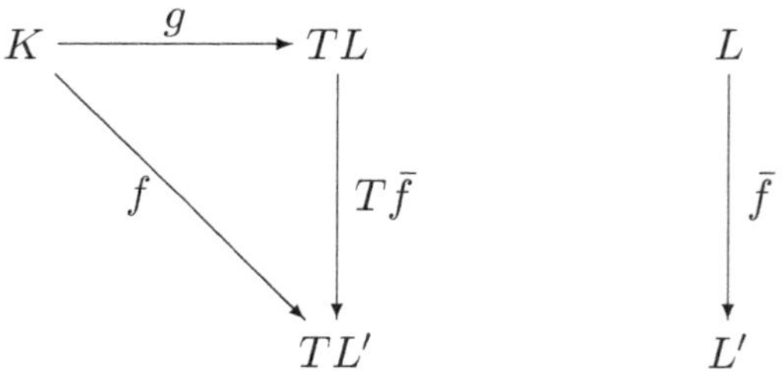

commutes.

Lemma 3.4 Consider the following commutative square of categories and functors

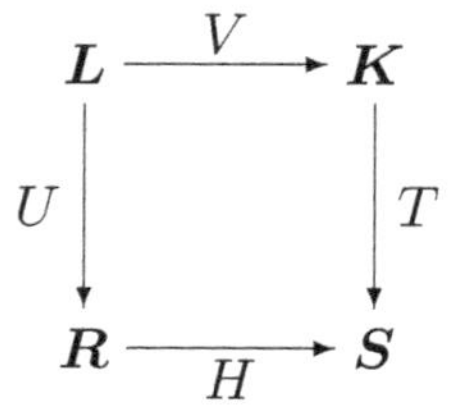

where $\boldsymbol{R}$ has (E, M) - factorizations, $\boldsymbol{L}$ has U-initial lifts for sources in M, M' is the class of all sources in $\boldsymbol{S}$ which have T-initial lifts, H has a left adjoint, H sends M-sources to M'-sources, and V sends M_U- sources to $(M')_T$- sources.

Then V has a left adjoint.

Moreover, if for each $\boldsymbol{S}$-object S there exists an $\boldsymbol{H}$-universal map (e, R) for S where $e : S \longrightarrow HR$ is an epimorphism in $\boldsymbol{S}$, then for each $\boldsymbol{K}$-object K there corresponds a V-universal map (f, L) for K such that $f : K \longrightarrow VL$ is an epimorphism in $\boldsymbol{K}$. ❙

Theorem 3.5 Suppose $\boldsymbol{A}$ and $\boldsymbol{B}$ are subcategories of $\boldsymbol{Alg(\Omega)}$.

(a) If $\boldsymbol{B}$ is an essentially algebraic subcategory of $\boldsymbol{Alg(\Omega)}$ that is reflective in a subcategory $\boldsymbol{A}$, then $\boldsymbol{X \diamond B}$ is a reflective subcategory of $\boldsymbol{X \diamond A}$.

(b) If $\boldsymbol{B}$ is further epireflective in $\boldsymbol{A}$, then $\boldsymbol{X \diamond B}$ is an epireflective subcategory of $\boldsymbol{X \diamond A}$.

Proof. Let $H : \boldsymbol{B} \longrightarrow \boldsymbol{A}$ be the inclusion map and define the concrete functor $\tilde{H} : \boldsymbol{X \diamond B} \longrightarrow \boldsymbol{X \diamond A}$ by $\tilde{H}(X, B) := (X, HB)$, for any $\boldsymbol{X \diamond B}$-object (X, B). We have the commutative square diagram

$$\begin{array}{ccc} \boldsymbol{X \diamond B} & \xrightarrow{\tilde{H}} & \boldsymbol{X \diamond A} \\ {\scriptstyle T}\downarrow & & \downarrow{\scriptstyle T'} \\ \boldsymbol{B} & \xrightarrow[H]{} & \boldsymbol{A} \end{array},$$

where T and T' are forgetful functors.

If $\boldsymbol{B}$ is an essentially algebraic subcategory of $\boldsymbol{Alg(\Omega)}$ that is reflective in a subcategory $\boldsymbol{A}$, then H is essentially algebraic and hence $\tilde{H}$ is essentially algebraic (see [5, Theorem 3.2 (c)]), which proves $\boldsymbol{X \diamond B}$ is a reflective subcategory of $\boldsymbol{X \diamond A}$.

The second part is a direct application of the Lemma 1. ■

4. Limits and Colimits

In [15]), Wyler shows, among other things, that if $\boldsymbol{X}$ is a topological category then all categorical limits and colimits can be lifted from a category $\boldsymbol{A}$ of algebras to the category $\boldsymbol{X \diamond A}$. In particular, if $\boldsymbol{X}$ is a topological category and $\boldsymbol{A}$ is complete and cocomplete then $\boldsymbol{X \diamond A}$ is complete and cocomplete. Since each monotopological category is an epireflective subcategory of a topological category, similar results are true if $\boldsymbol{X}$ is a monotopological category. In this section we intend to describe some limits and colimits in the paired category $\boldsymbol{X \diamond A}$ under the assumption that $\boldsymbol{X}$ is monotopological. We begin with a Theorem that is very useful in our work.

Theorem 4.1 Suppose $G : \boldsymbol{X} \longrightarrow \boldsymbol{Y}$ is a concrete functor which preserves concrete finite powers, X is an $\boldsymbol{X}$-object, $((X_i, A_i))_{i \in I}$ is a family of $\boldsymbol{X \diamond A}$-objects, $((Y, A), f_i : (Y, A) \longrightarrow (GX_i, A_i))_{i \in I}$ is a source in $\boldsymbol{Y \diamond A}$, and $(X, g_i : X \longrightarrow X_i)_{i \in I}$ is a G-initial source in $\boldsymbol{X}$. If $Y = GX$ and $Gg_i = f_i$ as $\boldsymbol{Y}$-morphisms for each $i \in I$, then (X, A) is an $\boldsymbol{X \diamond A}$-object.

Proof. Since $GX = Y$, we have that $|X| = |Y|(= |A|)$. Because X and A are objects in the categories $\boldsymbol{X}$ and $\boldsymbol{A}$ respectively, we only have to show that for each $j \in J$, the n_j-ary operation $\omega_{j,A}$ on A is a morphism in $\boldsymbol{X}$. Fix $j \in J$ and write n for n_j. For any $i \in I$, since $f_i : A \longrightarrow A_i$ is an $\boldsymbol{A}$-homomorphism we have the commutative diagram of $\boldsymbol{Y}$-morphisms:

$$\begin{array}{ccc} Y^n & \xrightarrow{f_i^n} & Y_i^n \\ {\scriptstyle \omega_{j,Y}}\downarrow & & \downarrow{\scriptstyle G\omega_{j,Y_i}} \\ Y & \xrightarrow[f_i]{} & Y_i \end{array}$$

where $Y_i = GX_i$. Since G preserves concrete finite powers and $GX = Y$, we have $Y^n = (GX)^n = G(X^n)$ and $G\omega_{j,X_i} \circ f_i = G\omega_{j,X_i} \circ (Gg_i)^n = G\omega_{j,X_i} \circ G(g_i) = G(\omega_{j,X_i} \circ g_i)$, hence the above diagram can be viewed as the following:

$$\begin{array}{ccc} G(X^n) & & \\ {\scriptstyle \omega_{j,Y_i}}\downarrow & \searrow^{G(\omega_{j,X_i} \circ g_i^n)} & \\ G(X) & \xrightarrow[Gg_i]{} & G(X_i) \end{array}$$

Since $(X, g_i : X \longrightarrow X_i)_{i \in I}$ is G-initial there exists a unique $\boldsymbol{X}$-morphism $\omega : X^n \longrightarrow X$ such that $G\omega = \omega_{j,Y}$ and the diagram

$$\begin{array}{ccc} X^n & & \\ {\scriptstyle \omega}\downarrow & \searrow^{\omega_{j,X_i} \circ g_i^n} & \\ X & \xrightarrow[g_i]{} & X_i \end{array}$$

commutes. Since G is concrete, ω and $\omega_{j,Y}$ have the same underlying functions. Thus $\omega_{j,A}$ is an $\boldsymbol{X}$-morphism. This shows that (X, A) is an $\boldsymbol{X}\diamond\boldsymbol{A}$-object. ■

As a consequence of this result we describe a construction of $\boldsymbol{X}\diamond\boldsymbol{A}$-objects from sources in the category $\boldsymbol{A}$.

Corollary 4.2 Let $((X_i, A_i))_{i \in I}$ be a family of objects in the category $\boldsymbol{X}\diamond\boldsymbol{A}$ and $(A, g_i : A \longrightarrow A_i)_{i \in I}$ be a source in $\boldsymbol{A}$. If X is an object in $\boldsymbol{X}$ having the same underlying set as A and is initial with respect to the source $(X, g_i : X \longrightarrow X_i)_{i \in I}$, then the pair (X, A) lies in $\boldsymbol{X}\diamond\boldsymbol{A}$.

In addition to the above hypothesis, if $\boldsymbol{A}$ is essentially algebraic and $(|A|, g_i : |A| \longrightarrow |A_i|)_{i \in I}$ is a monosource in $\boldsymbol{Set}$, then $((X, A), g_i : (X, A) \longrightarrow (X_i, A_i))_{i \in I}$ is initial in $\boldsymbol{X}\diamond\boldsymbol{A}$.

Proof. Since $G : \boldsymbol{X} \longrightarrow \boldsymbol{Set}$ is a concrete functor preserving concrete finite powers, the first part follows from the above Theorem.

In order to prove the second part, let us assume that $((X', A'), f_i : (X', A') \longrightarrow (X_i, A_i))_{i \in I}$ is an $\boldsymbol{X}\diamond\boldsymbol{A}$-source and $h : |A'| \longrightarrow |A|$ is a function such that the diagram

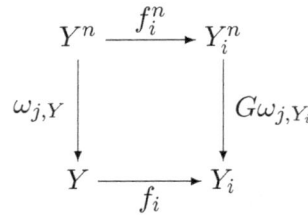

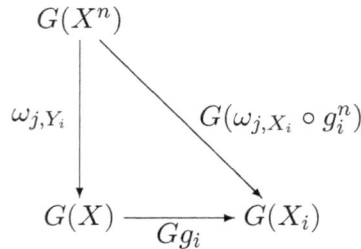

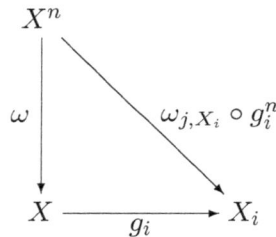

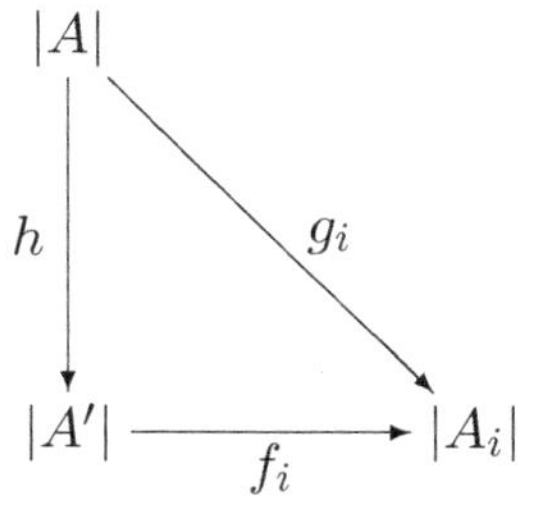

commutes for each $i \in I$. Since any source in $\boldsymbol{A}$ that is a monosource in $\boldsymbol{Set}$ is initial in $\boldsymbol{A}$ (see [1, 23.2(6)]), the source $(A, g_i : A \longrightarrow A_i)_{i \in I}$ is initial in $\boldsymbol{A}$ so that h is an $\boldsymbol{A}$-morphism. It is also an $\boldsymbol{X}$-morphism since $(X, g_i : X \longrightarrow X_i)_{i \in I}$ is G-initial. Hence h is an $\boldsymbol{X \diamond A}$-morphism. ■

The preceding corollary ensures, as shown in the following Theorem, the existence of products, subobjects, equalizers and intersections in the category $\boldsymbol{X \diamond A}$ which proves that $\boldsymbol{X \diamond A}$ is complete and is an (extremal epi, mono) - category, provided $\boldsymbol{X}$ is monotopological and $\boldsymbol{A}$ is an essentially algebraic subcategory of $\boldsymbol{Alg(\Omega)}$.

Theorem 4.3 Suppose $\boldsymbol{X}$ is a monotopological category and $\boldsymbol{A}$ is a subcategory of $\boldsymbol{Alg(\Omega)}$. Then the following hold.

(a) For any family $((X_i, A_i))_{i \in I}$ of objects in $\boldsymbol{X \diamond A}$ such that the concrete product A of $(A_i)_{i \in I}$ exists in $\boldsymbol{A}$, then (X, A), where X is given the initial structure in $\boldsymbol{X}$ with respect to the natural projections $\pi_i : A \longrightarrow A_i$, is the product in $\boldsymbol{X \diamond A}$.

(b) If (X, A) is an object in $\boldsymbol{X \diamond A}$, A' is an $\boldsymbol{A}$-object which is an Ω-subalgebra of A and X' is an $\boldsymbol{X}$-object having the same underlying set as A' with the initial structure with respect to the inclusion $u : X' \hookrightarrow X$, then (X', A') is an object in $\boldsymbol{X \diamond A}$ which is a subobject of (X, A).

(c) Suppose $f : (X, A) \longrightarrow (X', A')$ and $g : (X, A) \longrightarrow (X', A')$ are two morphisms such that the pair $f : A \longrightarrow A'$ and $g : A \longrightarrow A'$ admits an equalizer (E, e) in $\boldsymbol{A}$. Then, the $\boldsymbol{X}$-object Z, having the same underlying set as E, initial with respect to $e : E \longrightarrow X$ leads to an object (Z, E) in $\boldsymbol{X \diamond A}$ and $((Z, E), e)$ is an equalizer of the pair of morphisms f and g.

(d) If $((X_i, A_i))_{i \in I}$ is a family of $\boldsymbol{X \diamond A}$-subobjects of (X, A) such that the intersection $\cap A_i$ of $(A_i)_{i \in I}$ exists in $\boldsymbol{A}$, then the $\boldsymbol{X \diamond A}$-object $(\cap X_i, \cap A_i)$, where $\cap X_i$ has the initial structure with respect to inclusions into $X_i's$, is the intersection of $((X_i, A_i))_{i \in I}$ in $\boldsymbol{X \diamond A}$.

In particular,

(e) If $\boldsymbol{A}$ is an essentially algebraic subcategory of $\boldsymbol{Alg(\Omega)}$, then the category $\boldsymbol{X \diamond A}$ is complete and is an (extremal epi, mono) - category.

Proof. In order to prove (a), first note that the source $(|A|, \pi_i : |A| \longrightarrow |A_i|)_{i \in I}$ is point separating and hence is a monosource in $\boldsymbol{Set}$. Since $\boldsymbol{X}$ is monotopological, we can find an $\boldsymbol{X}$-object X, having the same underlying set as A, initial with respect to the source $(X, \pi_i : X \longrightarrow X_i)_{i \in I}$. By Corollary 9, (X, A) is an $\boldsymbol{X \diamond A}$-object and is the product of $((X_i, A_i))_{i \in I}$.

Analogously, statements $(b) - (d)$ follow from the observation that any inclusion is a monosource in $\boldsymbol{Set}$.

To establish statement (e), let us assume that $\boldsymbol{A}$ is an essentially algebraic subcategory of $\boldsymbol{Alg(\Omega)}$. Then, by Theorem 4 and the discussion following it, $\boldsymbol{A}$ is complete and wellpowered, and $\boldsymbol{A}$ has concrete products. It is straight forward to see that $\boldsymbol{X \diamond A}$

is wellpowered. Statements $(a), (c)$, and (d) imply that $\boldsymbol{X}\diamond\boldsymbol{A}$ has arbitrary products, equalizers, and intersections. Because any category having products and equalizers is complete (see [8, 23.8]) and because every wellpowered finitely complete category having intersections is an (extremal epi, mono) - category (see [8, 34.1]), $\boldsymbol{X}\diamond\boldsymbol{A}$ is complete and is an (extremal epi, mono) - category. ■

If (X, A) is an object in $\boldsymbol{X}\diamond\boldsymbol{A}$, X' is a quotient of X with respect to an $\boldsymbol{X}$-epimorphism $f : X \longrightarrow X'$ and A' is a quotient of A with respect to an $\boldsymbol{A}$-epimorphism $f : A \longrightarrow A'$ then, in general, (X', A') may not be a quotient of (X, A). In fact, it need not be an object in the category $\boldsymbol{X}\diamond\boldsymbol{A}$ as is seen in the following example (where $\boldsymbol{X}$ is the category ***Haus*** of Hausdorff topological spaces and $\boldsymbol{A}$ is the category ***Sgrp*** of semigroups) due to Lawson and Madison [10]. After presenting the example, in Theorem 11 we establish that an additional condition on the category $\boldsymbol{X}$ will eliminate the pathology.

***Example* 4.4** Let $\mathbb{R}^2 := \mathbb{R} \times \mathbb{R}$ be the two dimensional Euclidean space with the product topology inherited from the usual topology on the real line $\mathbb{R}$, $\mathbb{Q}$ be the set of all rational numbers, X be the subspace of $\mathbb{R}^2$ defined by

$$X := \{(x_1, x_2) \in \mathbb{R}^2 : x_1 \in \mathbb{Q} \text{ and } x_2 \geq 0\}$$

and

$$I := \{(x_1, x_2) \in \mathbb{R}^2 : x_1 \in \mathbb{Q} \text{ and } x_2 = 0\}.$$

Define a multiplication $\cdot$ on X by the formula

$$(x_1, x_2) \cdot (x_1', x_2') = (x_1 + x_1', \min(x_2, x_2')),$$

where $+$ is the usual addition of real numbers. Clearly X is a semigroup with multiplication $\cdot$ and I is a closed ideal of X (i.e., $X \cdot I \subseteq I$). Hence the so-called Rees congruence θ on X relative to the ideal I, given by $x\theta x'$ iff either $x = x'$ or I contains both x and x', leads to the semigroup X/θ under the multiplication defined by $(\pi(x), \pi(x')) \longrightarrow \pi(x \cdot x')$, where $\pi(z)$ is the θ-class containing z, i.e., $\pi(z) := I$ whenever $z \in I$ and $\pi(z) := \{z\}$ for any $z \notin I$. Thus $\pi : X \longrightarrow X/\theta$ is a quotient map in ***Sgrp*** (see [13, page 9]).

We give X/θ the quotient topology induced by the natural map $\pi : X \longrightarrow X/\theta$. It can be shown that that X/θ is not a topological semigroup. In fact, the multiplication on X/θ is not continuous at $(\pi(0,0), \pi(0,0))$ (see [10]).

Lemma 4.5 Suppose $\boldsymbol{X}$ has finitely productive quotients, (X, A) is an $\boldsymbol{X}\diamond\boldsymbol{A}$-object, $f : A \longrightarrow A'$ is an $\boldsymbol{A}$-morphism, and $f : X \longrightarrow X'$ is a quotient map in $\boldsymbol{X}$. Then (X', A') is an $\boldsymbol{X}\diamond\boldsymbol{A}$-object.

Proof. Let $j \in J$, $n = n_j$ and $\omega = \omega_{j,A}$, $\omega' = \omega_{j,A'}$ be n-ary operations on A and A' respectively. Since $\boldsymbol{X}$ has finitely productive quotients, X'^n is a quotient of X^n with respect to $f^n : X^n \longrightarrow X'^n$. Thus ω' is an $\boldsymbol{X}$-morphism iff $\omega' \circ f^n$ is an $\boldsymbol{X}$-morphism. However, because f is an Ω-homomorphism, we have the commutative diagram,

$$\begin{array}{ccc} |X^n| & \xrightarrow{\ \omega\ } & |X| \\ {\scriptstyle f^n}\downarrow & & \downarrow{\scriptstyle f} \\ |X'^n| & \xrightarrow[\ \omega'\ ]{} & |X'|, \end{array}$$

which shows that $\omega' \circ f^n$, being equal to $f \circ \omega$, is an $\boldsymbol{X}$-morphism. This being true for each $j \in J$, (X', A') is an $\boldsymbol{X \diamond A}$-object. ■

Theorem 4.6 If $\boldsymbol{X}$ has finitely productive quotients, (X, A) is an object in $\boldsymbol{X \diamond A}$, $f : X \longrightarrow X'$ is a quotient map in $\boldsymbol{X}$ and $f : A \longrightarrow A'$ is a quotient map in $\boldsymbol{A}$, then (X', A') is an object in $\boldsymbol{X \diamond A}$ and $f : (X, A) \longrightarrow (X', A')$ is a quotient map in $\boldsymbol{X \diamond A}$.

Proof. By the virtue of Lemma 2, it remains to show that $f : (X, A) \longrightarrow (X', A')$ is a quotient map in $\boldsymbol{X \diamond A}$. Let (X'', A'') be any object in $\boldsymbol{X \diamond A}$ and $g : |X'| \longrightarrow |X''|$ be a function between the two sets such that $g \circ f$ is an $\boldsymbol{X \diamond A}$-morphism. Then g is an $\boldsymbol{X}$-morphism between X' and X'' because $g \circ f$ is one such and $f : X \longrightarrow X'$ is a quotient map in $\boldsymbol{X}$. Similarly g is also an Ω-homomorphism. Thus $g : (X', A') \longrightarrow (X'', A'')$ is an $\boldsymbol{X \diamond A}$-morphism. ■

Thus we have proved that if $\boldsymbol{X}$ has finitely productive quotients then quotients of objects in $\boldsymbol{X}$ and $\boldsymbol{A}$ with the same underlying set resulting from the same underlying function can be paired to form quotients in the category $\boldsymbol{X \diamond A}$.

A similar construction is possible for coproducts under an additional condition that final epi sinks are finitely productive in $\boldsymbol{X}$ (in particular, under the assumption that $\boldsymbol{X}$ is well-fibred monotopological and cartesian closed, see [12]). As we will see later in Theorem 12, one can give other sufficient conditions that coproducts exist in $\boldsymbol{X \diamond A}$.

First we need the following definitions. We say that **final epi sinks are finitely productive** in $\boldsymbol{X}$ iff the product $(f_i \times g_k : X_i \times Y_k \longrightarrow X \times Y, X \times Y)_{i \in I, k \in K}$ of any two final epi sinks $(f_i : X_i \longrightarrow X, X)_{i \in I}$ and $(g_k : Y_k \longrightarrow Y, Y)_{k \in K}$ in $\boldsymbol{X}$ is final in $\boldsymbol{X}$. A class $\mathcal{F}$ of functions is said to be **Ω-admissible** to an Ω-algebra A iff each function in $\mathcal{F}$ has the codomain $|A|$, and for each $j \in J$, $n = n_j$, $f_1, \ldots, f_n \in \mathcal{F}$,

$$\omega_{j,A} \circ (f_1 \times \ldots \times f_n) \in \mathcal{F}.$$

Lemma 4.7 Suppose final epi sinks are finitely productive in $\boldsymbol{X}$, A is an $\boldsymbol{A}$-object, $(X_i)_{i \in I}$ is a family of $\boldsymbol{X}$-objects, and $(f_i : |X_i| \longrightarrow |A|)_{i \in I}$ is a class of functions Ω-admissible to A. If X is an $\boldsymbol{X}$-object with the same underlying set as A such that $(f_i : X_i \longrightarrow X, X)_{i \in I}$ is a final epi sink in $\boldsymbol{X}$, then (X, A) is an $\boldsymbol{X \diamond A}$-object.

Proof. Since X has the final structure with respect to $f_i : X_i \longrightarrow X$, for any positive integer n, X^n has the final structure with respect to $f_{i_1} \times \ldots \times f_{i_n}$, $i_1 \in I$, .., $i_n \in I$, by hypothesis. Let $j \in J$ and $n = n_j$. We have to show that $\omega_{j,A}$ is an $\boldsymbol{X}$-morphism. However, $\omega_{j,A} \circ (f_{i_1} \times \ldots \times f_{i_n})$, $i_1 \in I$, ... , $i_n \in I$, being one of the $f_i's$ as (f_i) is Ω-admissible to A, is an $\boldsymbol{X}$-morphism. Consequently, the n_j-ary operation $\omega_{j,A}$ on A is an $\boldsymbol{X}$-morphism $\omega_{j,x} : X^n \longrightarrow X$. This being true for each $j \in J$, (X, A) is an $\boldsymbol{X \diamond A}$-object. ■

Lemma 4.8 Suppose A is an Ω-algebra and $(C_i)_{i \in I}$ is a family of sets and $(f_i : C_i$

$\longrightarrow |A|)_{i\in I}$ is a class of functions. Then there exists the smallest class $\mathcal{F}$ of functions containing $(f_i)_{i\in I}$ that is Ω-admissible to A, i.e., there exists a class $\mathcal{F}$ of functions containing $(f_i)_{i\in I}$ such that $\mathcal{F}$ is Ω-admissible to A and any class of functions containing $(f_i)_{i\in I}$that is Ω-admissible to A contains $\mathcal{F}$.

Moreover, any member f of $\mathcal{F}$ has a domain of the form $C_{i_1} \times \ldots \times C_{i_n}$ $(i_1, \ldots, i_n \in I)$ and the codomain $|A|$.

Proof. Let $\mathcal{F}$ be the intersection of all classes of functions that are Ω-admissible to A and contain $(f_i)_{i\in I}$. Clearly $\mathcal{F}$ has the property described in the lemma. If $\mathcal{G}$ is the family of all functions of the form explained in the last statement of the lemma, $\mathcal{G}$ is certainly Ω-admissible to A and contains $(f_i)_{i\in I}$ so that $\mathcal{F} \subseteq \mathcal{G}$. ∎

For example, if A is an $\boldsymbol{Alg}(\mathbf{1})$ -object with the unary operation u and $(f_i : C_i \longrightarrow |A|)_{i\in I}$ is a class of functions, then $(u^n \circ f_i : C_i \longrightarrow |A|)_{i\in I, n\geq 0}$ is the smallest Ω-admissible class containing $(f_i : C_i \longrightarrow |A|)_{i\in I}$. If A is a group with the multiplication $\cdot$ and the inversion β_1 and $f : C \longrightarrow |A|$ is a function then write

$$f_{n,\mu,\sigma} := \left[(\beta_{\mu(1)} \circ f) \cdot (\beta_{\mu(2)} \circ f) \cdot \ldots \cdot (\beta_{\mu(n)} \circ f)\right] \circ r_\sigma$$

where n is a positive integer, μ is a function from $\mathbf{N}_n := \{1, 2, \ldots, n\}$ with values 0 and 1, σ is a bijection from $\mathbf{N}_n$ onto $\mathbf{N}_n$, r_σ is the function $r_\sigma : C^n \longrightarrow C^n$ given by $r_\sigma(c_1, \ldots, c_n) = (c_{\sigma(1)}, \ldots c_{\sigma(n)})$, and $\beta_0 :=\mathrm{id}_A$. The family $(f_{n,\mu,\sigma} : C^n \longrightarrow |A|)$ is the smallest Ω-admissible class containing $f : C \longrightarrow |A|$.

Theorem 4.9 Suppose final epi sinks are finitely productive in $\boldsymbol{X}$, $((X_i, A_i))_{i\in I}$ is a family of $\boldsymbol{X}\diamond\boldsymbol{A}$-objects, and $(f_i : A_i \longrightarrow A, A)_{i\in I}$ is a coproduct in $\boldsymbol{A}$.

Let $(g_k : |X'_k| \longrightarrow |A|)_{k\in K}$ be the smallest class of functions containing $(f_i)_{i\in I}$ that is Ω-admissible to A, where X'_k is of the form $X_{i_1} \times \ldots \times X_{i_n}$ $(i_1, \ldots, i_n \in I)$ for each $k \in K$, the product being formed in the category $\boldsymbol{X}$. Let X be an $\boldsymbol{X}$-object with the same underlying set as A such that $(g_k : X'_k \longrightarrow X, X)_{k\in K}$ is a final epi sink in $\boldsymbol{X}$. Then the pair (X, A) is an $\boldsymbol{X}\diamond\boldsymbol{A}$-object and the sink $(f_i : (X_i, A_i) \longrightarrow (X, A), (X, A))_{i\in I}$ is a coproduct in $\boldsymbol{X}\diamond\boldsymbol{A}$.

Proof. (X, A) is an $\boldsymbol{X}\diamond\boldsymbol{A}$-object by Lemma 3. Since $(g_k)_{k\in K}$ contains $(f_i)_{i\in I}$, $f_i : (X_i, A_i) \longrightarrow (X, A)$ is an $\boldsymbol{X}\diamond\boldsymbol{A}$-morphism for each $i \in I$. We now show that $(f_i)_{i\in I}$ is a coproduct in $\boldsymbol{X}\diamond\boldsymbol{A}$. Let (X', A') be any $\boldsymbol{X}\diamond\boldsymbol{A}$-object and $(f'_i : (X_i, A_i) \longrightarrow (X', A'))_{i\in I}$ be a family of $\boldsymbol{X}\diamond\boldsymbol{A}$-morphisms. Since $(f_i : A_i \longrightarrow A, A)_{i\in I}$ is a coproduct in $\boldsymbol{A}$, there exists a unique $\boldsymbol{A}$-morphism $\bar{f} : A \longrightarrow A'$ such that the diagram of $\boldsymbol{A}$-morphisms

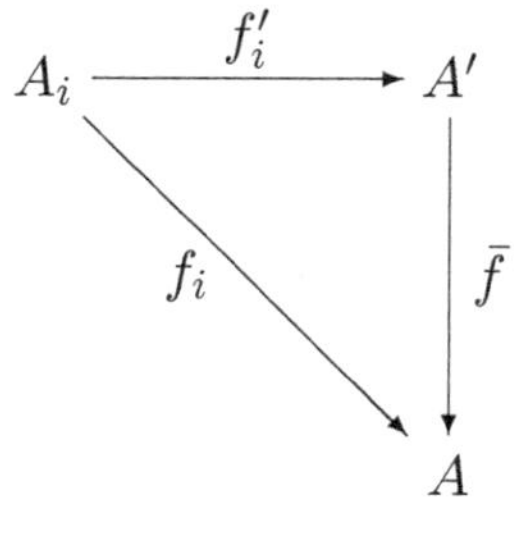

commutes for each $i \in I$. We show that $\bar{f} : X \longrightarrow X'$ is an $\boldsymbol{X}$-morphism. Let $\mathcal{H}$ be the family of all functions $h : |X_{i_1} \times \ldots \times X_{i_n}| \longrightarrow |A|$ $(i_1 \in I, \ldots, i_n \in I)$ such that $\bar{f} \circ h$ is an $\boldsymbol{X}$-morphism. $\mathcal{H}$ contains $(f_i)_{i\in I}$ because $\bar{f} \circ f_i = f'_i$ is an $\boldsymbol{X}$-morphism for each $i \in$I. Assume $j \in J, n = n_j$, and $h_1 \in \mathcal{H}, \ldots, h_n \in \mathcal{H}$. Then

$$\begin{aligned}
&\bar{f} \circ [\omega_{j,A} \circ (h_1 \times \ldots \times h_n)] \\
&\qquad = (\bar{f} \circ \omega_{j,A}) \circ (h_1 \times \ldots \times h_n) \\
&\qquad = (\omega_{j,A'} \circ \bar{f}^n) \circ (h_1 \times \ldots \times h_n), \text{ since } \bar{f} \text{ is an } \boldsymbol{A}\text{-morphism,} \\
&\qquad = \omega_{j,A'} \circ [\bar{f}^n \circ (h_1 \times \ldots \times h_n)] \\
&\qquad = \omega_{j,X'} \circ (\bar{f} \circ h_1 \times \ldots \times \bar{f} \circ h_n).
\end{aligned}$$

Since $\bar{f} \circ h_1, \ldots, \bar{f} \circ h_n$, and $\omega_{j,X'}$ are $\boldsymbol{X}$-morphisms, $\bar{f} \circ [\omega_{j,A} \circ (h_1 \times \ldots \times h_n)]$ is an $\boldsymbol{X}$-morphism. Thus

$$\omega_{j,A} \circ (h_1 \times \ldots \times h_n) \in \mathcal{H}.$$

This shows that $\mathcal{H}$ is Ω-admissible to A containing $(f_i)_{i \in I}$. Since $\mathcal{F}$ is the smallest such class, we conclude that $g_k \in \mathcal{H}$, i.e., $\bar{f} \circ g_k$ is an $\boldsymbol{X}$-morphism for each $k \in K$. Since $(g_k)_{k \in K}$ is final in $\boldsymbol{X}$, $\bar{f}$ is an $\boldsymbol{X}$-morphism. Thus we have the commutative diagram of $\boldsymbol{X} \diamond \boldsymbol{A}$-objects and $\boldsymbol{X} \diamond \boldsymbol{A}$-morphisms:

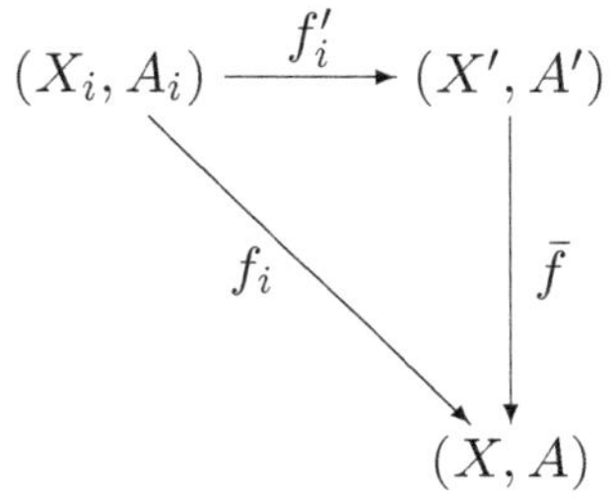

Of course, $\bar{f}$ is unique. ■

Acknowledgments

The author wishes to acknowledge thanks to Prof. H. Lamar Bentley for his helpful suggestions.

References

[1] J. Adámek, H. Herrlich and G. E. Strecker, Abstract and Concrete Categories, John Wiley & Sons, Inc., New York, 1990.
[2] H. L. Bentley, H. Herrlich and R. G. Ori, Zero sets and complete regularity for nearness spaces, In: Categorical Topology, World Scientific, Teaneck, New Jersey (1989), 446-461.
[3] P. M. Cohn, Universal Algebra, Harper and Row, Publishers, New York, 1965.
[4] T. H. Fay, An axiomatic approach to categories of topological algebras, Quaestiones Mathematicae 2 (1977), 113-137.
[5] V. L. Gompa, Essentially algebraic functors and topological algebra, Indian Journal of Mathematics, 35, (1993), 189-195.
[6] H. Herrlich, Essentially algebraic categories, Quaest. Math. 9 (1986), 245-262.
[7] Y. H. Hong, Studies on categories of universal topological algebras, Doctoral Dissertation, McMaster University, 1974.
[8] H. Herrlich and G. E. Strecker, Category Theory, Allyn and Bacon, Boston, 1973.
[9] J. Koslowski, Dual adjunctions and the compatibility of structures, In: Categorical Topology, Heldermann Verlag, Berlin (1984), 308-322.
[10] J. D. Lawson and B. L. Madison, On congruences and cones, Math. Zeit. 120 (1971), 18-24.

[11] L. D. Nel, Universal topological algebra needs closed topological categories, Topology and its Applications 12 (1981) 321-330.
[12] L. D. Nel, Initially structured categories and cartesian closedness, Canad. J. Math. 27 (1975) 1361-1377.
[13] M. Petrich, Lectures in Semigroups, John Wiley & Sons, New York, 1977.
[14] W. Tholen, On Wyler's taut lift theorem, General Topology and its Applications 8 (1978), 197-206.
[15] O. Wyler, On the categories of general topology and topological algebras, Arc. Math. (Basel) 22 (1971), 7-17.

12

On the energy of non-commuting graphs

M. Ghorbani[a*], Z. Gharavi-Alkhansari[a]

[a] *Department of Mathematics, Faculty of Science, Shahid Rajaee Teacher Training University, Tehran, 16785-136, Iran.*

Abstract. For given non-abelian group G, the non-commuting (NC)-graph $\Gamma(G)$ is a graph with the vertex set $G \setminus Z(G)$ and two distinct vertices $x, y \in V(\Gamma)$ are adjacent whenever $xy \neq yx$. The aim of this paper is to compute the spectra of some well-known NC-graphs.

Keywords: Non-commuting graph, characteristic polynomial, linear group.

1. Introduction

Paul Erdős was the first person who considered the non-commuting graph to answer a question on the size of the cliques of a graph in 1975, see [7]. In general, the non-commuting graph or briefly the $NC-$graph $\Gamma(G)$ associated to the non-abelian group G with center $Z(G)$, is a simple and undirected graph with the vertex set $G \setminus Z(G)$ in which two vertices join whenever they don't commute. For background materials about $NC-$graphs, we encourage the intrested reader to see references [1, 2, 5, 6].

For given graph Γ, its characteristic polynomial is defined as $\chi(\Gamma, \lambda) = det(\lambda I - A)$, where A is the adjacency matrix of Γ. The eigenvalues of Γ are the roots of this polynomial and the multi-set $\{\lambda_1, \cdots, \lambda_n\}$ of eigenvalues of A forms the spectrum of Γ. By this notation, the energy of Γ is defined as [4]:

$$\mathcal{E}(\Gamma) = \sum_{i=1}^{n} |\lambda_i|.$$

*Corresponding author.
E-mail address: mghorbani@srttu.edu (M. Ghorbani).

Here, in Section 2, we compute the spectrum of $NC-$graph of group G where $G \in \{QD_{2^n}, PSL(2,2^k), GL(2,q)\}$ and in Section 3, at first we compute the related Laplacian eigenvalues and then we compute the energy of these graphs.

2. Main Results

Here, we find the spectrum of NC-graph of three following groups: the projective special linear group $PSL(2,2^k)$, where $k \geqslant 2$, the general linear group $GL(2,q)$, where $q = p^n$ (p is a prime integer and $n \geqslant 4$) and the quasi-dihedral group QD_{2^n}, with the following presentations:

$$QD_{2^n} = \langle a, b : a^{2^{n-1}} = b^2 = 1,\ bab^{-1} = a^{2^{n-2}-1} \rangle.$$

Theorem 2.1 The spectrum of NC- graph of group QD_{2^n} is

$$\{[2^{n-2} - 1 - k]^1, [-2]^{2^{n-2}-1}, [0]^{3\times 2^{n-2}-3}, [2^{n-2} - 1 + k]^1\},$$

where $k = \sqrt{(5 \times 2^{n-2} - 1)(2^{n-2} - 1)}$.

Proof. The group QD_{2^n} is a non-abelian group of order 2^n and $Z(QD_{2^n}) = \{1, a^{2^{n-2}}\}$. For non-central elements, the centralizers are as follows

$$C_{QD_{2^n}}(a) = C_{QD_{2^n}}(a^i) = \langle a \rangle \text{ for } 1 \leqslant i \leqslant 2^{n-1} - 1, i \neq 2^{n-2}$$

and

$$C_{QD_{2^n}}(a^j b) = \{1, a^{2^{n-2}}, a^j b, a^{j+2^{n-2}} b\} \text{ for } 1 \leqslant j \leqslant 2^{n-2}.$$

Since the centralizers are abelian subgroups, QD_{2^n} is an AC-group and therefore $\Gamma(QD_{2^n})$ is a multipartite graph with parts V_i's $(0 \leqslant i \leqslant 2^{n-2})$ as follows:

$$V_0 = \{a, a^2, \cdots, a^{2^{n-2}-1}, a^{2^{n-2}+1}, \cdots, a^{2^{n-1}-1}\},$$
$$V_j = \{a^j b, a^{j+2^{n-2}} b\},\ 1 \leqslant j \leqslant 2^{n-2}.$$

By putting $p = 2^{n-1} - 2$ and $q = p/2$, the adjacency matrix of $\Gamma(QD_{2^n})$ is

$$M = \begin{pmatrix} 0_q & J_{q\times(q+1)} \\ J_{(q+1)\times q} & (J - I)_{q+1} \end{pmatrix} \otimes J_2 = A \otimes J_2.$$

The spectrum of J_2 is $\{[0]^1, [2]^1\}$ and by [3, Lemma 2], the spectrum of A is

$$\{[0]^{q-1}, [-1]^q, [\frac{q - \sqrt{q(5(q+1) - 1)}}{2}]^1, [\frac{q + \sqrt{q(5(q+1) - 1)}}{2}]^1\}.$$

Now by using [3, Theorem 1], the proof is completed. ∎

Theorem 2.2 The spectrum of NC-graph of the projective special linear group $PSL(2,2^k)$, where $k \geqslant 2$ is

$$\{[-u]^{t-1}, [-u+1]^u, [-u+2]^{s-1}, [0]^{(u+1)(u-2)+s(u-3)+t(u-1)}, [x_1]^1,\ [x_2]^1,\ [x_3]^1\},$$

where $u = 2^k$, $s = 2^{k-1}(2^k+1)$, $t = 2^{k-1}(2^k-1)$ and x_1, x_2, x_3 are the roots of the equation $x^3+(2^{k+2}-2^{3k}-2)x^2+(2^{3k+1}-2^{4k+1}+5\times 2^k(2^k-1))x+2^{4k+1}+2^{3k}-2^{5k}-2^{2k+1}$.

Proof. The center of non-abelian group $PSL(2,2^k)$ of order $2^k(2^{2k}-1)$ is trivial. By [1, Proposition 3.21], for non-central elements of $PSL(2,2^k)$ the set of centralizers is $\{gPg^{-1}, gAg^{-1}, gBg^{-1} : g \in PSL(2,2^k)\}$, where P is an elementary abelian 2-group of order 2^k and A, B are cyclic subgroups of order 2^k-1 and 2^k+1, respectively. Let $u = 2^k$, $s = 2^{k-1}(2^k+1)$ and $t = 2^{k-1}(2^k-1)$. The number of conjugates of P, A and B in $PSL(2,2^k)$ are $u+1, s$ and t respectively. All centralizers of $PSL(2,2^k)$ are abelian, since it is an AC-group. Also, it is not difficult to see that they are disjoint from each other. This yields $\Gamma(PSL(2,2^k))$ is a multipartite graph. Let $p = u-1$, $q = u-2$, then the adjacency matrix of $\Gamma(PSL(2,2^k))$ is the following block matrix:

$$M = \begin{pmatrix} 0_p & J_p & \cdots & J_p & J_{p\times q} & \cdots & J_{p\times q} & J_{p\times u} & \cdots & J_{p\times u} \\ J_p & & & & & & & & & \\ \vdots & & \ddots & & & & & & & \\ J_p & \cdots & J_p & 0_p & J_{p\times q} & \cdots & J_{p\times q} & J_{p\times u} & \cdots & J_{p\times u} \\ J_{q\times p} & \cdots & & J_{q\times p} & 0_q & & J_q & J_{q\times u} & \cdots & J_{q\times u} \\ \vdots & & & & & \ddots & & & & \\ J_{q\times p} & \cdots & & J_{q\times p} & J_q & \cdots & 0_q & J_{q\times u} & \cdots & J_{q\times u} \\ J_{u\times p} & \cdots & & J_{u\times p} & J_{u\times q} & \cdots & J_{u\times q} & 0_u & & J_u \\ \vdots & & & & & & & & \ddots & \\ J_{u\times p} & \cdots & & J_{u\times p} & J_{u\times q} & \cdots & J_{u\times q} & J_u & & 0_u \end{pmatrix}_{(u+1)+s+t}.$$

Let $N^r = xI + rJ$, then

$$det(xI - M) = \begin{vmatrix} A & B \\ C & D \end{vmatrix} = det(A)det(D - CA^{-1}B),$$

where

$$A = \begin{pmatrix} N_p^1 & 0 & \cdots & 0 & 0 & 0 & \cdots & 0 \\ 0 & N_p^1 & & 0 & 0 & & & 0 \\ \vdots & & \ddots & \vdots & \vdots & & & \vdots \\ 0 & 0 & \cdots & N_p^1 & 0 & \cdots & & 0 \\ 0 & 0 & \cdots & 0 & N_q^1 & 0 & \cdots & 0 \\ 0 & \cdots & & 0 & 0 & N_q^1 & & 0 \\ \vdots & & & \vdots & \vdots & & \ddots & \vdots \\ 0 & \cdots & & 0 & 0 & 0 & \cdots & N_q^1 \end{pmatrix}_{(u+1)+s},$$

$$B = \begin{pmatrix} -N^1_{p\times u} & 0 & \cdots & 0 \\ 0 & & & 0 \\ \vdots & & & \vdots \\ 0 & & \cdots & 0 \\ -N^1_{q\times u} & 0 & \cdots & 0 \\ 0 & & & 0 \\ \vdots & & & \vdots \\ 0 & & \cdots & 0 \end{pmatrix}_{((u+1)+s)\times t},$$

$$C = \begin{pmatrix} -(u+1)J_{u\times p} & -J_{u\times p} & \cdots & -J_{u\times p} & -sJ_{u\times q} & -J_{u\times q} & \cdots & -J_{u\times q} \\ 0 & & \cdots & 0 & 0 & & \cdots & 0 \\ \vdots & & & \vdots & \vdots & & & \vdots \\ 0 & & \cdots & 0 & 0 & & \cdots & 0 \end{pmatrix}_{t\times(u+s+1)}$$

and

$$D = \begin{pmatrix} N_u^{-t+1} & -J_u & \cdots & -J_u \\ 0 & N_u^1 & & 0 \\ \vdots & & \ddots & \vdots \\ 0 & 0 & \cdots & N_u^1 \end{pmatrix}_t .$$

We have

$$det(A) = det(N_p^1)^{u+1} det(N_q^1)^s = x^{(u+1)(u-2)+s(u-3)}(x+(u-1))^{u+1}(x+(u-2))^s$$

and so

$$det(D - CA^{-1}B) = \frac{(x-x_1)(x-x_2)(x-x_3)}{(x+(u-1))(x+(u-2))} x^{t(u-1)}(x+u)^{t-1},$$

where, $x_1,\ x_2,\ x_3,$ are the roots of the following equation

$$x^3 + (2^{k+2} - 2^{3k} - 2)x^2 + (2^{3k+1} - 2^{4k+1} + 5\times 2^k(2^k-1))x + 2^{4k+1} + 2^{3k} - 2^{5k} - 2^{2k+1}.$$

Thus

$$\begin{aligned} det(xI - M) &= det(A)det(D - CA^{-1}B) \\ &= x^{(u+1)(u-2)+s(u-3)+t(u-1)}(x+(u-1))^u(x+(u-2))^{s-1}(x+u)^{t-1} \\ &\quad (x-x_1)(x-x_2)(x-x_3). \end{aligned}$$

Hence, the result follows. ■

Theorem 2.3 The spectrum of NC- graph of the general linear group $GL(2,q)$, where $q = p^n > 2$ and p is prime integer is given by

$$\{[-2s]^{s-1}, [-k]^{t-1}, [-l]^{u-1}, [0]^{l(u-1)+s(2s-1)+t(k-1)}, [x_1]^1,\ [x_2]^1,\ [x_3]^1\},$$

where $u = \frac{q(q+1)}{2}$, $s = \frac{q(q-1)}{2}$, $t = q+1$, $l = (q-1)(q-2)$, $k = (q-1)^2$ and x_1, x_2, x_3 are the roots of the equation

$$x^3 + (-q^4 + q^3 + 4q^2 + 2)x^2 + (-2q^6 + 6q^5 - q^4 - 13q^3 + 15q^2 - 5q)x - q^8 + 5q^7 - 8q^6 + 2q^5 + 7q^4 - 7q^3 + 2q^2.$$

Proof. The order of non-abelian group $GL(2,q)$ is $(q^2-1)(q^2-q)$ and it is well-known that $|Z(GL(2,q))| = q-1$. By [1, Proposition 3.26], all non-central elements of $GL(2,q)$ have the set of centralizers as $\{gDg^{-1}, gIg^{-1}, gPZ(GL(2,q))g^{-1} : g \in GL(2,q)\}$, where D is the subgroup of $GL(2,q)$ consisting of all diagonal matrices, I is a cyclic subgroup $GL(2,q)$ having order q^2-1 and P is the Sylow p-subgroup of $GL(2,q)$ consisting of all upper triangular matrices whose diagonal entries are 1. The orders of D and $PZ(GL(2,q))$ are $(q-1)^2$ and $q(q-1)$ respectively. The number of conjugates of D, I and $PZ(GL(2,q))$ in $GL(2,q)$ are $\frac{q(q+1)}{2}$, $\frac{q(q-1)}{2}$ and $q+1$ respectively. Let $u = \frac{q(q+1)}{2}$, $s = \frac{q(q-1)}{2}$, $t = q+1$, $l = (q-1)(q-2)$ and $k = (q-1)^2$. Similar to the last case $\Gamma(GL(2,q))$ is a multipartite graph and the order of its parts are $|D| - |Z(GL(2,q))| = l$, $|I| - |Z(GL(2,q))| = 2s$ and $|PZ(GL(2,q))| - |Z(GL(2,q))| = k$. The adjacency matrix of $\Gamma(GL(2,q))$ has the following form:

$$M = \begin{pmatrix} 0_l & J_l & \cdots & J_l & J_{l\times k} & \cdots & J_{l\times k} & J_{l\times 2s} & \cdots & J_{l\times 2s} \\ J_l & & & & & & & & & \\ \vdots & & \ddots & & & & & & & \\ J_l & \cdots & J_l & 0_l & J_{l\times k} & \cdots & J_{l\times k} & J_{l\times 2s} & \cdots & J_{l\times 2s} \\ J_{k\times l} & \cdots & & J_{k\times l} & 0_k & & J_k & J_{k\times 2s} & \cdots & J_{k\times 2s} \\ \vdots & & & & & \ddots & & & & \\ J_{k\times l} & \cdots & & J_{k\times l} & J_k & \cdots & 0_k & J_{k\times 2s} & \cdots & J_{k\times 2s} \\ J_{2s\times l} & \cdots & & J_{2s\times l} & J_{2s\times k} & \cdots & J_{2s\times k} & 0_{2s} & & J_{2s} \\ \vdots & & & & & & & & \ddots & \\ J_{2s\times l} & \cdots & & J_{2s\times l} & J_{2s\times k} & \cdots & J_{2s\times k} & J_{2s} & & 0_{2s} \end{pmatrix}_{u+t+s}.$$

Let $N^r = xI + rJ$, similar to the proof of last theorem we have

$$det(xI - M) = det(A)det(D - CA^{-1}B),$$

where

$$A = \begin{pmatrix} N_l^1 & 0 & \cdots & 0 & 0 & 0 & \cdots & 0 \\ 0 & N_l^1 & & 0 & 0 & & & 0 \\ \vdots & & \ddots & \vdots & \vdots & & & \vdots \\ 0 & 0 & \cdots & N_l^1 & 0 & \cdots & & 0 \\ 0 & 0 & \cdots & 0 & N_k^1 & 0 & \cdots & 0 \\ 0 & \cdots & & 0 & 0 & N_k^1 & & 0 \\ \vdots & & & \vdots & \vdots & & \ddots & \vdots \\ 0 & \cdots & & 0 & 0 & 0 & \cdots & N_k^1 \end{pmatrix}_{u+t},$$

$$B = \begin{pmatrix} -N^1_{l\times 2s} & 0 & \cdots & 0 \\ 0 & & & 0 \\ \vdots & & & \vdots \\ 0 & & \cdots & 0 \\ -N^1_{k\times 2s} & 0 & \cdots & 0 \\ 0 & & & 0 \\ \vdots & & & \vdots \\ 0 & & \cdots & 0 \end{pmatrix}_{(u+t)\times s},$$

$$C = \begin{pmatrix} -uJ_{2s\times l} & -J_{2s\times l} & \cdots & -J_{2s\times l} & -tJ_{2s\times k} & -J_{2s\times k} & \cdots & -J_{2s\times k} \\ 0 & \cdots & & 0 & 0 & & \cdots & 0 \\ \vdots & & & \vdots & \vdots & & & \vdots \\ 0 & \cdots & & 0 & 0 & & \cdots & 0 \end{pmatrix}_{s\times(u+t)}$$

and

$$D = \begin{pmatrix} N^{-s+1}_{2s} & -J_{2s} & \cdots & -J_{2s} \\ 0 & N^1_{2s} & & 0 \\ \vdots & & \ddots & \vdots \\ 0 & 0 & \cdots & N^1_{2s} \end{pmatrix}_s.$$

We have

$$det(A) = det(N^1_l)^u det(N^1_k)^t = x^{u(l-1)+t(k-1)}(x+l)^u(x+k)^t$$

and hence

$$det(D - CA^{-1}B) = \frac{(x-x_1)(x-x_2)(x-x_3)}{(x+l)(x+k)} x^{s(2s-1)}(x+2s)^{s-1},$$

where, $x_1,\ x_2,\ x_3$ are the roots of the equation

$$x^3 + (-q^4 + q^3 + 4q^2 + 2)x^2 + (-2q^6 + 6q^5 - q^4 - 13q^3 + 15q^2 - 5q)x - q^8 + 5q^7 - 8q^6 + 2q^5 + 7q^4 - 7q^3 + 2q^2.$$

Thus

$$det(xI - M) = x^{l(u-1)+s(2s-1)+t(k-1)}(x+l)^{u-1}(x+k)^{t-1}(x+2s)^{s-1}(x-x_1)(x-x_2)(x-x_3).$$

■

3. Laplacian Eigenvalues

The aim of this section is to find the Laplacian spectrum of NC-graph of three groups $PSL(2,2^k)$, $GL(2,q)$ and QD_{2^n}.

Theorem 3.1 The Laplacian spectrum of NC-graph of the projective special linear group $PSL(2, 2^k)$, where $k \geqslant 2$, is given by

$$\{[0]^1, [u^3-2u-1]^{t(u-1)}, [u^3-2u]^{(u-2)(u+1)}, [u^3-2u+1]^{s(u-3)}, [u^3-u-1]^{t+s+u}\},$$

where $u = 2^k,\ s = 2^{k-1}(2^k+1),\ t = 2^{k-1}(2^k-1)$.

Proof. By Theorem 2, the degree of each element in the centralizer of the form xPx^{-1}, xAx^{-1} and xBx^{-1} is u^3-2u, u^3-2u+1 and u^3-2u-1, respectively. Let $z = u^3-2u,\ p = u-1,\ q = u-2,\ X = x-z,\ L = XI-J$ and $L' = (X-1)I-J$, then

$$det(xI-(\Delta-M)) = det(A)det(D-CA^{-1}B),$$

where

$$A = \begin{pmatrix} L_p & 0 & \cdots & 0 & 0 & 0 & \cdots & 0 \\ 0 & L_p & & 0 & 0 & & & 0 \\ \vdots & & \ddots & \vdots & \vdots & & & \vdots \\ 0 & 0 & \cdots & L_p & 0 & \cdots & & 0 \\ 0 & 0 & \cdots & 0 & L'_q & 0 & \cdots & 0 \\ 0 & \cdots & & 0 & 0 & L'_q & & 0 \\ \vdots & & & \vdots & \vdots & & \ddots & \vdots \\ 0 & \cdots & & 0 & 0 & 0 & \cdots & L'_q \end{pmatrix}_{(u+1)+s},$$

$$B = \begin{pmatrix} (J-(X+1)I)_{p\times u} & 0 & \cdots & 0 \\ 0 & & & 0 \\ \vdots & & & \vdots \\ 0 & & \cdots & 0 \\ (J-(X+1)I)_{q\times u} & 0 & \cdots & 0 \\ 0 & & & 0 \\ \vdots & & & \vdots \\ 0 & & \cdots & 0 \end{pmatrix}_{((u+1)+s)\times t},$$

$$C = \begin{pmatrix} (u+1)J_{u\times p} & J_{u\times p} & \cdots & J_{u\times p} & sJ_{u\times q} & J_{u\times q} & \cdots & J_{u\times q} \\ 0 & & \cdots & 0 & 0 & & \cdots & 0 \\ \vdots & & & \vdots & \vdots & & & \vdots \\ 0 & & \cdots & 0 & 0 & & \cdots & 0 \end{pmatrix}_{t\times(u+s+1)}$$

and

$$D = \begin{pmatrix} ((X+1)I+(t-1)J)_u & J_u & \cdots & J_u \\ 0 & ((X+1)-J)_u & & 0 \\ \vdots & & \ddots & \vdots \\ 0 & 0 & \cdots & ((X+1)-J)_u \end{pmatrix}_t.$$

We have

$$\begin{aligned}
det(A) &= det((XI-J)_p)^{u+1} det(((X-1)I-J)_q)^s \\
&= X^{(u+1)(u-2)}(X-(u-1))^{u+1}(x-1)^{s(u-3)}(X-1-(u-2))^s \\
&= (x-u^3+2u)^{(u-2)(u+1)}(x-u^3+2u-1)^{s(u-3)}(x-u^3+u+1)^{u+1+s}
\end{aligned}$$

and thus

$$\begin{aligned}
det(D-CA^{-1}B) &= x(x-u^3+2u+1)^{t(u-1)}(x-u^3+u+1)^{t-1} \\
&= x(x-u^3+2u)^{(u+1)(u-2)}(x-u^3+2u-1)^{s(u-3)} \\
&\quad (x-u^3+u+1)^{s+t+u}(x-u^3+2u+1)^{t(u-1)}.
\end{aligned}$$

■

Theorem 3.2 The Laplacian spectrum of the NC-graph of general linear group $GL(2,q)$, where $q = p^n > 2$ and p is prime integer is

$$\{[0]^1, [(q^3-2q-1)(q-1)]^{s(2s-1)}, [(q^3-2q)(q-1)]^{t(k-1)}, [(q^3-2q+1)(q-1)]^{u(l-1)}, [(q^3-q-1)(q-1)]^{u+t+s-1}\},$$

where $u = \frac{q(q+1)}{2}$, $s = \frac{q(q-1)}{2}$, $t = q+1$, $l = (q-1)(q-2)$, $k = (q-1)^2$.

Proof. By Theorem 3, the degree of each element in centralizer of the form gDg^{-1}, $gPZ(GL(2,q))g^{-1}$ and gIg^{-1} is $(q-1)(q^3-2q+1)$, $(q-1)(q^3-2q)$, $(q-1)(q^3-2q-1)$, respectively. Let $R = (q-1)(q^3-2q)$, $N^r = (R+r)I$, $X = x-R$, $L = (X-(q-1))I-J$ and $L' = XI-J$. In order to find the eigenvalues of $\Delta - M$, we compute $det(xI-(\Delta-M))$ which can easily comes to the following form

$$det(xI-(\Delta-M)) = \begin{vmatrix} A & B \\ C & D \end{vmatrix} = det(A)det(D-CA^{-1}B),$$

where

$$A = \begin{pmatrix} L_l & 0 & \cdots & 0 & 0 & 0 & \cdots & 0 \\ 0 & L_l & & 0 & 0 & & & 0 \\ \vdots & & \ddots & \vdots & \vdots & & & \vdots \\ 0 & 0 & \cdots & L_l & 0 & \cdots & & 0 \\ 0 & 0 & \cdots & 0 & L'_k & 0 & \cdots & 0 \\ 0 & \cdots & & 0 & 0 & L'_k & & 0 \\ \vdots & & & \vdots & \vdots & & \ddots & \vdots \\ 0 & \cdots & & 0 & 0 & 0 & \cdots & L'_k \end{pmatrix}_{u+t},$$

$$B = \begin{pmatrix} (J-(X+(q-1))I)_{l\times 2s} & 0 & \cdots & 0 \\ 0 & & & 0 \\ \vdots & & & \vdots \\ 0 & & \cdots & 0 \\ (J-(X+(q-1))I)_{k\times 2s} & 0 & \cdots & 0 \\ 0 & & & 0 \\ \vdots & & & \vdots \\ 0 & & \cdots & 0 \end{pmatrix}_{(u+t)\times s},$$

$$C = \begin{pmatrix} uJ_{2s\times l} & J_{2s\times l} & \cdots & J_{2s\times l} & tJ_{2s\times k} & J_{2s\times k} & \cdots & J_{2s\times k} \\ 0 & \cdots & & 0 & 0 & \cdots & & 0 \\ \vdots & & & \vdots & \vdots & & & \vdots \\ 0 & \cdots & & 0 & 0 & \cdots & & 0 \end{pmatrix}_{s\times(u+t)}$$

and

$$D = \begin{pmatrix} ((X+(q-1))I+(s-1)J)_{2s} & J_{2s} & \cdots & J_{2s} \\ \vdots & & \ddots & \vdots \\ 0 & 0 & \cdots & ((X+(q-1))I-J)_{2s} \end{pmatrix}_s.$$

We have

$$\begin{aligned} det(A) &= (det(L_l))^u (det(L'_k))^t \\ &= ((X-(q-1))^{l-1}(X-(q-1)-l))^u (X^{k-1}(X-k))^t \\ &= (x-(q-1)(q^3-2q+1))^{u(l-1)}(x-(q-1)(q^3-q-1))^{u+t} \\ &\quad (x-(q-1)(q^3-2q))^{(k-1)t}. \end{aligned}$$

It is not difficult to see that

$$\begin{aligned} det(D-CA^{-1}B) &= (X+R)(X+(q-1))^{s(2s-1)}(X+(q-1)-2s)^{s-1} \\ &= x(x-(q-1)(q^3-2q-1))^{s(2s-1)}(x-(q-1)(q^3-q-1))^{s-1} \end{aligned}$$

and the result follows. ■

Theorem 3.3 The Laplacian spectrum of the NC-graph of quasi-dihedral group QD_{2^n} is

$$\{[0]^1, [2^{n-1}]^{2^{n-1}-3}, [2^n-4]^{2^{n-2}}, [2^n-2]^{2^{n-2}}\}.$$

Proof. By Theorem 1, the NC-graph of group QD_{2^n} has the following parts

$$V_0 = \{a,\, a^2,\, \cdots,\, a^{2^{n-2}-1},\, a^{2^{n-2}+1},\, \cdots,\, a^{2^{n-1}-1}\}$$

and

$$V_j = \{a^j b, a^{j+2^{n-2}}b\} \text{ for } 1 \leqslant j \leqslant 2^{n-2}.$$

Each element in part V_0 has degree 2^{n-1} and every element in each V_j has degree 2^n-4. Let $X = x - 2^{n-1}$ and $X' = x - 2^n + 4$. In order to find the Laplacian spectrum of $\Gamma(QD_{2^n})$, we compute $det(xI - (\Delta - M))$, as follows:

$$det(xI-(\Delta-M)) = \begin{vmatrix} XI_2 & 0_2 & \cdots & 0_2 & J_2 & \cdots & J_2 \\ 0_2 & XI_2 & & 0_2 & J_2 & \cdots & J_2 \\ \vdots & & \ddots & \vdots & \vdots & \cdots & \vdots \\ 0_2 & \cdots & & XI_2 & J_2 & \cdots & J_2 \\ J_2 & \cdots & & J_2 & X'I_2 & & J_2 \\ \vdots & & & & & & \\ J_2 & \cdots & & J_2 & J_2 & \cdots & X'I_2 \end{vmatrix}.$$

It can be easily seen that:

$$det(xI-(\Delta-M)) = \begin{vmatrix} A & B \\ C & D \end{vmatrix} = det(A)det(D - CA^{-1}B),$$

where

$$A = \begin{pmatrix} XI_2 & 0_2 & \cdots & 0_2 \\ 0_2 & XI_2 & & 0_2 \\ \vdots & & \ddots & \vdots \\ 0_2 & \cdots & & XI_2 \end{pmatrix}_{2p},$$

$$B = \begin{pmatrix} 2^{n-1}J_2 & J_2 & \cdots & J_2 \\ 0_2 & & \cdots & 0_2 \\ \vdots & & & \vdots \\ 0_2 & & \cdots & 0_2 \end{pmatrix}_{2p\times\frac{p+2}{2}},$$

$$C = \begin{pmatrix} (2^{n-2}-1)J_2 & J_2 & \cdots & J_2 \\ 0_2 & & \cdots & 0_2 \\ \vdots & & & \vdots \\ 0_2 & & \cdots & 0_2 \end{pmatrix}_{\frac{p+2}{2}\times 2p}$$

and

$$D = \begin{pmatrix} X'I_2 + (2^{n-2}-1)J_2 & J_2 & \cdots & J_2 \\ 0_2 & X'I_2 - J_2 & & 0_2 \\ \vdots & & \ddots & \vdots \\ 0_2 & & \cdots & X'I_2 - J_2 \end{pmatrix}_{\frac{p+2}{2}}.$$

We have

$$det(A) = (det(XI_2))^{2^{n-2}-1} = ((x-2^{n-1})^2)^{2^{n-2}-1} = (x-2^{n-1})^{2^{n-1}-2}$$

and

$$det(D - CA^{-1}B) = \frac{x(x-2^n+2)^{2^{n-2}}(x-2^n+4)^{2^{n-2}-1}}{x-2^{n-1}}.$$

Therefore

$$\begin{aligned} det(xI - (\Delta - M)) &= det(A)det(D - CA^{-1}B) \\ &= x(x-2^{n-1})^{2^{n-1}-3}(x-2^n+4)^{2^{n-2}}(x-2^n+2)^{2^{n-2}}. \end{aligned}$$

Hence, the result follows. ∎

Here, we find the energy of NC-graph of three groups $PSL(2,2^k)$, $GL(2,q)$ and QD_{2^n}.

Theorem 3.4 The energy of NC-graph of group QD_{2^n} is

$$\mathcal{E}(\Gamma(QD_{2^n})) = 2^{n-1} - 2 + 2\sqrt{(5 \times 2^{n-2}-1)(2^{n-2}-1)}.$$

Proof. By using Theorem 1, we have

$$\begin{aligned} \mathcal{E}(\Gamma(QD_{2^n})) &= (2^{n-2}-1)|-2| + 1|2^{n-2}-1-k| + 1|2^{n-2}-1+k| \\ &= 2^{n-1} - 2 - 2^{n-2} + 1 + k + 2^{n-2} - 1 + k \\ &= 2^{n-1} - 2 + 2k = 2^{n-1} - 2 + 2\sqrt{(5 \times 2^{n-2}-1)(2^{n-2}-1)}. \end{aligned}$$

∎

Theorem 3.5

$$\mathcal{E}(\Gamma(PSL(2,2^k))) = 2^{3k} - 2^{k+2} + 2 + \alpha,$$

where $\alpha = |x_1| + |x_2| + |x_3|$ and x_1, x_2, x_3 are given in Theorem 2.

Theorem 3.6 The energy of the NC-graph of general linear group $GL(2,q)$, where $q = p^n > 2$ and p is prime integer is

$$\mathcal{E}(\Gamma(PSL(2,2^k))) = (q-1)(q^3 - 4q + 2) + \beta,$$

where $\beta = |x_1| + |x_2| + |x_3|$.

Proof. The proof follows from Theorem 3. ∎

References

[1] A. Abdollahi, S. Akbari, H. R. Maimani, Non-commuting graph of a group, J. Algebra 298 (2006) 468-492.
[2] M. R. Darafsheh, Groups with the same non-commuting graph, Discrete Appl. Math. 157 (2009) 833-837.
[3] M. Ghorbani, Z. Gharavi-AlKhansari, An algebraic study of non-commuting graphs, Filomat 31 (2017) 663-669.
[4] I. Gutman, The energy of a graph, Ber. Math. Statist. Sekt. Forschungsz Graz 103 (1978) 1-22.
[5] A. R. Moghaddamfar, W. J. Shi, W. Zhou, A. R. Zokayi, On the non-commuting graph associated with a finite group, Siberian Math. J. 46 (2005) 325-332.
[6] G. L. Morgan, C. W. Parker, The diameter of the commuting graph of a finite group with trivial centre, J. Algebra 393 (2013) 41-59.
[7] B. H. Neumann, A problem of Paul Erdős on groups, J. Austral. Math. Soc. Ser. A, 21 (1976) 467-472.

13

Random fixed point theorems with an application to a random nonlinear integral equation

R. A. Rashwan[a*], H. A. Hammad[b]

[a] *Department of Mathematics, Faculty of Science, Assuit University, Assuit 71516, Egypt.*
[b] *Department of Mathematics, Faculty of Science, Sohag University, Sohag 82524, Egypt.*

Abstract. In this paper, stochastic generalizations of some fixed point for operators satisfying random contractively generalized hybrid and some other contractive condition have been proved. We discuss also the existence of a solution to a nonlinear random integral equation.

Keywords: Random fixed point, nonlinear random integral equation, contractively generalized hybrid.

1. Introduction

Fixed point theory has the diverse applications in different branches of mathematics, statistics, engineering, and economics in dealing with the problems arising in approximation theory, potential theory, game theory, theory of differential equations, theory of integral equations, and others. Developments in the investigation on fixed points of non-expansive mappings, contractive mappings in different spaces like metric spaces, Banach spaces, Fuzzy metric spaces and cone metric spaces have almost been saturated. The study of random fixed point theorems was initiated by the Prague school of probabilistic in 1950's. The introduction of randomness leads to several new questions of measurability of solutions, probabilistic and statistical aspects of random solutions. Random fixed point theorems for random contraction mappings on separable complete metric spaces were first proved by Hanš [9] and Špaček [26]. The survey article

*Corresponding author.
E-mail address: rr_rashwan54@yahoo.com (R. A. Rashwan).

by Bharucha-Reid [7] in 1976 attracted the attention of several mathematicians and gave wings to this theory. The results of Špaček and Hanš in multi-valued contractive mappings was extended by Itoh [11]. By the same author random fixed point theorems with an application to random differential equations in Banach spaces are obtained. Mukherjee [16] gave a random version of Schauder's fixed point theorem on an atomic probability measure space. While Bharucha-Reid [6, 7] generalized Mukherjee's result on a general probability measure space. On the other hand, some authors [4, 17, 18, 23, 24] applied a random fixed point theorem to prove the existence of a solution in a Banach space of a random nonlinear integral equation. Sehgal and Waters [25] had obtained several random fixed point theorems including a random analogue of the classical results due to Rothe [20]. In some recent papers of Saha et al. [21, 22], some random fixed point theorems over separable Banach spaces and separable Hilbert spaces have been established.

On the other hand, the first fundamental fixed point theorem in deterministic form was due to S. Banach [3] in a metric space setting, this theorem runs as follows:

Theorem 1.1 (**Banach contraction principle**) Let (X, d) be a complete metric space, $c \in [0, 1)$ and let $T : X \to X$ be a mapping such that for each $x, y \in X$,

$$d(Tx, Ty) \leqslant cd(x, y).$$

Then T has a unique fixed point $z \in X$ such that for each $x \in X$, $\lim_{n\to\infty} T^n x = z$.

After this classical result, Kannan [13] gave a substantially new contractive mapping where the mapping T need not be continuous on X (but continuous at their fixed points, see [19]). He considered the contractive condition as follows: there exists a constant $b \in [0, \frac{1}{2})$ such that

$$d(Tx, Ty) \leqslant b[d(x, Tx) + d(y, Ty)],$$

for all $x, y \in X$. A mapping $T : X \to X$ is said to be contractively nonspreading [8, 27] if there exists $\beta \in [0, \frac{1}{2})$ such that

$$d(Tx, Ty) \leqslant \beta[d(x, Ty) + d(y, Tx)],$$

for all $x, y \in X$. A mapping $T : X \to X$ is called contractively hybrid [10] if there exists $\gamma \in [0, \frac{1}{2})$ such that

$$d(Tx, Ty) \leqslant \gamma[d(x, Ty) + d(y, Tx) + d(y, x)],$$

motivated by generalized hybrid mappings [14] in a Hilbert space, Takahashi et al. [10] introduced the concept of contractively generalized hybrid mappings on metric spaces and studied fixed point theorems for such mappings on complete metric spaces. Let (X, d) be a metric space, a mapping $T : X \to X$ is called contractively generalized hybrid [10] if there exist $\alpha, \beta \in \mathbb{R}$ and $\gamma \in [0, 1)$ such that

$$\alpha d(Tx, Ty) + (1 - \alpha)d(x, Ty) \leqslant \gamma\{\beta d(y, Tx) + (1 - \beta)d(y, x)\}, \tag{1.1}$$

for all $x, y \in X$, such a mapping T is also called contractively (α, β, γ)-generalized hybrid. For example, a contractively (α, β, γ)-generalized hybrid mapping is $r-$contractive for

$\alpha = 1$ and $\beta = 0$. It is contractively nonspreading for $\alpha = 1+r$ and $\beta = 1$, see Takahashi et al. [10].

2. Preliminaries

In order to prove our main results, we need to recall the following concepts and results.

Let$(X, \sum)$ be a separable Banach space where $\sum$ is a σ-algebra of Borel subsets of X and let $(\Omega, \sum, \mu)$ denote a complete probability measure space with measure μ and $\sum$ be a σ-algebra of subsets of Ω. For more details one can see Joshi and Bose [12].

Definition 2.1 A mapping $x : \Omega \to X$ is said to be an $X-$valued random variable, if the inverse image under the mapping x of every Borel set B of X belongs to $\sum$, that is, $x^{-1}(B) \in \sum$ for all $B \in \sum$.

Definition 2.2 A mapping $x : \Omega \to X$ is said to be a finitely valued random variable, if it is constant on each of a finite number of disjoint sets $A_i \in \sum$ and is equal to 0 on $\Omega - \left(\bigcup_{i=1}^{n} A_i\right)$. X is called a simple random variable if it is finitely valued and $\mu\{\omega : \|x(\omega)\| > 0\} < \infty$.

Definition 2.3 A mapping $x : \Omega \to X$ is said to be a strong random variable, if there exists a sequence $\{x_n(\omega)\}$ of simple random variables which converges to $x(\omega)$ almost surely, i.e., there exists a set $A_0 \in \sum$ with $(A_0) = 0$ such that $\lim_{n\to\infty} x_n(\omega) = x(\omega)$, $\omega \in \Omega - A_0$.

Definition 2.4 A mapping $x : \Omega \to X$ is said to be weak random variable, if the function $x^*(x(\omega))$ is a real valued random variables for each $x^* \in X^*$, the space X^* denoting the first normed dual space of X.

Remark 1

(1) In a separable Banach space X, the notions of strong and weak random variables $x : \Omega \to X$ coincide and respect of such a space X, x is termed as a random variable (see Joshi and Bose [12, Corollary 1]).

(2) If X is a separable Banach space then the $\sigma-$algebra generated by the class of all spherical neighourhoods of X is equal to the $\sigma-$algebra of Borel subsets of X. Hence every strong and also weak random variable is measurable in the sense of Definition 2.1.

Let Y be another Banach space. We also need the following definitions as cited in Joshi and Bose [12].

Definition 2.5 A mapping $F : \Omega \times X \to Y$ is said to be a random mapping if $F(\omega, x) = Y(\omega)$ is a $Y-$valued random variable for every $x \in X$.

Definition 2.6 A mapping $F : \Omega \times X \to Y$ is said to be a continuous random mapping if the set of all $\omega \in \Omega$ for which $F(\omega, x)$ is a continuous function of x has measure one.

Definition 2.7 An equation of the type $F(\omega, x(\omega)) = x(\omega)$, where $F : \Omega \times X \to Y$ is a random mapping, is called a random fixed point equation.

Definition 2.8 Any mapping $x : \Omega \to X$ which satisfies the random fixed point equation $F(\omega, x(\omega)) = x(\omega)$ almost surely is said to be a wide sense solution of the fixed point equation.

Definition 2.9 Any $X-$valued random variable $x(\omega)$ which satisfies $\mu\{\omega : F(\omega, x(\omega)) = x(\omega)\} = 1$ is said to be a random solution of the fixed point equation or a random fixed point of F.

Remark 2 *A random solution is a wide sense solution of the fixed point equation, but the converse is not necessarily true. This is evident from the following example as found under Joshi and Bose [12, Remark 1].*

Our main aim of this paper is to define the random analogue of a (α, β, γ)-generalized hybrid and thereby prove the stochastic version of the deterministic fixed point theorem in a separable Banach space. Also some more random fixed point theorems have been established in separable Banach space to investigate this relatively new field of research extensively with application.

Now, we define the random version of a $(\alpha, \beta, \gamma)-$generalized hybrid and then establish a random fixed point theorem for $(\alpha, \beta, \gamma)-$generalized hybrid.

3. Random analogue of (α, β, γ)-generalized hybrid

Definition 3.1 Let X be a separable Banach space and $(\Omega, \sum, \mu)$ a complete probability measure space. Then $T : \Omega \times X \to X$ is called a random contractively generalized hybrid if there exist a finitely real valued random variables $\alpha(\omega), \beta(\omega)$ and $\gamma(\omega)$ such that

$$\alpha(\omega) \|T(\omega, x_1) - T(\omega, x_2)\| \leqslant \gamma(\omega)\Big(\beta(\omega) \|x_2 - T(\omega, x_1)\| + (1 - \beta(\omega)) \|x_1 - x_2\| \Big) \\ - (1 - \alpha(\omega)) \|x_1 - T(\omega, x_2)\|, \tag{3.1}$$

for all $x_1, x_2 \in X$.

Theorem 3.2 Let X be a separable Banach space and $(\Omega, \sum, \mu)$ a complete probability measure space. Let $T : \Omega \times X \longrightarrow X$ be a continuous operator satisfying (3.1) almost surely, where $0 \leqslant \gamma(\omega) < 1$ is a real valued random variable and $\gamma(\omega).\beta(\omega) < \alpha(\omega)$ almost surely. Then there exists a random fixed point of T.

Proof. Let $A = \{\omega \in \Omega : T(\omega, x)$ is a continuous function of $x\}$,

$$B = \Big\{\omega \in \Omega : 0 \leqslant \gamma(\omega) < 1\Big\} \cap \Big\{\omega \in \Omega : \gamma(\omega).\beta(\omega) < \alpha(\omega)\Big\},$$

$$C_{x_1,x_2} = \Big\{\omega \in \Omega : \alpha(\omega) \|T(\omega, x_1) - T(\omega, x_2)\| \leqslant \gamma(\omega)\big(\beta(\omega) \|x_2 - T(\omega, x_1)\| \\ +(1 - \beta(\omega)) \|x_1 - x_2\| \big) - (1 - \alpha(\omega)) \|x_1 - T(\omega, x_2)\| \Big\}.$$

Let S be a countable dense subset of X, we now prove that

$$\bigcap_{x_1,x_2 \in X} (C_{x_1,x_2} \cap A \cap B) = \bigcap_{s_1,s_2 \in X} (C_{s_1,s_2} \cap A \cap B).$$

Let $\omega \in \bigcap_{s_1,s_2\in X} (C_{s_1,s_2} \cap A \cap B)$. Then for all $s_1, s_2 \in X$,

$$\begin{aligned}\alpha(\omega)\,\|T(\omega,s_1)-T(\omega,s_2)\| &\leqslant \gamma(\omega)\{\beta(\omega)\,\|s_2-T(\omega,s_1)\| \\ &\quad +(1-\beta(\omega))\,\|s_1-s_2\|\} - (1-\alpha(\omega))\,\|s_1-T(\omega,s_2)\|,\end{aligned}$$

note that for all $x_1, x_2 \in X$,

$$\|s_1-s_2\| \leqslant \|s_1-x_1\| + \|x_1-x_2\| + \|x_2-s_2\|, \tag{3.2}$$

$$\|s_1-T(\omega,s_1)\| \leqslant \|s_1-x_1\| + \|x_1-T(\omega,x_1)\| + \|T(\omega,x_1)-T(\omega,s_1)\|, \tag{3.3}$$

$$\|s_2-T(\omega,s_2)\| \leqslant \|s_2-x_2\| + \|x_2-T(\omega,x_2)\| + \|T(\omega,x_2)-T(\omega,s_2)\|, \tag{3.4}$$

$$\|s_1-T(\omega,s_2)\| \leqslant \|s_1-x_1\| + \|x_1-T(\omega,x_2)\| + \|T(\omega,x_2)-T(\omega,s_2)\|, \tag{3.5}$$

$$\|s_2-T(\omega,s_1)\| \leqslant \|s_2-x_2\| + \|x_2-T(\omega,x_1)\| + \|T(\omega,x_1)-T(\omega,s_1)\|. \tag{3.6}$$

Let $x_1, x_2 \in X$, we have

$$\begin{aligned}\alpha(\omega)\,\|T(\omega,x_1)-T(\omega,x_2)\| &\leqslant \alpha(\omega)\,\|T(\omega,x_1)-T(\omega,s_1)\| + \alpha(\omega)\,\|T(\omega,s_1)-T(\omega,s_2)\| \\ &\quad +\alpha(\omega)\,\|T(\omega,s_2)-T(\omega,x_2)\| \\ &\leqslant \alpha(\omega)\,\|T(\omega,x_1)-T(\omega,s_1)\| + \alpha(\omega)\,\|T(\omega,s_2)-T(\omega,x_2)\| \\ &\quad +\gamma(\omega)\{\beta(\omega)\,\|s_2-T(\omega,s_1)\| + (1-\beta(\omega))\,\|s_1-s_2\|\} \\ &\quad -(1-\alpha(\omega))\,\|s_1-T(\omega,s_2)\|,\end{aligned}$$

using (3.2), (3.5) and (3.6) we have

$$\begin{aligned}\alpha(\omega)\,\|T(\omega,x_1)-T(\omega,x_2)\| &\leqslant \alpha(\omega)\,\|T(\omega,x_1)-T(\omega,s_1)\| + \alpha(\omega)\,\|T(\omega,s_2)-T(\omega,x_2)\| \\ &\quad +\gamma(\omega).\beta(\omega)[\|s_2-x_2\| + \|x_2-T(\omega,x_1)\| + \|T(\omega,x_1)-T(\omega,s_1)\|] \\ &\quad +\gamma(\omega)(1-\beta(\omega))[\|s_1-x_1\| + \|x_1-x_2\| + \|x_2-s_2\|] \\ &\quad -(1-\alpha(\omega))[\|s_1-x_1\| + \|x_1-T(\omega,x_2)\| + \|T(\omega,x_2)-T(\omega,s_2)\|]\end{aligned}$$

$$\begin{aligned}&\leqslant (\alpha(\omega)+\gamma(\omega).\beta(\omega))\,\|T(\omega,x_1)-T(\omega,s_1)\| + (2\alpha(\omega)-1)\,\|T(\omega,s_2)-T(\omega,x_2)\| \\ &\quad +\gamma(\omega)\{\beta(\omega)\,\|x_2-T(\omega,x_1)\| + (1-\beta(\omega))\,\|x_1-x_2\|\} - (1-\alpha(\omega))\,\|x_1-T(\omega,x_2)\| \\ &\quad +[\gamma(\omega).\beta(\omega)+\gamma(\omega)(1-\beta(\omega))]\,\|x_2-s_2\| + [\gamma(\omega)(1-\beta(\omega))-(1-\alpha(\omega))]\,\|s_1-x_1\| \\ &< 2\alpha(\omega)\,\|T(\omega,x_1)-T(\omega,s_1)\| + (2\alpha(\omega)-1)\,\|T(\omega,s_2)-T(\omega,x_2)\| \\ &\quad +\gamma(\omega)\{\beta(\omega)\,\|x_2-T(\omega,x_1)\| + (1-\beta(\omega))\,\|x_1-x_2\|\} - (1-\alpha(\omega))\,\|x_1-T(\omega,x_2)\| \\ &\quad +\gamma(\omega)\,\|x_2-s_2\| + (\gamma(\omega)-1)\,\|s_1-x_1\|.\end{aligned}$$

Since for a particular $\omega \in \Omega$, $T(\omega, x)$ is a continuous function of x, so for any $\epsilon > 0$, there exists $\delta_i(x_i) > 0$ $(i = 1, 2)$ such that

$$\|T(\omega, x_1) - T(\omega, s_1)\| < \frac{\epsilon}{8\alpha(\omega))}, \text{ whenever } \|x_1 - s_1\| < \delta_1(x_1) = \frac{\epsilon}{4(\gamma(\omega) - 1)},$$

and

$$\|T(\omega, x_2) - T(\omega, s_2)\| < \frac{\epsilon}{4(2\alpha(\omega) - 1)}, \text{ whenever } \|x_2 - s_2\| < \delta_2(x_2) = \frac{\epsilon}{4\gamma(\omega)},$$

if we take $\rho_1 = \min(\delta_1, \frac{\varepsilon}{4})$ and $\rho_2 = \min(\delta_2, \frac{\varepsilon}{4})$, for such a choic of ρ_1 and ρ_2, we get

$$\alpha(\omega)\|T(\omega, x_1) - T(\omega, x_2)\| \leqslant \frac{\epsilon}{4} + \frac{\epsilon}{4} + \frac{\epsilon}{4} + \frac{\epsilon}{4} + \gamma(\omega)\{\beta(\omega)\|x_2 - T(\omega, x_1)\| + (1 - \beta(\omega))\|x_1 - x_2\|\} - (1 - \alpha(\omega))\|x_1 - T(\omega, x_2)\|.$$

As $\epsilon > 0$ is arbitrary, it follow that

$$\alpha(\omega)\|T(\omega, x_1) - T(\omega, x_2)\| \leqslant \gamma(\omega)\{\beta(\omega)\|x_2 - T(\omega, x_1)\| + (1 - \beta(\omega))\|x_1 - x_2\|\} - (1 - \alpha(\omega))\|x_1 - T(\omega, x_2)\|.$$

Thus $\omega \in \bigcap_{x_1, x_2 \in X} (C_{x_1, x_2} \cap A \cap B)$, which implies that

$$\bigcap_{s_1, s_2 \in X} (C_{s_1, s_2} \cap A \cap B) \subset \bigcap_{x_1, x_2 \in X} (C_{x_1, x_2} \cap A \cap B),$$

also, similar to above proof, we have

$$\bigcap_{x_1, x_2 \in X} (C_{x_1, x_2} \cap A \cap B) \subset \bigcap_{s_1, s_2 \in X} (C_{s_1, s_2} \cap A \cap B),$$

and so

$$\bigcap_{x_1, x_2 \in X} (C_{x_1, x_2} \cap A \cap B) = \bigcap_{s_1, s_2 \in X} (C_{s_1, s_2} \cap A \cap B).$$

Let $N = \bigcap_{s_1, s_2 \in X} (C_{s_1, s_2} \cap A \cap B)$, then $\mu(N) = 1$ and for each $\omega \in N$, $T(\omega, x)$ is a deterministic continuous operator satisfying the mapping referred to in [10] and hence, this has a wide sense solution $x(\omega)$. The randomness and measurability of $x(\omega)$ can be proved by generating an approximating sequence of random variable $x_n(\omega)$ as follows: Let $x_\circ(\omega)$ be a random variable, let $x_1(\omega) = T(\omega, x_\circ(\omega))$, then it follows that $x_1(\omega)$ is a random variable, then we consider $x_{n+1}(\omega) = T(\omega, x_n(\omega))$, by repeated iteration, it gives that $\{x_n(\omega)\}$ is a sequence of random variable convergence to $x(\omega)$. This implies that $x(\omega)$ is measurable and unique random fixed point of T. ■

In the following theorem, we prove the stochastic version of deterministic fixed point theorem for a general contractive mapping and some other related results.

Theorem 3.3 Let X be a separable Hilbert space and $(\Omega, \sum, \mu)$ be a complete propability measurable space. Let $T : \Omega \times X \longrightarrow X$ be a continuous operator such that for $\omega \in \Omega$, T satisfy the following condition:

$$\|T(\omega, x_1) - T(\omega, x_2)\| \leqslant \alpha(\omega) \ \max \left\{ \begin{array}{c} \|x_1 - x_2\|, \frac{\beta(\omega)}{2}[\|x_1 - T(\omega, x_1)\| + \|x_2 - T(\omega, x_2)\|], \\ \frac{\gamma(\omega)}{2}[\|x_1 - T(\omega, x_2)\| + \|x_2 - T(\omega, x_1)\|] \end{array} \right\}, \tag{3.7}$$

for all $x_1, x_2 \in X$ where $\alpha(\omega), \beta(\omega)$ and $\gamma(\omega)$ are nonnegative real valued random varibles such that $\beta(\omega), \gamma(\omega) \in (0, 1)$, $\alpha(\omega) > 0$ and $\alpha(\omega).\beta(\omega)$, $\alpha(\omega).\gamma(\omega) < \alpha(\omega)$ almost surely. Then T has a unique random fixed point in X.

Proof. Let $A = \{\omega \in \Omega : T(\omega, x) \text{ is a continuous function of } x\}$,

$$B = \{\omega \in \Omega : \alpha(\omega) > 0\} \cap \{\omega \in \Omega : 0 < \beta(\omega), \gamma(\omega) < 1\}$$
$$\cap \{\omega \in \Omega : \alpha(\omega).\beta(\omega), \alpha(\omega).\gamma(\omega) < \alpha(\omega)\},$$

$$C_{x_1,x_2} = \left\{ \begin{array}{c} \omega \in \Omega : \|T(\omega, x_1) - T(\omega, x_2)\| \\ \leqslant \alpha(\omega) \max \left\{ \begin{array}{c} \|x_1 - x_2\|, \frac{\beta(\omega)}{2}[\|x_1 - T(\omega, x_1)\| + \|x_2 - T(\omega, x_2)\|] \\ , \frac{\gamma(\omega)}{2}[\|x_1 - T(\omega, x_2)\| + \|x_2 - T(\omega, x_1)\|] \end{array} \right\} \end{array} \right\}.$$

Let S be a countable dense subset of X, we now prove that

$$\bigcap_{x_1,x_2 \in X} (C_{x_1,x_2} \cap A \cap B) = \bigcap_{s_1,s_2 \in X} (C_{s_1,s_2} \cap A \cap B).$$

Then for all $s_1, s_2 \in X$,

$$\|T(\omega, s_1) - T(\omega, s_2)\| \leqslant \alpha(\omega) \max \left\{ \begin{array}{c} \|s_1 - s_2\|, \frac{\beta(\omega)}{2}[\|s_1 - T(\omega, s_1)\| + \|s_2 - T(\omega, s_2)\|] \\ , \frac{\gamma(\omega)}{2}[\|s_1 - T(\omega, s_2)\| + \|s_2 - T(\omega, s_1)\|] \end{array} \right\}. \tag{3.8}$$

Now, we examine the following cases:

Case(i). Suppose

$$\|T(\omega, s_1) - T(\omega, s_2)\| = \alpha(\omega) \|s_1 - s_2\|,$$

now,

$$\|T(\omega, x_1) - T(\omega, x_2)\| \leqslant \|T(\omega, x_1) - T(\omega, s_1)\| + \|T(\omega, s_1) - T(\omega, s_2)\| + \|T(\omega, s_2) - T(\omega, x_2)\|$$

$$\leqslant \|T(\omega, x_1) - T(\omega, s_1)\| + \|T(\omega, s_2) - T(\omega, x_2)\| + \alpha(\omega) \|s_1 - s_2\|. \tag{3.9}$$

Using (3.2) and (3.9), we have

$$\|T(\omega, x_1) - T(\omega, x_2)\| \leqslant \|T(\omega, x_1) - T(\omega, s_1)\| + \|T(\omega, s_2) - T(\omega, x_2)\|$$

$$+\alpha(\omega)[\|s_1 - x_1\| + \|x_1 - x_2\| + \|x_2 - s_2\|], \tag{3.10}$$

since for a particular $\omega \in \Omega$, $T(\omega, x)$ is a continuous function of x, so for any $\epsilon > 0$, there exists $\delta_i(x_i) > 0$ $(i = 1, 2)$ such that

$$\|T(\omega, x_1) - T(\omega, s_1)\| < \frac{\epsilon}{4}, \text{ whenever } \|x_1 - s_1\| < \delta_1(x_1), \tag{3.11}$$

and

$$\|T(\omega, x_2) - T(\omega, s_2)\| < \frac{\epsilon}{4}, \text{ whenever } \|x_2 - s_2\| < \delta_2(x_2), \tag{3.12}$$

where

$$\delta = \delta_1(x_1) = \delta_2(x_2) = \frac{\epsilon}{4\alpha(\omega)}, \tag{3.13}$$

by choosing $\rho = \min(\delta, \frac{\epsilon}{4})$ then from (3.10), we have

$$\|T(\omega, x_1) - T(\omega, x_2)\| \leqslant \frac{\epsilon}{4} + \frac{\epsilon}{4} + \alpha(\omega)[\frac{\epsilon}{4\alpha(\omega)} + \|x_1 - x_2\| + \frac{\epsilon}{4\alpha(\omega)}]$$

$$= \frac{\epsilon}{4} + \frac{\epsilon}{4} + \frac{\epsilon}{4} + \frac{\epsilon}{4} + \alpha(\omega)\,\|x_1 - x_2\|\,.$$

As $\epsilon > 0$ is arbitrary, it follow that

$$\|T(\omega, x_1) - T(\omega, x_2)\| \leqslant \alpha(\omega)\,\|x_1 - x_2\|\,. \tag{3.14}$$

Case(ii). Suppose

$$\|T(\omega, s_1) - T(\omega, s_2)\| = \frac{\alpha(\omega).\beta(\omega)}{2}[\|s_1 - T(\omega, s_1)\| + \|s_2 - T(\omega, s_2)\|],$$

now,

$$\begin{aligned}\|T(\omega, x_1) - T(\omega, x_2)\| &\leqslant \|T(\omega, x_1) - T(\omega, s_1)\| + \|T(\omega, s_1) - T(\omega, s_2)\| + \|T(\omega, s_2) - T(\omega, x_2)\| \\ &\leqslant \|T(\omega, x_1) - T(\omega, s_1)\| + \|T(\omega, s_2) - T(\omega, x_2)\|\end{aligned}$$

$$+\frac{\alpha(\omega).\beta(\omega)}{2}[\|s_1 - T(\omega, s_1)\| + \|s_2 - T(\omega, s_2)\|]. \tag{3.15}$$

By using (3.3), (3.4) and (3.15), by routine calculation, we get

$$\begin{aligned}
\|T(\omega,x_1)-T(\omega,x_2)\| &\leqslant \|T(\omega,x_1)-T(\omega,s_1)\|+\|T(\omega,x_2)-T(\omega,s_2)\| \\
&\quad+\frac{\alpha(\omega).\beta(\omega)}{2}[\|s_1-x_1\|+\|x_1-T(\omega,x_1)\|+\|T(\omega,x_1)-T(\omega,s_1)\| \\
&\quad+\|s_2-x_2\|+\|x_2-T(\omega,x_2)\|+\|T(\omega,x_2)-T(\omega,s_2)\|] \\
&=(1+\frac{\alpha(\omega).\beta(\omega)}{2})[\|T(\omega,x_1)-T(\omega,s_1)\|+\|T(\omega,x_2)-T(\omega,s_2)\|] \\
&\quad+\frac{\alpha(\omega).\beta(\omega)}{2}[\|s_1-x_1\|+\|s_2-x_2\|] \\
&\quad+\frac{\alpha(\omega).\beta(\omega)}{2}[\|x_1-T(\omega,x_1)\|+\|x_2-T(\omega,x_2)\|] \\
&<(\frac{2+\alpha(\omega)}{2})[\|T(\omega,x_1)-T(\omega,s_1)\|+\|T(\omega,x_2)-T(\omega,s_2)\|] \\
&\quad+\frac{\alpha(\omega)}{2}[\|s_1-x_1\|+\|s_2-x_2\|]
\end{aligned}$$

$$+\frac{\alpha(\omega).\beta(\omega)}{2}[\|x_1-T(\omega,x_1)\|+\|x_2-T(\omega,x_2)\|], \tag{3.16}$$

since for a particular $\omega \in \Omega$, $T(\omega,x)$ is a continuous function of x, so for any $\epsilon>0$, there exists $\delta_i(x_i)>0$ $(i=1,2)$ such that

$$\|T(\omega,x_1)-T(\omega,s_1)\|<\frac{\epsilon}{2(2+\alpha(w))}, \text{ whenever } \|x_1-s_1\|<\delta_1(x_1),$$

and

$$\|T(\omega,x_2)-T(\omega,s_2)\|<\frac{\epsilon}{2(2+\alpha(w))}, \text{ whenever } \|x_2-s_2\|<\delta_2(x_2),$$

where

$$\delta=\delta_1(x_1)=\delta_2(x_2)=\frac{\epsilon}{2\alpha(\omega)},$$

by choosing $\rho=\min(\delta,\frac{\epsilon}{4})$ and from (3.16), we get

$$\begin{aligned}
\|T(\omega,x_1)-T(\omega,x_2)\| &\leqslant \frac{\epsilon}{4}+\frac{\epsilon}{4}+\frac{\epsilon}{4}+\frac{\epsilon}{4}+\frac{\alpha(\omega).\beta(\omega)}{2}[\|x_1-T(\omega,x_1)\|+\|x_2-T(\omega,x_2)\|] \\
&=\epsilon+\frac{\alpha(\omega).\beta(\omega)}{2}[\|x_1-T(\omega,x_1)\|+\|x_2-T(\omega,x_2)\|].
\end{aligned}$$

As $\epsilon>0$ is arbitrary, it follow that

$$\|T(\omega,x_1)-T(\omega,x_2)\| \leqslant \frac{\alpha(\omega).\beta(\omega)}{2}[\|x_1-T(\omega,x_1)\|+\|x_2-T(\omega,x_2)\|] \tag{3.17}$$

Case(iii). Suppose

$$\|T(\omega, s_1) - T(\omega, s_2)\| = \frac{\alpha(\omega).\gamma(\omega)}{2}[\|s_1 - T(\omega, s_2)\| + \|s_2 - T(\omega, s_1)\|],$$

now,

$$\begin{aligned}\|T(\omega, x_1) - T(\omega, x_2)\| &\leqslant \|T(\omega, x_1) - T(\omega, s_1)\| + \|T(\omega, s_1) - T(\omega, s_2)\| + \|T(\omega, s_2) - T(\omega, x_2)\| \\ &\leqslant \|T(\omega, x_1) - T(\omega, s_1)\| + \|T(\omega, s_2) - T(\omega, x_2)\| \\ &\quad + \frac{\alpha(\omega).\gamma(\omega)}{2}[\|s_1 - T(\omega, s_2)\| + \|s_2 - T(\omega, s_1)\|]. \end{aligned} \tag{3.18}$$

By using (3.5), (3.6) and (3.18), by routine check-up, we get

$$\begin{aligned}\|T(\omega, x_1) - T(\omega, x_2)\| &\leqslant \|T(\omega, x_1) - T(\omega, s_1)\| + \|T(\omega, s_2) - T(\omega, x_2)\| \\ &\quad + \frac{\alpha(\omega).\gamma(\omega)}{2}[\|s_1 - x_1\| + \|x_1 - T(\omega, x_2)\| + \|T(\omega, x_2) - T(\omega, s_2)\| \\ &\quad + \|s_2 - x_2\| + \|x_2 - T(\omega, x_1)\| + \|T(\omega, x_1) - T(\omega, s_1)\|] \\ &= (1 + \frac{\alpha(\omega).\gamma(\omega)}{2})[\|T(\omega, x_1) - T(\omega, s_1)\| + \|T(\omega, x_2) - T(\omega, s_2)\|] \\ &\quad + \frac{\alpha(\omega).\gamma(\omega)}{2}[\|s_1 - x_1\| + \|s_2 - x_2\|] \\ &\quad + \frac{\alpha(\omega).\gamma(\omega)}{2}[\|x_1 - T(\omega, x_2)\| + \|x_2 - T(\omega, x_1)\|] \\ &< (\frac{2 + \alpha(\omega))}{2})[\|T(\omega, x_1) - T(\omega, s_1)\| + \|T(\omega, x_2) - T(\omega, s_2)\|] \\ &\quad + \frac{\alpha(\omega)}{2}[\|s_1 - x_1\| + \|s_2 - x_2\|] \\ &\quad + \frac{\alpha(\omega).\gamma(\omega)}{2}[\|x_1 - T(\omega, x_2)\| + \|x_2 - T(\omega, x_1)\|],\end{aligned}$$

again choose $\rho = \min(\delta, \frac{\epsilon}{4})$ and by the same method of Case ii, we have

$$\begin{aligned}\|T(\omega, x_1) - T(\omega, x_2)\| &\leqslant \frac{\epsilon}{4} + \frac{\epsilon}{4} + \frac{\epsilon}{4} + \frac{\epsilon}{4} + \frac{\alpha(\omega).\gamma(\omega)}{2}[\|x_1 - T(\omega, x_2)\| + \|x_2 - T(\omega, x_1)\|] \\ &= \epsilon + \frac{\alpha(\omega).\gamma(\omega)}{2}[\|x_1 - T(\omega, x_2)\| + \|x_2 - T(\omega, x_1)\|].\end{aligned}$$

As $\epsilon > 0$ is arbitrary, it follow that

$$\|T(\omega, x_1) - T(\omega, x_2)\| \leqslant \frac{\alpha(\omega).\gamma(\omega)}{2}[\|x_1 - T(\omega, x_2)\| + \|x_2 - T(\omega, x_1)\|]. \tag{3.19}$$

Combining (3.14), (3.17) and (3.19), we get

$$\|T(\omega,x_1)-T(\omega,x_2)\| \leqslant \alpha(\omega)\ \max\left\{\begin{array}{c}\|x_1-x_2\|,\frac{k(\omega)}{2}[\|x_1-T(\omega,x_1)\|+\|x_2-T(\omega,x_2)\|],\\ \frac{\gamma(\omega)}{2}[\|x_1-T(\omega,x_2)\|+\|x_2-T(\omega,x_1)\|]\end{array}\right\}.$$

Thus $\omega \in \bigcap_{x_1,x_2\in X}(C_{x_1,x_2}\cap A\cap B)$, which implies that

$$\bigcap_{s_1,s_2\in X}(C_{s_1,s_2}\cap A\cap B)\subset \bigcap_{x_1,x_2\in X}(C_{x_1,x_2}\cap A\cap B),$$

also, similar to above proof, we have

$$\bigcap_{x_1,x_2\in X}(C_{x_1,x_2}\cap A\cap B)\subset \bigcap_{s_1,s_2\in X}(C_{s_1,s_2}\cap A\cap B),$$

and so

$$\bigcap_{x_1,x_2\in X}(C_{x_1,x_2}\cap A\cap B)= \bigcap_{s_1,s_2\in X}(C_{s_1,s_2}\cap A\cap B).$$

Let $N = \bigcap_{s_1,s_2\in X}(C_{s_1,s_2}\cap A\cap B)$, then $\mu(N) = 1$ and for each $\omega \in N$, $T(\omega,x)$ is a deterministic continuous operator satisfying the general contractive condition and hence, this has a wide sense solution $x(\omega)$. The randomness and measurability of $x(\omega)$ can be proved by generating an approximating sequence of random variable $x_n(\omega)$ as follows: Let $x_\circ(\omega)$ be a random variable, let $x_1(\omega) = T(\omega, x_\circ(\omega))$. Then it follows that $x_1(\omega)$ is a random variable, then we consider $x_{n+1}(\omega) = T(\omega, x_n(\omega))$. By repeated iteration, it gives that $\{x_n(\omega)\}$ is a sequence of random variable convergence to $x(\omega)$. This implies that $x(\omega)$ is measurable and unique random fixed point of T. ■

4. Application to a random nonlinear integral equation

In this section, we apply Theorem 3.3 to prove the existence of a solution in a Banach space of a random nonlinear integral equation of the form:

$$x(t;\omega) = h(t;\omega) + \lambda(\omega)\int_S k(t,s;\omega)f(s,x(s;\omega))d\mu_\circ(s), \tag{4.1}$$

where,
(i) S is a locally compact metric space with metric d on $S\times S$, $\mu_\circ$ is a complete $\sigma-$finite measure defined on the collection of Borel subsets of S,
(ii) $\omega \in \Omega$, where ω is a supporting set of the probability measure space $(\Omega, \sum, \mu)$,
(iii) $x(t;\omega)$ is an unknown vector-valued random variable for each $t \in S$,
(iv) $h(t;\omega)$ is the stochastic free term defined for $t \in S$,
(v) $k(t,s;\omega)$ is the stochastic kernel defined for t and s in S,
(vi) $f(t,x)$ is vector-valued function of $t \in S$ and x.
The integral equation (4.1) in stochastic version is a similar to Fredholm integral equation of the second kind in deterministic.

We shall further assume that S is the union of a decreasing sequence of countable family of compact sets $\{C_n\}$ having the properties that $C_1 \subset C_2 \subset ..$ and that for any other compact set S there is a C_i which contains it (see [2]).
we will the steps of Lee and Padjett [15] with necessary modification as required for the more general settings.

Definition 4.1 We define the space $C(S, L_2(\Omega, \sum, \mu))$ to be the space of all continuous functions from S into $L_2(\Omega, \sum, \mu)$ with the topology of uniform convergence on compacta i.e. for each fixed $t \in S$, $x(t;\omega)$ is a vector valued random variable such that

$$\|x(t;\omega)\|^2_{L_2(\Omega,\sum,\mu)} = \int\limits_{\Omega} |x(t;\omega)|^2 \, d\mu(\omega) < \infty.$$

It may be noted that $C(S, L_2(\Omega, \sum, \mu))$ is locally convex space (see [5]) whose topologies defined by a countable family of seminorms given by

$$\|x(t;\omega)\|_n = \sup_{t\in C_n} \|x(t;\omega)\|_{L_2(\Omega,\beta,\mu)}, n = 1, 2, ..$$

Moreover $C(S, L_2(\Omega, \sum, \mu))$ is complete relative to this topology since $L_2(\Omega, \sum, \mu)$ is complete. We will consider the function $h(t;\omega)$ and $f(t, x(t;\omega))$ to be in the space $C(S, L_2(\Omega, \sum, \mu))$ with respect to the stochastic kernel. We assume that for each pair (t, s), $k(t, s;\omega) \in L_\infty(\Omega, \beta, \mu)$ and denote the norm by

$$|||k(t, s;\omega)||| = \|k(t, s;\omega)\|_{L_\infty(\Omega,\sum,\mu)} = \mu - ess \sup_{\omega\in\Omega} |k(t, s;\omega)|.$$

Also we will suppose that $k(t, s;\omega)$ is such that $|||k(t, s;\omega)||| . \|x(t;\omega)\|_{L_2(\Omega,\sum,\mu)}$ is $\mu_\circ$-integrable with respect to s for each $t \in S$ and $x(s;\omega)$ in $C(S, L_2(\Omega, \sum, \mu))$, hence there exists a real valued function G defined $\mu_\circ-$a.e. on S, so that $G(S) \|x(s;\omega)\|_{L_2(\Omega,\sum,\mu)}$ is $\mu_\circ-$integrable so that for each pair $(t, s) \in S \times S$,

$$|||k(t, u;\omega) - k(s, u;\omega)||| . \|x(u;\omega)\|_{L_2(\Omega,\beta,\mu)} \leqslant G(u) \|x(u;\omega)\|_{L_2(\Omega,\beta,\mu)}$$

$\mu_\circ-$a.e. Further, for almost all $s \in S$, then $k(t, s;\omega) : S \to L_\infty(\Omega, \sum, \mu)$ will be continuous in t.

We now define the random integral operator T on $C(S, L_2(\Omega, \sum, \mu))$ by

$$(Tx)(t;\omega) = \lambda(\omega) \int_S k(t, s;\omega)x(s;\omega))d\mu_\circ(s), \quad |\lambda(\omega)| < 1 \tag{4.2}$$

where the integral is a Fredholm integral. Moreover, we have that for each $t \in S$, $(Tx)(t;\omega) \in L_2(\Omega, \sum, \mu)$ and that $(Tx)(t;\omega)$ is continuous in mean square by Lebesgue's dominated convergence theorem. So $(Tx)(t;\omega) \in C(S, L_2(\Omega, \sum, \mu))$.

Definition 4.2 [5] Let B and D be Banach spaces. The pair (B, D) is said to be admissible with respect to a random operator $T(\omega)$ if $T(\omega)(B) \subset D$.

Lemma 4.3 [15] The linear operator T defined by (4.2) is continuous from $C(S, L_2(\Omega, \sum, \mu))$ into itself.

Lemma 4.4 [15] If T is a continuous linear operator from $C(S, L_2(\Omega, \sum, \mu))$ into itself and $B, D \subset C(S, L_2(\Omega, \sum, \mu))$ are Banach spaces stronger than $C(S, L_2(\Omega, \sum, \mu))$ such that (B, D) is admissible with respect to T, then T is continuous from B into D.

***Remark** 3 [15] From Lemmas 4.3 and 4.4, it follows that:*
(1) The operator T defined by (4.2) is a bounded linear operator from B into D.
(2) By a random solution of the equation (4.1) we will mean a function $x(x;\omega)$ in $C(S, L_2(\Omega, \sum, \mu))$ which satisfies the equation (4.1) $\mu-a.e.$

Now we are in a position to prove theorem concerning the existence of a random solution of the equation (4.1) as the following:

Theorem 4.5 We consider the stochastic integral equation (4.1) subject to the following conditions:
(a) B and D are Banach spaces stronger than $C(S, L_2(\Omega, \sum, \mu))$ such that (B, D) is admissible with respect to the integral operator defined by (4.2),
(b) $h(t;\omega) \in D$,
(c) $x(t;\omega) \to f(t, x(t;\omega))$ is an operator from the set

$$Q(\rho) = \{x(t;\omega) : x(t;\omega) \in D, \|x(t;\omega)\|_D \leqslant \rho\}$$

into the space B satisfying

$$\|f(t, x_1(t;\omega)) - f(t, x_2(t;\omega))\|_B \leqslant \alpha(\omega) \max \left\{ \begin{array}{c} \|x_1(t;\omega) - x_2(t;\omega)\|_D, \\ \frac{\beta(\omega)}{2}[\|x_1(t;\omega) - f(t, x_1(t;\omega))\|_D \\ + \|x_2(t;\omega) - f(t, x_2(t;\omega))\|_D], \\ \frac{\gamma(\omega)}{2}[\|x_1(t;\omega) - f(t, x_2(t;\omega))\|_D \\ + \|x_2(t;\omega) - f(t, x_1(t;\omega))\|_D] \end{array} \right\}, \tag{4.3}$$

for $x_1(t;\omega), x_2(t;\omega) \in Q(\rho)$, where $\alpha(\omega), \beta(\omega)$ and $\gamma(\omega)$ are nonnegative real valued random variable such that $\beta(\omega), \gamma(\omega) \in (0,1)$, $\alpha(\omega) > 0$ almost surely.
Then there exists a unique random solution of (4.1) in $Q(\rho)$, provided $\alpha(\omega).\beta(\omega) < \alpha(\omega)$, $\alpha(\omega).\gamma(\omega) < \alpha(\omega)$ and

$$\|h(t;\omega)\|_D + (\frac{2+\alpha(\omega)}{2-\alpha(\omega)})c(\omega)\, \|f(t;0)\|_B \leqslant \rho(1 - \frac{c(\omega)\alpha(\omega)}{2-\alpha(\omega)}),$$

where $c(\omega)$ is the norm of the operator $T(\omega)$.

Proof. Define the operator $U(\omega)$ from $Q(\rho)$ into D by

$$(Ux)(t;\omega) = h(t;\omega) + \lambda(\omega) \int_S k(t,s;\omega) f(s, x(s;\omega)) d\mu_o(s).$$

So,

$$\begin{aligned} \|(Ux)(t;\omega)\|_D &\leqslant \|h(t;\omega)\|_D + c(\omega)\, \|f(t, x(t;\omega))\|_B \\ &\leqslant \|h(t;\omega)\|_D + c(\omega)\, \|f(t;0)\|_B + c(\omega)\, \|f(t, x(t;\omega)) - f(t,0)\|_B \,. \end{aligned}$$

Applying condition (4.3) of this theorem, we get

$$\|f(t,x(t;\omega))-f(t,0)\|_B \leqslant \alpha(\omega)\max\{\|x(t;\omega)\|_D, \frac{\beta(\omega)}{2}[\|x(t;\omega)-f(t,x(t;\omega))\|_D+\|f(t,0)\|_D],$$
$$\frac{\gamma(\omega)}{2}[\|x(t;\omega)-f(t,0))\|_D+\|f(t,x(t;\omega))\|_D]\}.$$

By the same manner of three cases of Theorem 3.2 and by simple proof, we have

$$\|(Ux)(t;\omega)\|_D \leqslant \|h(t;\omega)\|_D + c(\omega)\,\|f(t;0)\|_B + c(\omega)\alpha(\omega)\rho < \rho \text{ from case (i)}, \tag{4.4}$$

also,

$$\|(Ux)(t;\omega)\|_D \leqslant \|h(t;\omega)\|_D + \frac{c(\omega)\alpha(\omega)}{2-\alpha(\omega)}\rho + (\frac{2+\alpha(\omega)}{2-\alpha(\omega)})c(\omega)\,\|f(t,0)\|_B < \rho \text{ from cases (ii), (iii)}. \tag{4.5}$$

Then by (4.4) and (4.5), we have $(Ux)(t;\omega) \in Q(\rho)$, then for $x_1(t;\omega)$, $x_2(t;\omega) \in Q(\rho)$, we have by condition (c)

$$\|(Ux_1)(t;\omega)-(Ux_2)(t;\omega)\|_D = |\lambda(\omega)|\left\|\int_S k(t,s;\omega)[f(s,x_1(s;\omega)-f(s,x_2(s;\omega))]d\mu_\circ(s)\right\|_D$$
$$\leqslant \left\|\int_S k(t,s;\omega)[f(s,x_1(s;\omega)-f(s,x_2(s;\omega))]d\mu_\circ(s)\right\|_D \quad \text{since } |\lambda(\omega)|<1$$
$$< c(\omega)\,\|f(t,x_1(t;\omega)-f(t,x_2(t;\omega))\|_B$$
$$\leqslant \alpha(\omega)\ \max\{\|x_1(t;\omega)-x_2(t;\omega)\|_D,$$
$$\frac{\beta(\omega)}{2}[\|x_1(t;\omega)-(Ux_1)(t;\omega))\|_D+\|x_2(t;\omega)-(Ux_2)(t;\omega))\|,$$
$$\frac{\gamma(\omega)}{2}[\|x_1(t;\omega)-(Ux_2)(t;\omega))\|_D+\|x_2(t;\omega)-(Ux_1)(t;\omega))\|_D]\}.$$

Therefore $U(\omega)$ is a random contractive nonlinear operator on $Q(\rho)$ hence, by Theorem 3.3 there exists a random fixed point of $U(\omega)$, which is the random solution of equation (4.1). ■

Remark 4 *If we take $h(t;\omega)=0$ and $\lambda(\omega)=1$ in Equation (4.1), we have Fredholm equation of the first kind. By a similar method of Theorem 4.5 there exists a random fixed point of $U(\omega)$, which is the random solution of it.*

Acknowledgements

The authors are thankful to the editor-in-chief and referees for giving the valuable suggestions to improve the presentation of the paper.

References

[1] J. Achari, On a pair of random generalized non-linear contractions. Int. J. Math. Math. Sci., 6 (3) (1983), 467-475.

[2] R. F. Arens, A topology for spaces of transformations. Annals Math., 47 (2) (1946), 480-495.
[3] S. Banach, Sur les opérations dans les ensembles abstraits et leur application aux équations intégrals. Fundam. Math., 3 (1922), 133-181.
[4] I. Beg, D. Dey and M. Saha, Converegence and stability of two random iteration algorithms. J. Nonlinear Funct. Anal., 2014 (2014), 1-15.
[5] V. Berinde, Approximating fixed points of weak contractions using Picard iteration. Nonlinear Anal. Forum, 9 (1) (2004), 43-53.
[6] A. T. Bharucha-Reid, Random integral equations. Academic Press, New York, 1972.
[7] A. T. Bharucha-Reid, Fixed point theorems in probabilistic analysis. Bull. Amer. Math. Soc., 82 (5) (1976), 641-657.
[8] S. K. Chatterjea, Fixed point theorems. C. R. Acad. Bulgare Sci., 25 (1972), 727-730.
[9] O. Hanš, Reduzierende zufällige transformationen. Czechoslov. Math. J., 7 (82) (1957), 154-158.
[10] K. Hasegawa, T. Komiya and W. Takahashi, Fixed point theorems for general contractive mappings in metric spaces and estimating expressions. Sci. Math. Jpn., 74 (2011), 15-27.
[11] S. Itoh, Random fixed point theorems with an application to random differential equations in Banach spaces. J. Math. Anal. Appl., 67 (2) (1979), 261-273.
[12] M. C. Joshi and R. K. Bose, Some topics in nonlinear functional analysis. Wiley Eastern Ltd., New Delhi, 1984.
[13] R. Kannan, Some results on fixed points. Bull. Cal. Math. Soc., 60 (1968), 71-76.
[14] P. Kocourek, W. Takahashi and J. C. Yao, Fixed point theorems and weak convergence theorems for generalized hybrid mappings in Hilbert spaces. Taiw. J. Math., 14 (2010), 2497-2511.
[15] A. C. H. Lee and W. J. Padgett, On random nonlinear contraction. Math. Systems Theory, ii (1977), 77-84.
[16] A. Mukherjee, Transformation aleatoires separable theorem all point fixed aleatoire. C. R. Acad. Sci. Paris, Ser. A-B, 263 (1966), 393-395.
[17] W. J. Padgett, On a nonlinear stochastic integral equation of the Hammerstein type. Proc. Amer. Math. Soc., 38 (1) (1973), 625-631.
[18] R. A. Rashwan and D. M. Albaqeri, A common random fixed point theorem and application to random integral equations. Int. J. Appl. Math. Reser., 3 (1) (2014), 71-80.
[19] B. E. Rhoades, Fixed point iterations using infinite matrices. Trans. Amer. Math. Soc., 196 (1974), 161-176.
[20] E. Rothe, Zur theorie der topologische ordnung und der vektorfelder in Banachschen Rau-men. Composito Math., 5 (1937), 177-197.
[21] M. Saha, On some random fixed point of mappings over a Banach space with a probability measure. Proc. Nat. Acad. Sci., 76 (III) (2006), 219-224.
[22] M. Saha and L. Debnath, Random fixed point of mappings over a Hilbert space with a probability measure Adv. Stud. Contemp. Math., 1 (2007), 79-84.
[23] M. Saha and A. Ganguly, Random fixed point theorem on a Ćirić-type contractive mapping and its consequence. Fixed Point Theory and Appl., 2012 (2012), 1-18.
[24] M. Saha and D. Dey, Some random fixed point theorems for (θ, L)-weak contractions. Hacett. J. Math. Statist., 41 (6) (2012), 795-812.
[25] V. M. Sehgal and C. Waters, Some random fixed point theorems for condensing operators. Proc. Amer. Math. Soc., 90 (1) (1984), 425-429.
[26] A. Špaček, Zufällige Gleichungen. Czechoslovak Math. J., 5 (80) (1955), 462-466.
[27] T. Zamfirescu, Fixd point theorems in metric spaces. Arch. Math. (Basel), 23 (1972), 292-298.

14

Second order linear differential equations with generalized trapezoidal intuitionistic Fuzzy boundary value

S. P. Mondal[a*], T. K. Roy[b]

[a]*Department of Mathematics, National Institute of Technology, Agartala, Jirania-799046, Tripura, India;*

[b]*Department of Mathematics, Indian Institute of Engineering Science and Technology, Shibpur, Howrah-711103, West Bengal, India.*

Abstract. In this paper the solution of a second order linear differential equations with intuitionistic fuzzy boundary value is described. It is discussed for two different cases: coefficient is positive crisp number and coefficient is negative crisp number. Here fuzzy numbers are taken as generalized trapezoidal intutionistic fuzzy numbers (GTrIFNs). Further a numerical example is illustrated.

Keywords: Fuzzy set, fuzzy differential equation, generalized trapezoidal intutionistic fuzzy number.

1. Introduction

1.1 *Fuzzy and intutionistic fuzzy set theory*

Zadeh [1] and Dubois and Parade [2] were the first who introduced the conception based on fuzzy number and fuzzy arithmetic. The generalizations of fuzzy sets theory [1] is considered as Intuitionistic fuzzy set (IFS). Out of several higher-order fuzzy sets, IFS was first introduced by Atanassov [3] and have been found to be suitable to deal with unexplored areas.The fuzzy set considers only the degree of belongingness and non belongingness. Fuzzy set theory does not incorporate the degree of hesitation (i.e.,degree

*Corresponding author.
E-mail address: sankar.res07@gmail.com (S. P. Mondal).

of non-determinacy defined as, 1- sum of membership function and non-membership function).To handle such situations, Atanassov [4] explored the concept of fuzzy set theory by intuitionistic fuzzy set (IFS) theory.The degree of acceptance is only considered in Fuzzy Sets but IFS is characterized by a membership function and a non-membership function so that the sum of both values is less than one [4].

1.2 *Fuzzy derivative and fuzzy differential equation*

The topic "fuzzy differential equation" (FDE) has been rapidly developing in recent years. The appliance of fuzzy differential equations is a inherent way to model dynamic systems under possibilistic uncertainty [5]. The concept of the fuzzy derivative was first introduced by Chang and Zadeh [6] and it was followed up by Dubois and Prade [7]. Other methods have been discussed by Puri and Ralescu [8] and Goetschel and Voxman [9]. The concept of differential equations in a fuzzy environment was first formulated by Kaleva [10]. In fuzzy differential equation all derivative is deliberated as either Hukuhara or generalized derivatives. The Hukuhara differentiability has a imperfection (see [12], [19]). The solution turns fuzzier as time goes by. Bede exhibited that a large class of BVPs has no solution if the Hukuhara derivative is used [11]. To overcome this difficulty, the concept of a generalized derivative was developed [[12], [13]) and fuzzy differential equations were discussed using this concept (see [14], [15], [16], [17]). Khastan and Nieto found solutions for a large enough class of boundary value problems using the generalized derivative [18].Bede in [26] discussed the generalized differentiability for fuzzy valued functions. Pointedly the disadvantage of strongly generalized differentiability of a function in comparison H-differentiability is that, a fuzzy differential equation has no unique solution [20]. Recently, Stefanini and Bede by the concept of generalization of the Hukuhara difference for compact convex set [21], introduced generalized Hukuhara differentiability [22] for fuzzy valued function and they demonstrated that, this concept of differentiability have relationships with weakly generalized differentiability and strongly generalized differentiability. Recently Gasilov et. all. [25] solve the fuzzy initial value problem by a new technique where Barros et. all. [33] solve fuzzy differential equation via fuzzification of the derivative operator. Recently the research on fuzzy boundary value is grew attention among all. Armand and Gouyandeh [23] solve two point fuzzy boundary value problems using variational iteration method. Gasilov et. all. [24] solve linear differential equation with fuzzy boundary value. Lopez [27] find the existence of solutions to periodic boundary value problems for linear differential equation.

1.3 *Intutionistic fuzzy differential equation*

Intutionistic FDE is very rare. Melliani and Chadli [28] solve partial differential equation with intutionistic fuzzy number. Abbasbandy and Allahviranloo [29] discussed numerical Solution of fuzzy differential equations by runge-kutta method and in the intuitionistic fuzzy environment.. Lata and Kumar [30] worked on time-dependent intuitionistic fuzzy differential equation and its application to analyze the intutionistic fuzzy reliability of industrial system. First order homogeneous ordinary differential equation with initial value as triangular intuitionistic fuzzy number is described by Mondal and Roy [31]. System of differential equation with initial value as triangular intuitionistic fuzzy number and its application is solved by Mondal and Roy [32].

1.4 *Novelties*

In spite of above mentioned developments, following lacunas are still exists in the formulation and solution of the fuzzy boundary value problem, which are summarized below:

(i) Though there are some articles of fuzzy boundary value problem was solved but till now none has solve boundary value problem with intuitionistic fuzzy number.
(ii) Here also second order differential equation is solved with intuitionistic fuzzy number.
(iii) The intutionistic fuzzy number is also taken as generalized trapezoidal intuitionistic fuzzy number.

1.5 *Structure of the paper*

The paper is organized as follows. In Section 2, the basic concept on fuzzy number and fuzzy derivative are discussed. Also basic definition and properties of generalized trapezoidal intuitionistic fuzzy number are defined. In Section 3 we solve the intutionistic fuzzy boundary value problem. In Section 4 the proposed method is illustrated by an example. The conclusion and future research scope is drawn in the last section 5.

2. Preliminary concept

Definition 2.1:Fuzzy Set: A fuzzy set $\tilde{A}$ is defined by $\tilde{A} = \{(x, \mu_{\tilde{A}}(x)) : x\epsilon A, \mu_{\tilde{A}}(x)\epsilon[0,1]\}$. In the pair $(x, \mu_{\tilde{A}}(x))$ the first element belongs to the classical set A, the second element $\mu_{\tilde{A}}(x)$, belongs to the interval $[0,1]$, called membership function.

Definition 2.2:α-cut of a fuzzy set: The α-level set (or, interval of confidence at level α or α-cut) of the fuzzy set $\tilde{A}$ of X that have membership values in A greater than or equal to α i.e., $\tilde{A} = \{x : \mu_{\tilde{A}}(x) \geqslant \alpha, x\epsilon X, \alpha\epsilon[0,1]\}$.

Definition 2.3: Fuzzy number:A fuzzy number is an extension of a regular number in the sense that it does not refer to one single value but rather to a connected set of possible values, where each possible value has its own weight between 0 and 1.This weight is called membership function. Thus a fuzzy number is a convex and normal fuzzy set.

Definition 2.4:Intuitionistic Fuzzy set: Let a set X be fixed. An IFS $\tilde{A^i}$ in X is an object having the form $\tilde{A^i} = \{< x, \mu_{\tilde{A^i}}(x), \vartheta_{\tilde{A^i}}(x) > 0 : x\epsilon X\}$, where the $\mu_{\tilde{A^i}}(x) : X \to [0,1]$ and $\vartheta_{\tilde{A^i}}(x) : X \to [0,1]$ define the degree of membership and degree of non-membership respectively, of the element $x\epsilon X$ to the set $\tilde{A^i}$, which is a subset of X, for every element of $x\epsilon X, 0 < \mu_{\tilde{A^i}}(x) + \vartheta_{\tilde{A^i}}(x) \leqslant 1$.

Definition 2.5:Intuitionistic Fuzzy number: An IFN $\tilde{A^i}$ is defined as follows
(i) an intuitionistic fuzzy subject of real line.
(ii) normal. i.e., there is any $x_0\epsilon R$ such that $\mu_{\tilde{A^i}}(x_0) = 1$(so $\vartheta_{\tilde{A^i}}(x_0) = 0$)
(iii) a convex set for the membership function $\mu_{\tilde{A^i}}(x)$, i.e.,
$\mu_{\tilde{A^i}}(\lambda x_1 + (1-\lambda)x_2) \geqslant min(\mu_{\tilde{A^i}}(x_1), \mu_{\tilde{A^i}}(x_2))\forall x_1, x_2\epsilon R, \lambda\epsilon[0,1]$
(iv)a concave set for the non-membership function $\vartheta_{\tilde{A^i}}(x)$, i.e.,
$\vartheta_{\tilde{A^i}}(\lambda x_1 + (1-\lambda)x_2) \leqslant min(\vartheta_{\tilde{A^i}}(x_1), \vartheta_{\tilde{A^i}}(x_2))\forall x_1, x_2\epsilon R, \lambda\epsilon[0,1]$

Definition 2.6: Trapezoidal Intuitionistic Fuzzy number: A TrIFN $\tilde{A}^i$ is a subset of IFN in R with following membership function and non membership function as follows:

$$\mu_{\tilde{A}_i}(x) = \begin{cases} \frac{x-a_1}{a_2-a_1} & \text{if } a_1 \leqslant x < a_2 \\ 1 & \text{if } a_2 \leqslant x \leqslant a_3 \\ \frac{a_4-x}{a_4-a_3} & \text{if } a_3 < x \leqslant a_4 \\ 0 & \text{otherwise} \end{cases}$$

and

$$\vartheta_{\tilde{A}_i}(x) = \begin{cases} \frac{a_2-x}{a_2-a_1^{'}} & \text{if } a_1^{'} \leqslant x < a_2 \\ 0 & \text{if } a_2 \leqslant x \leqslant a_3 \\ \frac{x-a_3}{a_4^{'}-a_3} & \text{if } a_3 < x \leqslant a_4^{'} \\ 1 & \text{otherwise} \end{cases}$$

where $a_1^{'} \leqslant a_2 \leqslant a_3 \leqslant a_4^{'}$, $a_1 \leqslant a_2 \leqslant a_3 \leqslant a_4$ and TrIFN is denoted by $\tilde{A}_{TrIFN} = (a_1, a_2, a_3, a_4; a_1^{'}, a_2, a_3, a_4^{'})$

Note 2.1: Here $\mu_{\tilde{A}_i}(x)$ increases with constant rate for $x\epsilon[a_1, a_2]$ and decreases with constant rate for $x\epsilon[a_3, a_4]$ but $\vartheta_{\tilde{A}_i}(x)$ decreases with constant rate for $x\epsilon[a_1^{'}, a_2]$ and increases with constant rate for $x\epsilon[a_3, a_4^{'}]$

Definition 2.7: Generalized Intuitionistic Fuzzy Number: An IFN $\tilde{A}^i$ is defined as follows

(i) an intuitionistic fuzzy subject of real line.

(ii) normal. i.e., there is any $x_0\epsilon R$ such that $\mu_{\tilde{A}^i}(x_0) = \omega$(so $\vartheta_{\tilde{A}^i}(x_0) = \sigma$) for $0 < \omega + \sigma \leqslant 1$.

(iii) a convex set for the membership function $\mu_{\tilde{A}^i}(x)$, i.e., $\mu_{\tilde{A}^i}(\lambda x_1 + (1-\lambda)x_2) \geqslant min(\mu_{\tilde{A}^i}(x_1), \mu_{\tilde{A}^i}(x_2))\forall x_1, x_2\epsilon R, \lambda\epsilon[0, \omega]$

(iv)a concave set for the non-membership function $\vartheta_{\tilde{A}^i}(x)$, i.e., $\vartheta_{\tilde{A}^i}(\lambda x_1 + (1-\lambda)x_2) \leqslant min(\vartheta_{\tilde{A}^i}(x_1), \vartheta_{\tilde{A}^i}(x_2))\forall x_1, x_2\epsilon R, \lambda\epsilon[\sigma, 1]$

(v) $\mu_{\tilde{A}^i}$ and $\vartheta_{\tilde{A}^i}$ is continuous mapping from R to the closed interval $[0, \omega]$ and $[\sigma, 1]$ respectively and $x_0\epsilon R$, the relation $0 \leqslant \mu_{\tilde{A}^i} + \vartheta_{\tilde{A}^i} \leqslant 1$ holds.

Definition 2.8: Generalized Trapezoidal Intuitionistic Fuzzy number: A GTrIFN $\tilde{A}^i$ is a subset of IFN in R with following membership function and non membership function as follows:

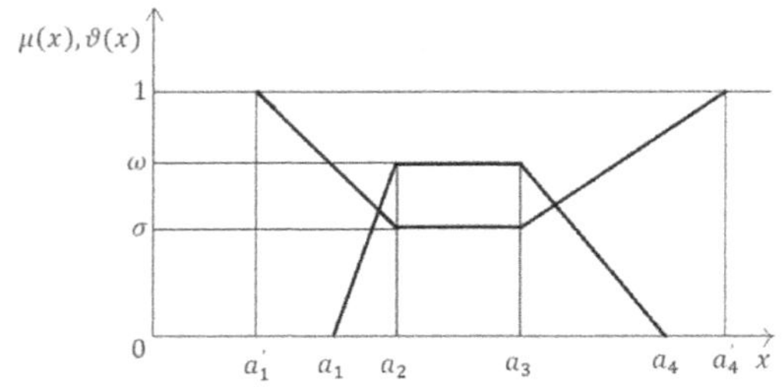

Figure 1. Generalized trapezoidal intutionistic fuzzy number

$$\mu_{\tilde{A}_i}(x) = \begin{cases} \omega\frac{x-a_1}{a_2-a_1} & \text{if } a_1 \leqslant x < a_2 \\ \omega & \text{if } a_2 \leqslant x \leqslant a_3 \\ \omega\frac{a_4-x}{a_3-a_3} & \text{if } a_3 < x \leqslant a_4 \\ 0 & \text{otherwise} \end{cases}$$

and

$$\vartheta_{\tilde{A}_i}(x) = \begin{cases} \sigma\frac{a_2-x}{a_2-a_1^{'}} & \text{if } a_1^{'} \leqslant x < a_2 \\ \sigma & \text{if } a_2 \leqslant x \leqslant a_3 \\ \sigma\frac{x-a_3}{a_4^{'}-a_3} & \text{if } a_3 < x \leqslant a_4^{'} \\ 0 & \text{otherwise} \end{cases}$$

where $a_1^{'} \leqslant a_2 \leqslant a_3 \leqslant a_4^{'}$, $a_1 \leqslant a_2 \leqslant a_3 \leqslant a_4$ and GTrIFN is denoted by $\tilde{A}_{GTrIFN} = ((a_1, a_2, a_3, a_4; \omega), (a_1^{'}, a_2, a_3, a_4^{'}; \sigma))$

Definition 2.9:Non negative GTrIFN: A GTrIFN $\widetilde{A}^i_{GTrIFN} = ((a_1, a_2, a_3, a_4; \omega), (a_1^{'}, a_2, a_3, a_4^{'}; \sigma))$ iff $a_1^{'} \geqslant 0$.

Definition 2.10:Equality of two GTrIFN: A GTrIFN $\widetilde{A}^i_{GTIFN} = ((a_1, a_2, a_3, a_4; \omega_1), (a_1^{'}, a_2, a_3, a_4^{'}; \sigma_1))$ and $\widetilde{B}^i_{GTIFN} = ((b_1, b_2, b_3, b_4; \omega_2), (b_1^{'}, b_2, b_4, b_4^{'}; \sigma_2))$ are said to be equal iff $a_1 = b_1$,$a_2 = b_2$,$a_3 = b_3$,$a_4 = b_4$,$a_1^{'} = b_1^{'}$,$a_4^{'} = b_4^{'}$,$\omega_1 = \omega_2$ and $\sigma_1 = \sigma_2$.

Definition 2.11: α-cut set: α-cut set of a GTrIFN $\widetilde{A}^i_{GTrIFN} = ((a_1, a_2, a_3, a_4; \omega), (a_1^{'}, a_2, a_3, a_4^{'}; \sigma))$ is a crisp subset of R which is defined as follows

$A_\alpha = \{x : \mu_{\tilde{A}_i}(x) \geqslant \alpha\} = [A_1(\alpha), A_2(\alpha)] = [a_1 + \frac{\alpha}{\omega}(a_2 - a_1), a_4 - \frac{\alpha}{\omega}(a_4 - a_3)]$

Definition 2.12: β-cut set: β-cut set of a GTrIFN $\widetilde{A}^i_{GTIFN} = ((a_1, a_2, a_3, a_4; \omega), (a_1^{'}, a_2, a_3, a_4^{'}; \sigma))$ is a crisp subset of R which is defined as follows

$A_\beta = \{x : \vartheta_{\tilde{A}_i}(x) \leqslant \beta\} = [A_1^{'}(\beta), A_2^{'}(\beta)] = [a_2 - \beta(a_2 - a_1^{'}), a_3 + \beta(a_4^{'} - a_3)]$

Definition 2.13: (α,β)-cut set: (α,β)-cut set of a GTrIFN $\widetilde{A}^i_{GTrIFN} = ((a_1,a_2,a_3,a_4;\omega),(a_1^{'},a_2,a_3,a_4^{'};\sigma))$ is a crisp subset of R which is defined as follows
$A_{\alpha,\beta} = \{[A_1(\alpha),A_2(\alpha)];[A_1^{'}(\beta),A_2^{'}(\beta)]\}, 0 < \alpha+\beta \leqslant \omega,\sigma, \alpha\epsilon[0,\omega], \beta\epsilon[\sigma,1]$

Definition 2.14: Addition of two GTrIFN: Let two $\widetilde{A}^i_{GTrIFN} = ((a_1,a_2,a_3,a_4;\omega_1),(a_1^{'},a_2,a_3,a_4^{'};\sigma_1))$ and $\widetilde{B}^i_{GTrIFN} = ((b_1,b_2,b_3,b_4;\omega_2),(b_1^{'},b_2,b_3,b_4^{'};\sigma_2))$ be GTrIFN, then the addition of two GTrIFN is given by

$\widetilde{A}^i_{GTrIFN} \bigoplus \widetilde{B}^i_{GTrIFN} = ((a_1+b_1,a_2+b_2,a_3+b_3,a_4+b_4;\omega),(a_1^{'}+b_1^{'},a_2+b_2,a_3+b_3,a_4^{'}+b_4^{'};\sigma))$
where $\omega = min\{\omega_1,\omega_2\}$ and $\sigma_1 = min\{\sigma_1,\sigma_2\}$.

Definition 2.15: Subtraction of two GTrIFN: Let two $\widetilde{A}^i_{GTrIFN} = ((a_1,a_2,a_3m,a_4;\omega_1),(a_1^{'},a_2,a_3,a_3^{'};\sigma_1))$ and
$\widetilde{B}^i_{GTrIFN} = ((b_1,b_2,b_3,b_4;\omega_2),(b_1^{'},b_2,b_3,b_4^{'};\sigma_2))$ be GTrIFN, then the subtraction of two GTrIFN is given by

$\widetilde{A}^i_{GTrIFN} \ominus \widetilde{B}^i_{GTrIFN} = ((a_1-b_4,a_2-b_3,a_3-b_2,a_4-b_1;\omega),(a_1^{'}-b_4^{'},a_2-b_3,a_3-b_2,a_4^{'}-b_1^{'};\sigma))$
where $\omega = \min\{\omega_1,\omega_2\}$ and $\sigma_1 = \min\{\sigma_1,\sigma_2\}$.

Definition 2.16: Multiplication by a scalar: Let $\widetilde{A}^i_{GTrIFN} = ((a_1,a_2,a_3;\omega),(a_1^{'},a_2,a_3^{'};\sigma))$ and k is a scalar then $k\widetilde{A}^i_{GTrIFN}$ is also a GTrIFN and is defined as

$$k\widetilde{A}^i_{GTrIFN} = \begin{cases} ((ka_1,ka_2,ka_3;\omega),(ka_1^{'},ka_2,ka_3,ka_4^{'};\sigma)) \text{ if } k > 0 \\ ((ka_3,ka_2,ka_1;\omega),(ka_4^{'},ka_3,ka_2,ka_1^{'};\sigma)) \text{ if } k < 0 \end{cases}$$

where $0 < \omega,\sigma \leqslant 1$.

Definition 2.17: Multiplication of two GTIFN: Let two $\widetilde{A}^i_{GTrIFN} = ((a_1,a_2,a_3,a_4;\omega_1),(a_1^{'},a_2,a_3,a_4^{'};\sigma_1))$ and $\widetilde{B}^i_{GTrIFN} = ((b_1,b_2,b_3,b_4;\omega_2),(b_1^{'},b_2,b_3,b_3^{'};\sigma_2))$ be GTrIFN, then the multiplication of two GTIFN is given by
$\widetilde{A}^i_{GTrIFN} \bigotimes \widetilde{B}^i_{GTrIFN} = ((a_1b_1,a_2b_2,a_3b_3,a_4b_4;\omega),(a_1^{'}b_1^{'},a_2b_2,a_3b_3,a_4^{'}b_4^{'};\sigma))$
where $\omega = \min\{\omega_1,\omega_2\}$ and $\sigma_1 = \min\{\sigma_1,\sigma_2\}$.

Definition 2.18: Division of two GTrIFN: Let two $\widetilde{A}^i_{GTrIFN} = ((a_1,a_2,a_3,a_4;\omega_1),(a_1^{'},a_2,a_3,a_4^{'};\sigma_1))$ and $\widetilde{B}^i_{GTrIFN} = ((b_1,b_2,b_3,b_4;\omega_2),(b_1^{'},b_2,b_3,b_4^{'};\sigma_2))$ be GTrIFN, then the multiplication of two GTIFN is given by
$\widetilde{A}^i_{GTrIFN}/\widetilde{B}^i_{GTrIFN} = ((\frac{a_1}{b_4},\frac{a_2}{b_3},\frac{a_3}{b_2},\frac{a_4}{b_1};\omega),(\frac{a_1^{'}}{b_4^{'}},\frac{a_2}{b_3},\frac{a_3}{b_2},\frac{a_4^{'}}{b_1^{'}};\sigma))$
where $\omega = \min\{\omega_1,\omega_2\}$ and $\sigma_1 = \min\{\sigma_1,\sigma_2\}$.

Theorem 2.1: [34] If ω and σ represent the maximum degree of membership and minimum degree of non-membership function respectively then they satisfy the conditions $0 \leqslant \omega \leqslant 1$, $0 \leqslant \sigma \leqslant 1$ and $0 < \omega+\sigma \leqslant 1$.

Definition 2.19: Generalized Hukuhara difference: [20] The generalized Hukuhara difference of two fuzzy number $u, v \in \Re_F$ is defines as follows

$u \ominus_{gH} v = w$ is equivalent to $\begin{cases} (i)u = v \oplus w \\ (ii)v = u \oplus (-1)w \end{cases}$

Consider $[w]_\alpha = [w_1(\alpha), w_2(\alpha)]$, then $w_1(\alpha) = min\{u_1(\alpha) - v_1(\alpha), u_2(\alpha) - v_2(\alpha)\}$ and $w_2(\alpha) = max\{u_1(\alpha) - v_1(\alpha), u_2(\alpha) - v_2(\alpha)\}$

Here the parametric representation of a fuzzy valued function $f : [a, b] \to \Re_F$ is expressed by $[f(t)]_\alpha = [f_1(t, \alpha), f_2(t, \alpha)]$, $t \in [a, b], \alpha \in [0, 1]$.

Definition 2.20: Generalized Hukuhara derivative for first order: [20] The generalized Hukuhara derivative of a fuzzy valued function $f : (a, b) \to \Re_F$ at t_0 is defined as $f'(t_0) = lim_{h \to 0} \frac{f(t_0+h) \ominus_{gH} f(t_0)}{h}$

If $f'(t_0) \in \Re_F$ satisfying (2.1) exists, we say that f is generalized Hukuhara differentiable at t_0.

Also we say that $f(t)$ is (i)-gH differentiable at t_0 if

$$[f'(t_0)]_\alpha = [f_1'(t_0, \alpha), f_2'(t_0, \alpha)]$$

and $f(t)$ is (ii)-gH differentiable at t_0 if

$$[f'(t_0)]_\alpha = [f_2'(t_0, \alpha), f_1'(t_0, \alpha)]$$

Definition 2.21: Generalized Hukuhara derivative for second order: [23] The second order generalized Hukuhara derivative of a fuzzy valued function $f : (a, b) \to \Re_F$ at t_0 is defined as

$$f''(t_0) = lim_{h \to 0} \frac{f'(t_0+h) \ominus_{gH} f'(t_0)}{h}$$

If $f'' \in \Re_F$, we say that $f'(t_0)$ is generalized Hukuhara at t_0.

Also we say that $f'(t_0)$ is (i)-gH differentiable at t_0 if

$$f''(t_0, \alpha) = \begin{cases} (f_1'(t_0, \alpha), f_2'(t_0, \alpha)) \text{ if f be (i)-gH differentiable on (a,b)} \\ (f_2'(t_0, \alpha), f_1'(t_0, \alpha)) \text{ if f be (ii)-gH differentiable on (a,b)} \end{cases}$$

for all $\alpha \in [0, 1]$, and that $f'(t_0)$ is (ii)-gH differentiable at t_0 if

$$f''(t_0, \alpha) = \begin{cases} (f_2'(t_0, \alpha), f_1'(t_0, \alpha)) \text{ if f be (i)-gH differentiable on (a,b)} \\ (f_1'(t_0, \alpha), f_2'(t_0, \alpha)) \text{ if f be (ii)-gH differentiable on (a,b)} \end{cases}$$

for all $\alpha \in [0, 1]$.

Definition 2.22: Strong and week solution: If the solution of intutionistic fuzzy differential equation is of the form $[x_1(t, \alpha), x_2(t, \alpha); x_1'(t, \beta), x_2'(t, \beta)]$, the solution is called strong solution when

(i) $\frac{dx_1(t,\alpha)}{d\alpha} > 0$, $\frac{dx_2(t,\alpha)}{d\alpha} < 0 \; \forall \alpha \in [0, \omega], x_1(t, \omega) \leqslant x_2(t, \omega)$

(ii) $\frac{dx_1'(t,\beta)}{d\beta} < 0$, $\frac{dx_2(t,\beta)}{d\beta} > 0 \; \forall \beta \in [\sigma, 1], x_1'(t, \sigma) \leqslant x_2'(t, \sigma)$

Definition 2.23: (α, β)-cuts to Intutionistic fuzzy number: Let $[A_1(\alpha), A_2(\alpha); A_1'(\beta), A_2'(\beta)]$ be the (α, β)-cuts of a trapezoidal intutionistic fuzzy number $\widetilde{A}$ and ω, σ be the gradation of membership and non membership function respectively then the intutionistic fuzzy number is given by

$\widetilde{A} = ((A_1(\alpha = 0), A_1(\alpha = \omega), A_2(\alpha = \omega), A_2(\alpha = 0); \omega); (A_1^{'}(\beta = \sigma), A_1^{'}(\beta = 0), A_2^{'}(\beta = 0), A_2^{'}(\beta = \sigma); \sigma))$

3. Second order intutionistic fuzzy boundary value problem

Consider the differential equation $\frac{d^2x(t)}{dt^2} = kx(t)$, with boundary condition $x(0) = \widetilde{a}$ and $x(L) = \widetilde{b}$. Where $\widetilde{a}, \widetilde{b}$ are generalized trapezoidal intutionistic fuzzy number.

Let $\widetilde{a} = ((a_1, a_2, a_3, a_4; \omega_1), (a_1^{'}, a_2, a_3, a_4^{'}; \sigma_1))$ and $\widetilde{b} = ((b_1, b_2, b_3, b_4; \omega_2), (b_1^{'}, b_2, b_3, b_4^{'}; \sigma_2))$

3.1 *Second order intutionistic fuzzy boundary value problem with coefficients is positive constant i.e., k > 0*

Here two cases arise

Case 3.1.1: When $x(t)$ is (i)-gH differentiable and $\frac{dx(t)}{dt}$ is (i)-gH then we have

$\frac{d^2x_1(t,\alpha)}{dt^2} = kx_1(t, \alpha)$

$\frac{d^2x_2(t,\alpha)}{dt^2} = kx_2(t, \alpha)$

$\frac{d^2x_1^{'}(t,\beta)}{dt^2} = kx_1^{'}(t, \beta)$

$\frac{d^2x_2^{'}(t,\beta)}{dt^2} = kx_2^{'}(t, \beta)$

With boundary conditions
$x_1(0, \alpha) = a_1 + \frac{\alpha l_{\widetilde{a}}}{\omega}$, $x_2(0, \alpha) = a_4 - \frac{\alpha r_{\widetilde{a}}}{\omega}$, $x_1^{'}(0, \beta) = a_2 - \frac{\beta l_{\widetilde{a}}^{'}}{\sigma}$, $x_2^{'}(0, \beta) = a_3 + \frac{\beta r_{\widetilde{a}}^{'}}{\sigma}$

and $x_1(L, \alpha) = b_1 + \frac{\alpha l_{\widetilde{b}}}{\omega}$, $x_2(L, \alpha) = b_4 - \frac{\alpha r_{\widetilde{b}}}{\omega}$, $x_1^{'}(L, \beta) = b_2 - \frac{\beta l_{\widetilde{b}}^{'}}{\sigma}$, $x_2^{'}(L, \beta) = b_3 + \frac{\beta r_{\widetilde{b}}^{'}}{\sigma}$

Where, $l_{\widetilde{a}} = a_2 - a_1$,$r_{\widetilde{a}} = a_4 - a_3$, $l_{\widetilde{a}}^{'} = a_2 - a_1^{'}$, $r_{\widetilde{a}}^{'} = a_4^{'} - a_3$
and $l_{\widetilde{b}} = b_2 - b_1$,$r_{\widetilde{b}} = b_4 - b_3$, $l_{\widetilde{b}}^{'} = b_2 - b_1^{'}$, $r_{\widetilde{b}}^{'} = b_4^{'} - b_3$
and $\omega = min\{\omega_1, \omega_2\}$, $\sigma = min\{\sigma_1, \sigma_2\}$.

Solution: The general solution of first equation is
$x_1(t, \alpha) = c_1 e^{\sqrt{k}t} + c_2 e^{-\sqrt{k}t}$

Using initial condition we have

$c_1 + c_2 = a_1 + \frac{\alpha l_{\widetilde{a}}}{\omega}$ and $c_1 e^{\sqrt{k}L} + c_2 e^{-\sqrt{k}L} = b_1 + \frac{\alpha l_{\widetilde{b}}}{\omega}$

Solving we get

$c_1 = \frac{1}{e^{\sqrt{k}L} - e^{-\sqrt{k}L}}\{(b_1 + \frac{\alpha l_{\widetilde{b}}}{\omega}) - (a_1 + \frac{\alpha l_{\widetilde{a}}}{\omega})e^{-\sqrt{k}L}\}$

and

$c_2 = -\frac{1}{e^{\sqrt{k}L}-e^{-\sqrt{k}L}}\{(b_1 + \frac{\alpha l_{\tilde{b}}}{\omega}) - (a_1 + \frac{\alpha l_{\tilde{a}}}{\omega})e^{\sqrt{k}L}\}$

Therefore the solution is

$x_1(t,\alpha) = \frac{1}{e^{\sqrt{k}L}-e^{-\sqrt{k}L}}[\{(b_1 + \frac{\alpha l_{\tilde{b}}}{\omega}) - (a_1 + \frac{\alpha l_{\tilde{a}}}{\omega})e^{-\sqrt{k}L}\}e^{\sqrt{k}t} - \{(b_1 + \frac{\alpha l_{\tilde{b}}}{\omega}) - (a_1 + \frac{\alpha l_{\tilde{a}}}{\omega})e^{\sqrt{k}L}\}e^{-\sqrt{k}t}]$

Similarly

$x_2(t,\alpha) = \frac{1}{e^{\sqrt{k}L}-e^{-\sqrt{k}L}}[\{(b_4 - \frac{\alpha r_{\tilde{b}}}{\omega}) - (a_4 - \frac{\alpha r_{\tilde{a}}}{\omega})e^{-\sqrt{k}L}\}e^{\sqrt{k}t} - \{(b_4 - \frac{\alpha r_{\tilde{b}}}{\omega}) - (a_4 - \frac{\alpha r_{\tilde{a}}}{\omega})e^{\sqrt{k}L}\}e^{-\sqrt{k}t}]$

$x_1'(t,\beta) = \frac{1}{e^{\sqrt{k}L}-e^{-\sqrt{k}L}}[\{(b_2 - \frac{\beta l_{\tilde{b}}'}{\sigma}) - (a_2 - \frac{\beta l_{\tilde{a}}'}{\sigma})e^{-\sqrt{k}L}\}e^{\sqrt{k}t} - \{(b_2 - \frac{\beta l_{\tilde{b}}'}{\sigma}) - (a_2 - \frac{\beta l_{\tilde{a}}'}{\sigma})e^{\sqrt{k}L}\}e^{-\sqrt{k}t}]$

$x_2'(t,\beta) = \frac{1}{e^{\sqrt{k}L}-e^{-\sqrt{k}L}}[\{(b_3 + \frac{\beta l_{\tilde{b}}'}{\sigma}) - (a_3 + \frac{\beta l_{\tilde{a}}'}{\sigma})e^{-\sqrt{k}L}\}e^{\sqrt{k}t} - \{(b_3 + \frac{\beta l_{\tilde{b}}'}{\sigma}) - (a_3 + \frac{\beta l_{\tilde{a}}'}{\sigma})e^{\sqrt{k}L}\}e^{-\sqrt{k}t}]$

Case 3.1.2: When $x(t)$ is (ii)-gH differentiable and $\frac{dx(t)}{dt}$ is (i)-gH then we have

$\frac{d^2x_2(t,\alpha)}{dt^2} = kx_1(t,\alpha)$

$\frac{d^2x_1(t,\alpha)}{dt^2} = kx_2(t,\alpha)$

$\frac{d^2x_2'(t,\beta)}{dt^2} = kx_1'(t,\beta)$

$\frac{d^2x_1'(t,\beta)}{dt^2} = kx_2'(t,\beta)$

With same boundary conditions.

Solution: The general solution is given by

$x_1(t,\alpha) = c_1e^{\sqrt{k}t} + c_2e^{-\sqrt{k}t} + c_3\cos\sqrt{k}t + c_4\sin\sqrt{k}t$

$x_2(t,\alpha) = c_1e^{\sqrt{k}t} + c_2e^{-\sqrt{k}t} - c_3\cos\sqrt{k}t - c_4\sin\sqrt{k}t$

$x_1'(t,\beta) = d_1e^{\sqrt{k}t} + d_2e^{-\sqrt{k}t} + d_3\cos\sqrt{k}t + d_4\sin\sqrt{k}t$

$x_2'(t,\beta) = d_1e^{\sqrt{k}t} + d_2e^{-\sqrt{k}t} - d_3\cos\sqrt{k}t - d_4\sin\sqrt{k}t$

Where, $c_1 = \frac{1}{e^{\sqrt{k}L}-e^{-\sqrt{k}L}}[\{\frac{b_1+b_4}{2} + \frac{\alpha(l_{\tilde{b}}-r_{\tilde{b}})}{2\omega}\} - \{\frac{a_1+a_4}{2} + \frac{\alpha(l_{\tilde{a}}-r_{\tilde{a}})}{2\omega}\}e^{-\sqrt{k}L}]$

$c_2 = -\frac{1}{e^{\sqrt{k}L}-e^{-\sqrt{k}L}}[\{\frac{b_1+b_4}{2} + \frac{\alpha(l_{\tilde{b}}-r_{\tilde{b}})}{2\omega}\} - \{\frac{a_1+a_4}{2} + \frac{\alpha(l_{\tilde{a}}-r_{\tilde{a}})}{2\omega}\}e^{\sqrt{k}L}]$

$c_3 = \frac{a_1-a_4}{2} + \frac{\alpha(l_{\tilde{b}}+r_{\tilde{b}})}{2\omega}$

$c_4 = \frac{1}{\sin\sqrt{k}L}[\{\frac{b_1+b_4}{2} + \frac{\alpha(l_{\tilde{b}}-r_{\tilde{b}})}{2\omega}\} - \{\frac{a_1+a_4}{2} + \frac{\alpha(l_{\tilde{a}}-r_{\tilde{a}})}{2\omega}\}\cos\sqrt{k}L]$

$d_1 = \frac{1}{e^{\sqrt{k}L}-e^{-\sqrt{k}L}}[\{\frac{b_2+b_3}{2} - \frac{\beta(l'_{\tilde{b}}-r'_{\tilde{b}})}{2\sigma}\} - \{\frac{a_2+a_3}{2} - \frac{\beta(l'_{\tilde{a}}-r'_{\tilde{a}})}{2\sigma}\}e^{-\sqrt{k}L}]$

$d_2 = -\frac{1}{e^{\sqrt{k}L}-e^{-\sqrt{k}L}}[\{\frac{b_2+b_3}{2} - \frac{\beta(l'_{\tilde{b}}-r'_{\tilde{b}})}{2\sigma}\} - \{\frac{a_2+a_3}{2} - \frac{\beta(l'_{\tilde{a}}-r'_{\tilde{a}})}{2\sigma}\}e^{\sqrt{k}L}]$

$d_3 = \frac{a_2-a_3}{2} - \frac{\beta(l'_{\tilde{a}}+r'_{\tilde{a}})}{2\sigma}$

$d_4 = \frac{1}{\sin\sqrt{k}L}[\{\frac{b_2-b_3}{2} - \frac{\beta(l'_{\tilde{b}}-r'_{\tilde{b}})}{2\sigma}\} - \{\frac{a_2-a_3}{2} - \frac{\beta(l'_{\tilde{a}}+r'_{\tilde{a}})}{2\sigma}\}\cos\sqrt{k}L]$

Case 3.1.3: When $x(t)$ is (i)-gH differentiable and $\frac{dx(t)}{dt}$ is (ii)-gH then we have

$\frac{d^2x_2(t,\alpha)}{dt^2} = kx_1(t,\alpha)$

$\frac{d^2x_1(t,\alpha)}{dt^2} = kx_2(t,\alpha)$

$\frac{d^2x'_2(t,\beta)}{dt^2} = kx'_1(t,\beta)$

$\frac{d^2x'_1(t,\beta)}{dt^2} = kx'_2(t,\beta)$

With same boundary conditions.

Solution: The result is same as Case 3.1.2.

Case 3.1.4: When $x(t)$ is (ii)-gH differentiable and $\frac{dx(t)}{dt}$ is (ii)-gH then we have

$\frac{d^2x_1(t,\alpha)}{dt^2} = kx_1(t,\alpha)$

$\frac{d^2x_2(t,\alpha)}{dt^2} = kx_2(t,\alpha)$

$\frac{d^2x'_1(t,\beta)}{dt^2} = kx'_1(t,\beta)$

$\frac{d^2x'_2(t,\beta)}{dt^2} = kx'_2(t,\beta)$

With same boundary conditions.

Solution: The result is same as Case 3.1.1.

3.2 *Second order intutionistic fuzzy boundary value problem with coefficients is negative constant i.e., $k < 0$, Consider $k = -m$, $m > 0$:*

In this problem four cases arise

Case 3.2.1 When $x(t)$ is (i)-gH differentiable and $\frac{dx(t)}{dt}$ is (i)-gH then we have

$\frac{d^2x_1(t,\alpha)}{dt^2} = -mx_2(t,\alpha)$

$\frac{d^2x_2(t,\alpha)}{dt^2} = -mx_1(t,\alpha)$

$\frac{d^2x_1'(t,\beta)}{dt^2} = -mx_2'(t,\beta)$

$\frac{d^2x_2'(t,\beta)}{dt^2} = -mx_1'(t,\beta)$

With boundary conditions

$x_1(0,\alpha) = a_1 + \frac{\alpha l_{\tilde{a}}}{\omega}$, $x_2(0,\alpha) = a_4 - \frac{\alpha r_{\tilde{a}}}{\omega}$, $x_1'(0,\beta) = a_2 - \frac{\beta l_{\tilde{a}}'}{\sigma}$, $x_2'(0,\beta) = a_3 + \frac{\beta r_{\tilde{a}}'}{\sigma}$

and $x_1(L,\alpha) = b_1 + \frac{\alpha l_{\tilde{b}}}{\omega}$, $x_2(L,\alpha) = b_4 - \frac{\alpha r_{\tilde{b}}}{\omega}$, $x_1'(L,\beta) = b_2 - \frac{\beta l_{\tilde{b}}'}{\sigma}$, $x_2'(L,\beta) = b_3 + \frac{\beta r_{\tilde{b}}'}{\sigma}$

Where, $l_{\tilde{a}} = a_2 - a_1$,$r_{\tilde{a}} = a_4 - a_3$, $l_{\tilde{a}}' = a_2 - a_1'$, $r_{\tilde{a}}' = a_4' - a_3$
and $l_{\tilde{b}} = b_2 - b_1$,$r_{\tilde{b}} = b_4 - b_3$, $l_{\tilde{b}}' = b_2 - b_1'$, $r_{\tilde{b}}' = b_4' - b_3$
and $\omega = min\{\omega_1, \omega_2\}$, $\sigma = min\{\sigma_1, \sigma_2\}$.

Solution: The general solution of this equation may be written as
$x_1(t,\alpha) = c_1e^{\sqrt{m}t} + c_2e^{-\sqrt{m}t} + c_3\cos\sqrt{m}t + c_4\sin\sqrt{m}t$

$x_2(t,\alpha) = -c_1e^{\sqrt{m}t} - c_2e^{-\sqrt{m}t} + c_3\cos\sqrt{m}t + c_4\sin\sqrt{m}t$

$x_1'(t,\beta) = d_1e^{\sqrt{m}t} + d_2e^{-\sqrt{m}t} + d_3\cos\sqrt{m}t + d_4\sin\sqrt{m}t$

$x_2'(t,\beta) = -d_1e^{\sqrt{m}t} - d_2e^{-\sqrt{m}t} + d_3\cos\sqrt{m}t + d_4\sin\sqrt{m}t$

Where, $c_1 = \frac{1}{e^{\sqrt{m}L} - e^{-\sqrt{m}L}}[\{\frac{b_1-b_2}{2} + \frac{\alpha(l_{\tilde{b}}+r_{\tilde{b}})}{2\omega}\} - \{\frac{a_1-a_4}{2} + \frac{\alpha(l_{\tilde{a}}+r_{\tilde{a}})}{2\omega}\}e^{-\sqrt{m}L}]$

$c_1 = -\frac{1}{e^{\sqrt{m}L} - e^{-\sqrt{m}L}}[\{\frac{b_1-b_4}{2} + \frac{\alpha(l_{\tilde{b}}+r_{\tilde{b}})}{2\omega}\} - \{\frac{a_1-a_4}{2} + \frac{\alpha(l_{\tilde{a}}+r_{\tilde{a}})}{2\omega}\}e^{\sqrt{m}L}]$

$c_3 = \frac{a_1+a_4}{2} + \frac{\alpha(l_{\tilde{b}}-r_{\tilde{b}})}{2\omega}$

$c_4 = \frac{1}{\sin\sqrt{m}L}[\{\frac{b_1+b_4}{2} + \frac{\alpha(l_{\tilde{b}}-r_{\tilde{b}})}{2\omega}\} - \{\frac{a_1+a_4}{2} + \frac{\alpha(l_{\tilde{a}}-r_{\tilde{a}})}{2\omega}\}\cos\sqrt{m}L]$

$d_1 = \frac{1}{e^{\sqrt{m}L} - e^{-\sqrt{m}L}}[\{\frac{b_2-b_3}{2} - \frac{\beta(l_{\tilde{b}}'+r_{\tilde{b}}')}{2\sigma}\} - \{\frac{a_2-a_3}{2} - \frac{\beta(l_{\tilde{a}}'+r_{\tilde{a}}')}{2\sigma}\}e^{-\sqrt{m}L}]$

$d_2 = -\frac{1}{e^{\sqrt{m}L} - e^{-\sqrt{m}L}}[\{\frac{b_2-b_3}{2} - \frac{\beta(l_{\tilde{b}}'+r_{\tilde{b}}')}{2\sigma}\} - \{\frac{a_2-a_3}{2} - \frac{\beta(l_{\tilde{a}}'+r_{\tilde{a}}')}{2\sigma}\}e^{\sqrt{m}L}]$

$d_3 = \frac{a_2+a_3}{2} - \frac{\beta(l_{\tilde{a}}'-r_{\tilde{a}}')}{2\sigma}$

$d_4 = \frac{1}{\sin\sqrt{m}L}[\{\frac{b_2+b_3}{2} - \frac{\beta(l_{\tilde{b}}'-r_{\tilde{b}}')}{2\sigma}\} - \{\frac{a_2+a_3}{2} - \frac{\beta(l_{\tilde{a}}'-r_{\tilde{a}}')}{2\sigma}\}\cos\sqrt{m}L]$

Case 3.2.2: When $x(t)$ is (ii)-gH differentiable and $\frac{dx(t)}{dt}$ is (i)-gH then we have

$\frac{d^2x_2(t,\alpha)}{dt^2} = -mx_2(t,\alpha)$

$\frac{d^2x_1(t,\alpha)}{dt^2} = -mx_1(t,\alpha)$

$\frac{d^2x_2^{'}(t,\beta)}{dt^2} = -mx_2^{'}(t,\beta)$

$\frac{d^2x_1^{'}(t,\beta)}{dt^2} = -mx_1^{'}(t,\beta)$

With same boundary conditions.

Solution: The solution is written as

$$x_1(t,\alpha) = (a_1 + \tfrac{\alpha l_{\tilde{a}}}{\omega})\cos\sqrt{m}t + \tfrac{1}{\sin\sqrt{m}L}\{(b_1 + \tfrac{\alpha l_{\tilde{b}}}{\omega}) - (a_1 + \tfrac{\alpha l_{\tilde{a}}}{\omega})\cos\sqrt{m}L\}\sin\sqrt{m}t$$

$$x_2(t,\alpha) = (a_4 - \tfrac{\alpha r_{\tilde{a}}}{\omega})\cos\sqrt{m}t + \tfrac{1}{\sin\sqrt{m}L}\{(b_4 - \tfrac{\alpha r_{\tilde{b}}}{\omega}) - (a_4 - \tfrac{\alpha r_{\tilde{a}}}{\omega})\cos\sqrt{m}L\}\sin\sqrt{m}t$$

$$x_1^{'}(t,\beta) = (a_3 + \tfrac{\beta r_{\tilde{a}}^{'}}{\sigma})\cos\sqrt{m}t + \tfrac{1}{\sin\sqrt{m}L}\{(b_2 - \tfrac{\beta l_{\tilde{b}}^{'}}{\sigma}) - (a_3 + \tfrac{\beta r_{\tilde{a}}^{'}}{\sigma})\cos\sqrt{m}L\}\sin\sqrt{m}t$$

$$x_2^{'}(t,\beta) = (a_2 - \tfrac{\beta l_{\tilde{a}}^{'}}{\sigma})\cos\sqrt{m}t + \tfrac{1}{\sin\sqrt{m}L}\{(b_3 + \tfrac{\beta r_{\tilde{b}}^{'}}{\sigma}) - (a_2 - \tfrac{\beta l_{\tilde{a}}^{'}}{\sigma})\cos\sqrt{m}L\}\sin\sqrt{m}t$$

Case 3.2.3: When $x(t)$ is (i)-gH differentiable and $\frac{dx(t)}{dt}$ is (ii)-gH then we have

$\frac{d^2x_2(t,\alpha)}{dt^2} = -mx_2(t,\alpha)$

$\frac{d^2x_1(t,\alpha)}{dt^2} = -mx_1(t,\alpha)$

$\frac{d^2x_2^{'}(t,\beta)}{dt^2} = -mx_2^{'}(t,\beta)$

$\frac{d^2x_1^{'}(t,\beta)}{dt^2} = -mx_1^{'}(t,\beta)$

With same boundary conditions.

Solution: The result is same as Case 3.2.2.

Case 3.2.4: When $x(t)$ is (ii)-gH differentiable and $\frac{dx(t)}{dt}$ is (ii)-gH differentiable then we have

$\frac{d^2x_1(t,\alpha)}{dt^2} = -mx_2(t,\alpha)$

$\frac{d^2x_2(t,\alpha)}{dt^2} = -mx_1(t,\alpha)$

$\frac{d^2x_1^{'}(t,\beta)}{dt^2} = -mx_2^{'}(t,\beta)$

$\frac{d^2x_2^{'}(t,\beta)}{dt^2} = -mx_1^{'}(t,\beta)$

With boundary conditions

Solution: The result is same as Case 3.2.1.

Table 1. Solutions for t=1

α	$x_{(t,\alpha)}$	$x_2(t,\alpha)$	β	$x_1^{'}(t,\beta)$	$x_2^{'}(t,\beta)$
0	1.9271	2.7245	0.2	1.7942	2.8574
0.1	1.9461	2.7055	0.3	1.6613	2.9903
0.2	1.9650	2.6865	0.4	1.5284	3.1232
0.3	1.9840	2.6675	0.5	1.3955	3.2561
0.4	2.0030	2.6485	0.6	1.2626	3.3890
0.5	2.0220	2.6295	0.7	1.1297	3.5219
0.6	2.0410	2.6106	0.8	0.9968	3.6548
0.7	2.0600	2.5916	0.9	0.8639	3.7877
			1	0.7310	3.9206

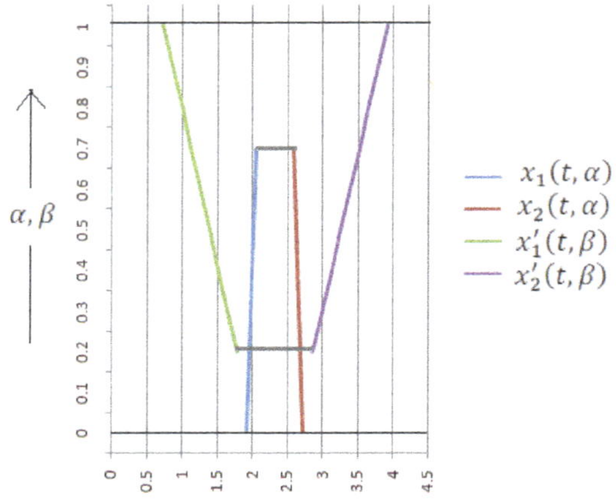

Figure 2. Graph of solutions for t=1

4. Numerical example:

Consider the differential equation $\frac{d^2x(t)}{dt^2} = 4x(t)$, with boundary conditions $x(0) = ((4.5, 5, 7, 7.5; 0.7), (4, 5, 7, 8; 0.2))$

and $x(2) = ((10, 10.5, 12.5, 13; 0.7), (9.5, 10.5, 12.5, 13.5; 0.2))$

Find the solution when $x(t)$ and $\frac{dx(t)}{dt}$ is (i)-gH differentiable.

Solution: The solution is given by

$$x_1(t,\alpha)=\frac{1}{e^4-e^{-4}}[\{(10+\tfrac{5\alpha}{7})-(4.5+\tfrac{5\alpha}{7})e^{-4}\}e^{2t}-\{(10+\tfrac{5\alpha}{7})-(4.5+\tfrac{5\alpha}{7})e^{4}\}e^{-2t}]$$

$$x_2(t,\alpha)=\frac{1}{e^4-e^{-4}}[\{(13-\tfrac{5\alpha}{7})-(7.5-\tfrac{5\alpha}{7})e^{-4}\}e^{2t}-\{(13-\tfrac{5\alpha}{7})-(7.5-\tfrac{5\alpha}{7})e^{4}\}e^{-2t}]$$

$$x_1^{'}(t,\beta)=\frac{1}{e^4-e^{-4}}[\{(10.5-5\beta)-(5-5\beta)e^{-4}\}e^{2t}-\{(10.5-5\beta)-(5-5\beta)e^{4}\}e^{-2t}]$$

$$x_2^{'}(t,\beta)=\frac{1}{e^4-e^{-4}}[\{(12.5+5\beta)-(7+5\beta)e^{-4}\}e^{2t}-\{(12.5+5\beta)-(7+5\beta)e^{4}\}e^{-2t}]$$

Remarks: From the graph and table we conclude that $x_1(t,\alpha)$ is increasing and $x_2(t,\alpha)$ is decreasing whereas $x_1^{'}(t,\beta)$ is decreasing and $x_2^{'}(t,\beta)$ is increasing function for $t = 1$. Thus in this case the solution is a strong solution.

5. Conclusion:

In this paper second order intutionistic fuzzy boundary value problem is solved. The intutionistic fuzzy number is taken as generalized trapezoidal intuitionistic fuzzy number. The coefficient is positive crisp number and negative crisp number are taken in this paper and solved the problem analytically. Finally a numerical example is illustrated. In future we can apply this procedure in n-th order intutionistic fuzzy boundary value problem and use the results in different applications in the field of sciences and engineering.

Acknowledgement

We are grateful to the reviewers for their careful reading, valuable comments and helpful suggestions which enable to improve the presentation of this paper significantly.

References

[1] L. A. Zadeh, Fuzzy sets, Information and Control, 8, (1965) 338-353.
[2] D.Dubois, H.Parade, Operation on Fuzzy Number, International Journal of Fuzzy system, 9, (1978) 613-626.
[3] K. T. Atanassov, Intuitionistic fuzzy sets, VII ITKRs Session, Sofia, Bulgarian, 1983.
[4] K. T. Atanassov, Intuitionistic fuzzy sets, Fuzzy Sets and Systems, Vol. 20, (1986) 8796.
[5] L. A. Zadeh, Toward a generalized theory of uncertainty (GTU) an outline, Information Sciences 172, (2005) 140.
[6] S. L. Chang, L. A. Zadeh, On fuzzy mapping and control, IEEE Transaction on Systems Man Cybernetics 2, (1972) 3034.
[7] D. Dubois, H. Prade, Towards fuzzy differential calculus: Part 3, Differentiation, Fuzzy Sets and Systems 8, (1982) 225233.
[8] M. L. Puri, D. A. Ralescu, Differentials of fuzzy functions, Journal of Mathematical Analysis and Application 91, (1983) 552558.
[9] R. Goetschel, W. Voxman, Elementary fuzzy calculus, Fuzzy Sets and Systems 18, (1986) 3143.
[10] O. Kaleva, Fuzzy differential equations, Fuzzy Sets and Systems 24, (1987) 301317.
[11] B. Bede, A note on two-point boundary value problems associated with non-linear fuzzy differential equations, Fuzzy Sets. Syst.157, (2006) 986989.
[12] B. Bede,S. G. Gal, Generalizations of the differentiability of fuzzy-number-valued functions with applications to fuzzy differential equations, Fuzzy Sets Syst.151, (2005) 581599.
[13] Y. Chalco-Cano, H. Romn-Flores, On the new solution of fuzzy differential equations, Chaos Solitons Fractals 38, (2008) 112119.
[14] B. Bede,I. J. Rudas and A. L. Bencsik, First order linear fuzzy differential equations under generalized differentiability, Inf. Sci. 177, (2007) 16481662.
[15] Y. Chalco-Cano, M.A.Rojas-Medar,H.Romn-Flores, Sobre ecuaciones diferencial esdifusas, Bol. Soc. Esp. Mat. Apl. 41, (2007) 9199.
[16] Y. Chalco-Cano, H. Romn-Flores and M. A. Rojas-Medar, Fuzzy differential equations with generalized derivative, in:Proceedings of the 27th North American Fuzzy Information Processing Society International Conference, IEEE, 2008.
[17] L. Stefanini, B. Bede, Generalized Hukuhara differentiability of interval-valued functions and interval differential equations, Nonlinear Anal. 71, (2009) 13111328.
[18] A. Khastan,J. J. Nieto, A boundary value problem for second-order fuzzy differential equations, Nonlinear Anal. 72, (2010) 35833593.
[19] P. Diamond, P. Kloeden, Metric Spaces of Fuzzy Sets, World Scientific, Singapore, 1994.
[20] B. Bede, S. G. Gal, Generalizations of the differentiability of fuzzy-number-valued functions with applications to fuzzy differential equations, Fuzzy Set Systems, 151 (2005) 581-599.
[21] L. Stefanini, A generalization of Hukuhara difference for interval and fuzzy arithmetic, in: D. Dubois, M.A. Lubiano, H. Prade, M. A. Gil, P. Grzegorzewski, O. Hryniewicz (Eds.), Soft Methods for Handling Variability and Imprecision, in: Series on Advances in Soft Computing, 48 (2008).
[22] L. Stefanini, B. Bede, Generalized Hukuhara differentiability of interval-valued functions and interval differential equations, Nonlinear Analysis, 71 (2009) 1311-1328.
[23] A. Armand, Z. Gouyandeh, Solving two-point fuzzy boundary value problem using the variational iteration method, Communications on Advanced Computational Science with Applications, Vol. 2013, (2013) 1-10.
[24] N. Gasilov, S. E. Amrahov, A. G. Fatullayev, Solution of linear differential equations with fuzzy boundary values, Fuzzy Sets and Systems 257, (2014) 169-183.

[25] N. Gasilov, S. E. Amrahov, A. G. Fatullayev, A. Khastan, A new approach to fuzzy initial value problem ,18 (2), (2014) 217-225.
[26] B. Bede, L. Stefanini, Generalized differentiability of fuzzy-valued functions, Fuzzy Sets and Systems, 230, (2013) 119-141.
[27] R. Rodrguez-Lpez, On the existence of solutions to periodic boundary value problems for fuzzy linear differential equations, Fuzzy Sets and Systems, 219, (2013) 1-26.
[28] S. Melliani, L. S. Chadli, Introduction to intuitionistic fuzzy partial differential Equations, Fifth Int. Conf. on IFSs, Sofia, 22-23 Sept. 2001.
[29] S. Abbasbandy, T. Allahviranloo, Numerical Solution of Fuzzy Differential Equations by Runge-Kutta and the Intuitionistic Treatment, Journal of Notes on Intuitionistic Fuzzy Sets, Vol. 8, No. 3, (2002) 43-53.
[30] S. Lata, A.Kumar, A new method to solve time-dependent intuitionistic fuzzy differential equation and its application to analyze the intutionistic fuzzy reliability of industrial system, Concurrent Engineering: Research and Applications, (2012) 1-8.
[31] S. P. Mondal and T. K. Roy, First order homogeneous ordinary differential equation with initial value as triangular intuitionistic fuzzy number, Journal of Uncertainty in Mathematics Science (2014) 1-17.
[32] S. P. Mondal. and T. K. Roy, System of Differential Equation with Initial Value as Triangular Intuitionistic Fuzzy Number and its Application, Int. J. Appl. Comput. Math, (2010).
[33] L. C. Barros, L. T. Gomes, P. A Tonelli, Fuzzy differential equations: An approach via fuzzification of the derivative operator, Fuzzy Sets and Systems, 230, (2013) 39-52.
[34] M.R.Seikh, P.K.Nayak and M.Pal, Generalized Triangular Fuzzy Numbers In Intuitionistic Fuzzy Environment, International Journal of Engineering Research and Development, Volume 5, Issue 1 (2012) 08-13.
[35] H.J.Zimmerman, Fuzzy set theory and its applications, Kluwer Academi Publishers, Dordrecht (1991).

15

Existence and multiplicity of positive solutions for a class of semilinear elliptic system with nonlinear boundary conditions

F. M. Yaghoobi[a*], J. Shamshiri[b]

[a] *Department of Mathemetics, College of Science, Hamedan Branch, Islamic Azad University, Hamedan, Iran.*

[b] *Department of Mathematics, Mashhad Branch, Islamic Azad University, Mashhad, Iran.*

Abstract. This study concerns the existence and multiplicity of positive weak solutions for a class of semilinear elliptic systems with nonlinear boundary conditions. Our results is depending on the local minimization method on the Nehari manifold and some variational techniques. Also by using Mountain Pass Lemma, we establish the existence of at least one solution with positive energy.

Keywords: Critical point, semilinear elliptic system, nonlinear boundary value problem, Fibering map, Nehari manifold.

1. Introduction

The class of elliptic partial differential problems involving the nonlinear boundary conditions arise from many branches of science, for example in mechanics, geometry and other sciences. The attention of many authors have been attracted toward the existence and multiplicity of solutions for semilinear elliptic problems with nonlinear boundary conditions. The aim of this paper is to prove existence and multiplicity results of nontrivial

*Corresponding author.
E-mail address: yaghoobi@iauh.ac.ir (F. M. Yaghoobi).

nonnegative solutions for the semilinear elliptic system:

$$\begin{cases} \Delta u + m_1(x)u = \dfrac{1}{p}\lambda f_u(x,u,v) + g_u(x,u,v) & x \in \Omega, \\ \Delta v + m_2(x)v = \dfrac{1}{p}\lambda f_v(x,u,v) + g_v(x,u,v) & x \in \Omega, \\ \dfrac{\partial u}{\partial n} = \dfrac{1}{q}\mu h_u(x,u,v) + \dfrac{1}{r}j_u(x,u,v) & x \in \partial\Omega, \\ \dfrac{\partial v}{\partial n} = \dfrac{1}{q}\mu h_v(x,u,v) + \dfrac{1}{r}j_v(x,u,v) & x \in \partial\Omega, \end{cases} \tag{1}$$

where $\lambda, \mu > 0$, $1 < q, r < 2 < p < 2^*$, where 2^* is the critical Sobolev exponent ($2^* = \frac{2N}{N-2}$ if $N > 2$, $2^* = \infty$ if $N \leqslant 2$), $\frac{\partial}{\partial n}$ is the outer normal derivative, $\Omega \subset \mathbb{R}^N$ is a bounded domain with smooth boundary $\partial\Omega$ and $m_1, m_2 \in C(\bar{\Omega})$ are positive bounded functions. Also f, g, h and j are C^1-positively homogenous functions of degrees p, 1, q and r respectively such that $f(x,0,0) = g(x,0,0) = h(x,0,0) = j(x,0,0) = 0$.

We say $\psi(x,u,v)$ is a positively homogeneous function of degree α whenever $\psi(x,tu,tv) = t^\alpha \psi(x,u,v)$ for every $t > 0$. It is clear that if $\alpha \geqslant 0$ and $\psi(x,u,v)$ is an $\alpha-$homogeneous C^1-function, then $\psi(x,u,v) \leqslant K_\psi(|u|^\alpha + |v|^\alpha)$, where $K_\psi = \max\{\psi(x,u,v) : (x,u,v) \in \overline{\Omega} \times \mathbb{R}^2,\ |u|^\alpha + |v|^\alpha = 1\}$. So from assumptions over f, g, h and j we conclude that there exist positive constants K_f, K_g, K_h and K_j such that

$$\begin{cases} f(x,u,v) \leqslant K_f(|u|^p + |v|^p), & g(x,u,v) \leqslant K_g(|u| + |v|), \\ h(x,u,v) \leqslant K_h(|u|^q + |v|^q), & j(x,u,v) \leqslant K_j(|u|^r + |v|^r). \end{cases} \tag{2}$$

Over the last years, many authors have studied the existence of solutions for the following elliptic system

$$\begin{cases} -\Delta u + m_1(x)|u|^{p-2}u = F_u(x,u,v) + G_u(x,u,v) & x \in \Omega, \\ -\Delta v + m_2(x)|v|^{p-2}v = F_v(x,u,v) + G_v(x,u,v) & x \in \Omega, \\ \dfrac{\partial u}{\partial n} = H_u(x,u,v) + J_u(x,u,v), \quad \dfrac{\partial v}{\partial n} = H_v(x,u,v) + J_v(x,u,v) & x \in \partial\Omega, \end{cases}$$

where Ω is a bounded region in $\mathbb{R}^N (N > 2)$ with smooth boundary $\partial\Omega$ and F, G, H and J are positively homogenous functions of different degrees. For instance, Brown and Wu [6] considered the case $m_1 = m_2 = 0, G(x,u,v) = J(x,u,v) = 0$ and

$$\begin{cases} F(x,u,v) = \dfrac{1}{\alpha+\beta} f(x) u^\alpha v^\beta \\ H(x,u,v) = \lambda \dfrac{1}{q} g(x) u^q + \mu \dfrac{1}{q} h(x) v^q \end{cases}$$

where $\alpha > 1, \beta > 1, 2 < \alpha + \beta < 2^*$ and the weight functions f, g, h satisfy the following conditions:

- $f \in C(\overline{\Omega})$ with $\|f\|_\infty = 1$ and $f^+ = max\{f, 0\} \not\equiv 0$.
- g, $h \in C(\partial\Omega)$ with $\|g\|_\infty = \|h\|_\infty = 1$ and $g^\pm = max\{\pm g, 0\} \not\equiv 0$ and $h^\pm = max\{\pm h, 0\} \not\equiv 0$.

They found that the above problem has at least two nonnegative solutions when the pair of the parameters (λ, μ) belongs to a certain subset of $\mathbb{R}^2$. Also in [17], Wu considered the case $F(x,u,v) = \frac{1}{q}\lambda f(x)u^q + \frac{1}{q}\mu g(x)v^q, G(x,u,v) = J(x,u,v) = 0$ and $H(x,u,v) = \frac{2}{\alpha+\beta}h(x)u^\alpha v^\beta$, where $1 < q < 2, \alpha > 1, \beta > 1$ satisfy $2 < \alpha + \beta < 2^*$ and the weights f, g, h satisfy some suitable conditions. The author showed this problem has at least two solutions when (λ, μ) belongs to a certain subset of R^2.
In [14], Feng-Yun Lu proved the existence at least two nontrivial nonnegative solutions in the case $m_1 = m_2 = 1, F(x,u,v) = \frac{1}{q}\lambda f(x)u^q + \frac{1}{q}\mu g(x)v^q, G(x,u,v) = J(x,u,v) = 0$ and $H(x,u,v) = \frac{1}{\alpha+\beta}h(x)u^\alpha v^\beta$, where $\alpha > 1, \beta > 1$ and $2 < \alpha + \beta < 2^*$. Recently, Fan [11] studied the case $F(x,u,v) = \frac{1}{r}\lambda u^r + \frac{1}{r}\mu v^r, H(x,u,v) = J(x,u,v) = 0$ and $G(x,u,v) = \frac{2}{\alpha+\beta}u^\alpha v^\beta$ for $1 < r < p < 2^*$. By using the Nehari manifold and the Lusternik-Schnirelman category, the author proved the problem admits at least $\mathrm{cat}(\Omega)+1$ positive solutions. Moreover, equations involving positively homogeneous functions have been considered in many papers, such as [2–4, 8, 12, 13, 15] and the references cited therein.

In this paper, at first, by exploiting the relationship between the Nehari manifold, fibering maps and extraction of the palais-smale sequences in the Nehari manifold and using the Rellich-Kondrachov Theorem [5] we establish the existence of local minimizers for Euler functional associated with the equation and so we prove the existence of nonnegative solutions of system (1). Then by using Mountain Pass Lemma [16], we establish the existence of at least one solution with positive energy.

Problem (1) is posed in the framework of the Sobolev space $W = W^{1,2}(\Omega) \times W^{1,2}(\Omega)$ with the norm

$$\|(u,v)\|_W = \left(\int_\Omega (|\nabla u|^2 + m_1(x)|u|^2)dx + \int_\Omega (|\nabla v|^2 + m_2(x)|v|^2)dx\right)^{\frac{1}{2}},$$

which is equivalent to the standard one and we use the standard $\mathrm{L}^r(\Omega)$ spaces whose norms denoted by $\|u\|_r$. Throughout this paper, we denote by S_r and $\bar{S}_r$ the best Sobolev and the best Sobolev trace constant for the embeddings of $W^{1,2}(\Omega)$ into $\mathrm{L}^r(\Omega)$ and $W^{1,2}(\Omega)$ into $\mathrm{L}^r(\partial\Omega)$, respectively. So we have

$$\frac{(\|(u,v)\|_W^2)^r}{(\int_{\partial\Omega}(|u|^r + |v|^r)dx)^2} \geqslant \frac{1}{4\bar{S}_r^{2r}} \quad \text{and} \quad \frac{(\|(u,v)\|_W^2)^r}{(\int_\Omega(|u|^r + |v|^r)dx)^2} \geqslant \frac{1}{4S_r^{2r}}. \tag{3}$$

Now we will show the existence and multiplicity results of nontrivial solutions of system (1) by looking for critical points of the associated Euler functional

$$\ell_{\lambda,\mu}(u,v) = \frac{1}{2}M(u,v) - \frac{1}{p}\lambda F(u,v) - G(u,v) - \frac{1}{q}\mu H(,u,v) - \frac{1}{r}J(u,v), \tag{4}$$

where

$$M(u,v) = \int_\Omega (|\nabla u|^2 + m_1(x)|u|^2)dx + \int_\Omega (|\nabla v|^2 + m_2(x)|v|^2)dx$$

and

$$\begin{cases} F(u,v) = \int_{\Omega} f(x,|u|,|v|)dx, & G(u,v) = \int_{\Omega} g(x,|u|,|v|)dx, \\ H(u,v) = \int_{\partial\Omega} h(x,|u|,|v|)d\sigma, & J(u,v) = \int_{\partial\Omega} j(x,|u|,|v|)d\sigma. \end{cases} \tag{5}$$

Moreover, a pair of functions $(u,v) \in W$ is said to be a weak solution of the problem (1), if $\langle \ell'_{\lambda,\mu}(u,v), (\varphi_1,\varphi_2)\rangle = 0$, i.e.

$$\begin{aligned} \int_{\Omega} (\nabla u.\nabla\varphi_1 + m_1(x)u\varphi_1)dx + \int_{\Omega} (\nabla v.\nabla\varphi_2 + m_2(x)v\varphi_2)dx \\ = \frac{1}{p}\lambda \int_{\Omega} (f_u\varphi_1 + f_v\varphi_2)dx + \int_{\Omega} (g_u\varphi_1 + g_v\varphi_2)dx \\ + \frac{1}{q}\mu \int_{\partial\Omega} (h_u\varphi_1 + h_v\varphi_2)d\sigma + \frac{1}{r}\int_{\partial\Omega} (j_u\varphi_1 + j_v\varphi_2)d\sigma, \end{aligned}$$

for all $(\varphi_1,\varphi_2) \in W$.

To get the solutions of system (1) we look for minimizers of the energy functional $\ell_{\lambda,\mu}$. But $\ell_{\lambda,\mu}$ is not bounded neither above nor below on W, so we introduce the Nehari manifold

$$\mathcal{N}_{\lambda,\mu}(\Omega) = \{(u,v) \in W \setminus \{(0,0)\} : \langle \ell'_{\lambda,\mu}(u,v),(u,v)\rangle = 0\},$$

where $\langle , \rangle$ denotes the usual duality between W and W^{-1}, where W^{-1} is the dual space of the Sobolev space W. We recall that any nonzero solution of problem (1) belongs to $\mathcal{N}_{\lambda,\mu}(\Omega)$. Moreover, by definition, we have that $(u,v) \in \mathcal{N}_{\lambda,\mu}(\Omega)$ if and only if

$$M(u,v) - \lambda F(u,v) - G(u,v) - \mu H(u,v) - J(u,v) = 0. \tag{6}$$

The following result concerns the behavior of $\ell_{\lambda,\mu}$ on $\mathcal{N}_{\lambda,\mu}(\Omega)$.

Lemma 1.1 $\ell_{\lambda,\mu}$ is coercive and bounded from below on $\mathcal{N}_{\lambda,\mu}(\Omega)$.

Proof. Let $(u,v) \in \mathcal{N}_{\lambda,\mu}(\Omega)$ be an arbitrary. Then by (2)–(6) we get

$$\begin{aligned} \ell_{\lambda,\mu}(u,v) &= \frac{p-2}{2p}M(u,v) - \mu\frac{q-p}{qq}H(u,v) - \frac{p-1}{p}G(u,v) - \frac{p-r}{rp}J(u,v) \\ &\geqslant \frac{p-2}{2p}M(u,v) - \mu\frac{p-q}{pq}K_h\int_{\partial\Omega}(|u|^q + |v|^q)d\sigma \\ &\quad - \frac{p-1}{p}K_g\int_{\Omega}(|u| + |v|)dx - \frac{p-r}{pr}K_j\int_{\partial\Omega}(|u|^r + |v|^r)d\sigma \\ &\geqslant \frac{p-2}{2p}M(u,v) - 2\mu\bar{S}_q^q\frac{p-q}{pq}K_h(M(u,v))^{q/2} \\ &\quad - 2S_1\frac{p-1}{p}K_g(M(u,v))^{\frac{1}{2}} - 2\bar{S}_r^r\frac{p-r}{pr}K_j(M(u,v))^{r/2}. \end{aligned}$$

Thus, $\ell_{\lambda,\mu}$ is coercive and bounded from below on $\mathcal{N}_{\lambda,\mu}(\Omega)$.

The Nehari manifold is closely linked to the behavior of functions of the form $\phi_{u,v}$: $t \mapsto \ell_{\lambda,\mu}(tu, tv)$ $(t > 0)$. Such maps are known as fibering maps. They were introduced by Drabek and Pohozaev in [9] and also were discussed in Brown and Zhang [7]. So for $(u, v) \in W$, we have

$$\begin{aligned} \phi_{u,v}(t) &= \ell_{\lambda,\mu}(tu, tv) \\ &= \frac{t^2}{2}M(u,v) - \frac{t^p}{p}\lambda F(u,v) - tG(u,v) - \frac{t^q}{q}\mu H(u,v) - \frac{t^r}{r}J(u,v), \\ \phi'_{u,v}(t) &= \langle \ell'_{\lambda,\mu}(tu, tv), (u,v)\rangle \\ &= tM(u,v) - \lambda t^{p-1}F(u,v) - G(u,v) - \mu t^{q-1}H(u,v) - t^{r-1}J(u,v). \end{aligned} \tag{7}$$

It is easy to see that $\phi'_{u,v}(t) = 0$ if and only if $(tu, tv) \in \mathcal{N}_{\lambda,\mu}(\Omega)$. In particular, $(u, v) \in \mathcal{N}_{\lambda,\mu}(\Omega)$ if and only if $\phi'_{u,v}(1) = 0$, i.e. elements in $\mathcal{N}_{\lambda,\mu}(\Omega)$ correspond to stationary points of fibering maps. Thus, it is natural to split $\mathcal{N}_{\lambda,\mu}$ into three parts corresponding to local minima, local maxima and points of inflection and so we define

$$\begin{aligned} \mathcal{N}^+_{\lambda,\mu} &= \{(tu, tv) \in W : \phi'_{u,v}(t) = 0, \phi''_{u,v}(t) > 0\}, \\ \mathcal{N}^-_{\lambda,\mu} &= \{(tu, tv) \in W : \phi'_{u,v}(t) = 0, \phi''_{u,v}(t) < 0\}, \\ \mathcal{N}^0_{\lambda,\mu} &= \{(tu, tv) \in W : \phi'_{u,v}(t) = 0, \phi''_{u,v}(t) = 0\}. \end{aligned} \tag{8}$$

The following lemma shows that minimizers for $\ell_{\lambda,\mu}(u, v)$ on $N_{\lambda,\mu}(\Omega)$ are usually critical points for $\ell_{\lambda,\mu}$, as proved by Brown and Zhang in [7] or in Aghajani et al. [1].

Lemma 1.2 Let (u_0, v_0) be a local minimizer for $\ell_{\lambda,\mu}(u, v)$ on $\mathcal{N}_{\lambda,\mu}(\Omega)$, if $(u_0, v_0) \notin \mathcal{N}^0_{\lambda,\mu}(\Omega)$, then (u_0, v_0) is a critical point of $\ell_{\lambda,\mu}$.

The purpose of this paper is to prove the following results.

Theorem 1.3 If $q \leqslant r$, $G(u, v) > 0$ and $J(u, v) \leqslant 0$, then there exists $\Lambda^* \subset (\mathbb{R}^+)^2$ such that for $(\lambda, \mu) \in \Lambda^*$, system (1) has at least two positive distinct solutions.

Theorem 1.4 If $G(u, v) > 0$ and $F(u, v) > 0$, then there exists Λ^{**} $\left(\Lambda^* \subseteq \Lambda^{**} \subsetneq (\mathbb{R}^+)^2\right)$ such that for $(\lambda, \mu) \in \Lambda^{**}$, system (1) has at least one nontrivial solution with positive energy.

This paper is organized as follows. In section 2 we point out some notations and preliminaries and give a fairly complete description of the Nehari manifold and fibering map. Finally Theorem 1.3 and Theorem 1.4 are proved in section 3.

2. Preliminaries and auxiliary results

In this section some properties of the Nehari manifold and fibering map will be perused. First, motivated by Lemma 1.2, we will get conditions for $\mathcal{N}^0_{\lambda,\mu} = \emptyset$.

Lemma 2.1 If $q \leqslant r$, $G(u, v) > 0$ and $J(u, v) \leqslant 0$, then there exists $\Lambda_0 \subset (\mathbb{R}^+)^2$ such that for $(\lambda, \mu) \in \Lambda_0$ and $q \leqslant r$, we have $\mathcal{N}^0_{\lambda,\mu} = \emptyset$.

proof. Suppose otherwise, let $(u,v) \in \mathcal{N}^0_{\lambda,\mu}$ be an arbitrary, then by (7) and (8) we have

$$\phi''_{u,v}(1) = M(u,v) - \lambda(p-1)F(u,v) - \mu(q-1)H(u,v) - (r-1)J(u,v) = 0, \quad (9)$$

and

$$\phi'_{u,v}(1) = M(u,v) - \lambda F(u,v) - G(u,v) - \mu H(u,v) - J(u,v) = 0. \quad (10)$$

Using (2), (3), (5), (9) and (10) we obtain

$$\begin{aligned}(2-q)M(u,v) &= (p-q)\lambda F(u,v) + (1-q)G(u,v) + (r-q)J(u,v)\\ &< (p-q)\lambda K_f \int_\Omega (|u|^p + |v|^p)dx \leqslant 2(p-q)(S_p^p \lambda K_f)(M(u,v))^{\frac{p}{2}},\end{aligned}$$

which concludes

$$M(u,v) > \left(\frac{2-q}{2(p-q)S_p^p \lambda K_f}\right)^{\frac{2}{p-2}}. \quad (11)$$

On the other hand, by relations (9), (10), (2), (3), (5) and Young inequality we get

$$\begin{aligned}(p-2)M(u,v) &= \mu(p-q)H(u,v) + (p-1)G(u,v) + (p-r)J(u,v)\\ &\leqslant \mu(p-q)K_h \int_{\partial\Omega} (|u|^q + |v|^q)d\sigma + (p-1)K_g \int_\Omega (|u| + |v|)dx\\ &\leqslant 2\mu \bar{S}_q^q (p-q)K_h(M(u,v))^{\frac{q}{2}} + 2S_1(p-1)K_g(M(u,v))^{\frac{1}{2}}\\ &\leqslant \frac{2(p-2)}{3q}\left(\frac{2-q}{2}\left(\frac{3q(p-q)}{(p-2)}\mu \bar{S}_q^q K_h\right)^{\frac{2}{2-q}} + \frac{q}{2}M(u,v)\right)\\ &+ \frac{2(p-2)}{3}\left(\frac{1}{2}\left(\frac{3(p-1)}{(p-2)}S_1K_g\right)^2 + \frac{1}{2}M(u,v)\right),\end{aligned}$$

so we have

$$M(u,v) \leqslant L + L', \quad (12)$$

where $L = \frac{(2-q)}{q}\left(\frac{3q(p-q)}{(p-2)}\mu \bar{S}_q^q K_h\right)^{\frac{2}{2-q}}$ and $L' = \left(\frac{3(p-1)}{(p-2)}S_1K_g\right)^2$. Now by (12) and (11) we must have

$$\left(\frac{2-q}{2(p-q)S_p^p \lambda K_f}\right)^{\frac{2}{p-2}} < L + L',$$

which is a contradiction for λ and μ sufficiently small. So there exists $\Lambda_0 \subset (\mathbb{R}^+)^2$ such that for $(\lambda,\mu) \in \Lambda_0$, $\mathcal{N}^0_{\lambda,\mu} = \emptyset$.

Lemma 2.2 If $(u,v) \in N^-_{\lambda,\mu}$, $q \leqslant r$, $G(u,v) > 0$ and $J(u,v) \leqslant 0$, then $F(u,v) > 0$.

Proof. By using (7) and (8) for $(u,v) \in N^-_{\lambda,\mu}$ we have

$$F(u,v) \geqslant \frac{2-q}{\lambda(p-q)} M(u,v) + \frac{q-1}{\lambda(p-q)} G(u,v) + \frac{q-r}{\lambda(p-q)} J(u,v) > 0. \tag{13}$$

As it was mentioned in previous section we have, $\phi'_{u,v}(t) = 0$ if and only if $(tu,tv) \in \mathcal{N}_{\lambda,\mu}(\Omega)$. Therefore our purpose is to describe the nature of the derivative of the fibering maps for all possible signs of $F(u,v)$, to do this, at first we define the following functions

$$\begin{cases} R_\lambda(t) := \frac{1}{2} t^2 M(u,v) - \lambda \frac{1}{p} t^p F(u,v), \\ S_\mu(t) := \mu \frac{1}{q} t^q H(u,v) + tG(u,v) + \frac{1}{r} t^r J(u,v), \end{cases} \tag{14}$$

follows from (7) that, $\phi_{u,v}(t) = R_\lambda(t) - S_\mu(t)$ and in particular $\phi'_{u,v}(t) = 0$ if and only if $R'_\lambda(t) = S'_\mu(t)$, where

$$\begin{cases} R'_\lambda(t) = tM(u,v) - t^{p-1}\lambda F(u,v), \\ S'_\mu(t) = \mu t^{q-1} H(u,v) + G(u,v) + t^{r-1} J(u,v). \end{cases} \tag{15}$$

In the next result we see that, $\phi_{u,v}$ has positive values for all nonzero $(u,v) \in W$ whenever, λ and μ are sufficiently small.

Lemma 2.3 There exists $\Lambda_1 \subset (\mathbb{R}^+)^2$ such that $\phi_{u,v}(t) = R_\lambda(t) - S_\mu(t)$ takes on positive values for all non-zero $(u,v) \in W$, whenever $(\lambda,\mu) \in \Lambda_1$.

Proof. If $F(u,,v) \leqslant 0$, then $R_\lambda(t) > S_\mu(t)$ for t sufficiently large and so $\phi_{u,v}(t) > 0$. Otherwise, suppose there exists $(u,v) \in W$ such that $F(u,v) > 0$. By elementary calculus, we infer that $R_\lambda(t)$ takes a maximum at

$$t_{max} = \left(\frac{\|(u,v)\|^2_W}{\lambda F(u,v)} \right)^{\frac{1}{p-2}}, \tag{16}$$

then follow by (14), (16), (2), (3) and (5)

$$\begin{aligned} R_\lambda(t_{max}) &= \frac{p-2}{2p} \left(\frac{(\|(u,v)\|^2_W)^p}{(\lambda F(u,v))^2} \right)^{\frac{1}{p-2}} \\ &\geqslant \frac{p-2}{2p(\lambda K_f)^{\frac{2}{p-2}}} \left(\frac{(\|(u,v)\|^2_W)^p}{\left(\int_\Omega (|u|^p + |v|^p)dx\right)^2} \right)^{\frac{1}{p-2}} \\ &\geqslant \frac{p-2}{2p(\lambda K_f)^{\frac{2}{p-2}}} \left(\frac{1}{4S_p^{2p}} \right)^{\frac{1}{p-2}} \geqslant \frac{\delta}{\lambda^{\frac{2}{p-2}}}, \end{aligned} \tag{17}$$

where δ is independent of (u,v). Now, we are going to prove that there exists $\Lambda_1 \subset (\mathbb{R}^+)^2$ such that for all non-zero $(u,v) \in W$, $\phi_{u,v}(t_{max}) > 0$, provided that $(\lambda,\mu) \in \Lambda_1$. To do this, first note that from (16), (17) and (3) for $1 \leqslant \alpha < 2^*$

$$\begin{aligned}(t_{max})^\alpha \int_\Omega (|u|^\alpha + |v|^\alpha)dx &\leqslant 2S_\alpha^\alpha \left(\frac{\|(u,v)\|_W^2}{\lambda F(u,v)} \right)^{\frac{\alpha}{p-2}} (\|(u,v)\|_W^2)^{\frac{\alpha}{2}} \\ &= 2S_\alpha^\alpha \left(\frac{(\|(u,v)\|_W^2)^p}{(\lambda F(u,v))^2} \right)^{\frac{\alpha}{2(p-2)}} \\ &= 2S_\alpha^\alpha (\frac{2p}{p-2})^{\frac{\alpha}{2}} (R_\lambda(t_{max}))^{\frac{\alpha}{2}} = c_1 (R_\lambda(t_{max}))^{\frac{\alpha}{2}},\end{aligned} \tag{18}$$

similarly, $(t_{max})^\alpha \int_{\partial\Omega} (|u|^\alpha + |v|^\alpha)dx = \bar{c}_1 (R_\lambda(t_{max}))^{\frac{\alpha}{2}}$. By computing (2), (5), (14) and (18) we find

$$\begin{aligned}S_\mu(t_{max}) &= \frac{1}{q}\mu(t_{max})^q H(u,v) + t_{max}G(u,v) + \frac{1}{r}(t_{max})^r J(u,v) \\ &\leqslant \frac{1}{q}\mu K_h (t_{max})^q \int_{\partial\Omega} (|u|^q + |v|^q)d\sigma \\ &+ K_g t_{max} \int_\Omega (|u| + |v|)dx + \frac{1}{r} K_j (t_{max})^r \int_{\partial\Omega} (|u|^r + |v|^r)d\sigma \\ &\leqslant \mu C_1 \big(R_\lambda(t_{max})\big)^{\frac{q}{2}} + C_2 \big(R_\lambda(t_{max})\big)^{\frac{1}{2}} + C_3 \big(R_\lambda(t_{max})\big)^{\frac{r}{2}},\end{aligned} \tag{19}$$

where C_1, C_2 and C_3 are positive constants and independent of (u,v). Hence using (7), (17) and (19) we observe that

$$\begin{aligned}\phi_{u,v}(t_{max}) &= R_\lambda(t_{max}) - S_\mu(t_{max}) \\ &\geqslant \mathfrak{R}\Big(1 - \mu C_1 \mathfrak{R}^{\frac{q-2}{2}} - C_2 \mathfrak{R}^{\frac{-1}{2}} - C_3 \mathfrak{R}^{\frac{r-2}{2}}\Big) \\ &\geqslant \frac{\delta}{\lambda^{\frac{2}{p-2}}} \Big(1 - \mu C_1 \delta^{\frac{q-2}{2}} \lambda^{\frac{2-q}{p-2}} - C_2 \delta^{\frac{-1}{2}} \lambda^{\frac{1}{p-2}} - C_3 \delta^{\frac{r-2}{2}} \lambda^{\frac{2-r}{p-2}}\Big),\end{aligned}$$

where $\mathfrak{R} = R_\lambda(t_{max})$. Since $1 < q, r < 2 < p$, so there exist $\Lambda_1 \subset (\mathbb{R}^+)^2$ and $\epsilon > 0$ such that if $(\lambda, \mu) \in \Lambda_1$, then $\phi_{u,v}(t_{max}) > \epsilon > 0$ for all nonzero (u,v) and this completes the proof.

Corollary 2.4 If $(\lambda, \mu) \in \Lambda_1$, then $\ell_{\lambda,\mu}(u,v) > \epsilon > 0$ for all $(u,v) \in \mathcal{N}_{\lambda,\mu}^-$.

Proof. Since $(u,v) \in N_{\lambda,\mu}^-$ thus, $\phi_{u,v}$ has a positive global maximum at $t = 1$, i.e.

$$\ell_{\lambda,\mu}(u,v) = \phi_{u,v}(1) \geqslant \phi_{u,v}(t_{max}) > \epsilon > 0.$$

Corollary 2.5 For $(\lambda, \mu) \in \Lambda_1$, $\phi'_{u,v}(t) = R'_\lambda(t) - S'_\mu(t)$ takes on positive values for all non-zero $(u,v) \in W$.

Proof. Let $(u,v) \in W$, using (7) $\phi_{u,v}(0) = 0$, and by Lemma 2.3, $\phi_{u,v}(t_{max}) > 0$. So there exists $0 < \tau < t_{max}$, such that $\phi'_{u,v}(\tau) > 0$.

To state our main results, we now present some important properties of $N_{\lambda,\mu}^-$ and $N_{\lambda,\mu}^+$.

Corollary 2.6 If $G(u,v)>0$, then for $(u,v)\in W\setminus\{(0,0)\}$ and $(\lambda,\mu)\in\Lambda_1$, we have
(i) there exists $t_1>0$ such that $(t_1u,t_1v)\in N^+_{\lambda,\mu}$ and $\phi_{u,v}(t_1)<0$.
(ii) if $F(u,v)>0$, then there exists $0<t_1<t_2$ such that $(t_1u,t_1v)\in N^+_{\lambda,\mu}$, $(t_2u,t_2v)\in N^-_{\lambda,\mu}$ and $\phi_{u,v}(t_1)<0$.

Proof. (i) From the definition of $\phi'_{u,v}(t)$, we know $\phi'_{u,v}(0)<0$ and by Corollary 2.5, we obtain that $\phi'_{u,v}(\tau)>0$ for suitable $\tau>0$, so exists $0<t_1<\tau$ such that $\phi'_{u,v}(t_1)=0$ and $\phi''_{u,v}(t_1)>0$. Therefore, we conclude that $(t_1u,t_1v)\in N^+_{\lambda,\mu}$ and $\phi_{u,v}(t_1)<\phi_{u,v}(0)=0$.
Proof. (ii) As in the proof of (i), we have that $\phi'_{u,v}(0)<0$ and $\phi'_{u,v}(\tau)>0$. Moreover $\lim_{t\to\infty}\phi'_{u,v}(t)=-\infty$, so there exist t_1,t_2 such that $0<t_1<\tau<t_2$ and $\phi'_{u,v}(t_1)=\phi'_{u,v}(t_2)=0$. Furthermore $(t_1u,t_1v)\in N^+_{\lambda,\mu}$, $(t_2u,t_2v)\in N^-_{\lambda,\mu}$ and $\phi_{u,v}(t_1)<\phi_{u,v}(0)=0$.

3. Existence of solutions

In order to prove of Theorem 1.3, we need to show the existence of local minimum for $\ell_{\lambda,\mu}$ on $N^+_{\lambda,\mu}$ and $N^-_{\lambda,\mu}$, for this, we need the following remark:

Remark 1 *By using relation (2) we have* $f(x,u,v)\leqslant K_f(|u|^p+|v|^p)$, $g(x,u,v)\leqslant K_g(|u|+|v|)$, $h(x,u,v)\leqslant K_h(|u|^q+|v|^q)$ *and* $j(x,u,v)\leqslant K_j(|u|^r+|v|^r)$ *for* $1<q,\ r<2<p<2^*$. *Hence from compactness of the embeddings* $W^{1,2}(\Omega)\hookrightarrow L^\alpha(\Omega)$ *and* $W^{1,2}(\Omega)\hookrightarrow L^\alpha(\partial\Omega)$ *for* $1\leqslant\alpha<2^*$ *(Rellich-Kondrachov Theorem [5]) and the fact that the functions* $f(x,u,v)$, $g(x,u,v)$, $h(x,u,v)$ *and* $j(x,u,v)$ *are continuous, we conclude that the functionals* $I_1(u,v)=\int_\Omega f(x,u,v)dx$, $I_2(u,v)=\int_\Omega g(x,u,v)dx$, $I_3(u,v)=\int_{\partial\Omega}h(x,u,v)d\sigma$ *and* $I_4(u,v)=\int_{\partial\Omega}j(x,u,v)d\sigma$ *are weakly continuous, i.e. if* $(u_n,v_n)\rightharpoonup(u,v)$, *then* $I_i(u_n,v_n)\to I_i(u,v)$, *i=1,2,3,4.*

Definition 3.1 A sequence $y_n=(u_n,v_n)\subset W$ is called a Palais-Smale sequence ((PS)-sequence) if $\{\ell_{\lambda,\mu}(y_n)\}$ is bounded and $\ell'_{\lambda,\mu}(y_n)\to 0$ as $n\to\infty$. It is said that the functional $\ell_{\lambda,\mu}$ satisfies the Palais-Smale condition if each Palais-Smale sequence has a convergent subsequence.

Now we prove the boundedness of Palais-Smale sequence.

Lemma 3.2 If $\{(u_n,v_n)\}$ is a $(PS)-$ sequence for $\ell_{\lambda,\mu}$, then $\{(u_n,v_n)\}$ is bounded in W.

Proof. By using Young inequality and from (2), (3), (5) and (7) we get

$$
\begin{aligned}
&\ell_{\lambda,\mu}(u_n, v_n) - \frac{1}{p}\langle \ell'_{\lambda,\mu}(u_n, v_n), (u_n, v_n)\rangle \\
&= \frac{p-2}{2p}M(u_n, v_n) - \mu\frac{p-q}{pq}H(u_n, v_n) \\
&- \frac{p-1}{p}G(u_n, v_n) - \frac{p-r}{pr}J(u_n, v_n) \\
&\geqslant \frac{p-2}{2p}M(u_n, v_n) - \frac{p-q}{pq}2\mu\bar{S}_q^q K_h M(u_n, v_n)^{\frac{q}{2}} \\
&- \frac{p-1}{p}2S_1 K_g M(u_n, v_n)^{\frac{1}{2}} - \frac{p-r}{pr}2\bar{S}_r^r K_j M(u_n, v_n)^{\frac{r}{2}} \\
&\geqslant \frac{p-2}{2p}M(u_n, v_n) - \frac{(p-2)}{4pq}\left(\frac{2-q}{2}\left(\frac{4(p-q)}{(p-2)}2\mu\bar{S}_q^q K_h\right)^{\frac{2}{2-q}} + \frac{q}{2}M(u, v)\right) \\
&- \frac{(p-2)}{4p}\left(\frac{1}{2}\left(\frac{4(p-1)}{(p-2)}2S_1 K_g\right)^2 + \frac{1}{2}M(u_n, v_n)\right) \\
&- \frac{(p-2)}{4pr}\left(\frac{2-r}{2}\left(\frac{4(p-r)}{(p-2)}2\bar{S}_r^r K_j\right)^{\frac{2}{2-r}} + \frac{r}{2}M(u, v)\right) \geqslant \frac{p-2}{8p}\|(u_n, v_n)\|_W^2 - L,
\end{aligned}
$$

where

$$
\begin{aligned}
L = &\frac{(p-2)(2-q)}{8pq}\left(\frac{4(p-q)}{(p-2)}2\mu\bar{S}_q^q K_h\right)^{\frac{2}{2-q}} \\
&+ \frac{(p-2)}{8p}\left(\frac{4(p-1)}{(p-2)}2S_1 K_g\right)^2 + \frac{(p-2)(2-r)}{8pr}\left(\frac{4(p-r)}{(p-2)}2\bar{S}_r^r K_j\right)^{\frac{2}{2-r}},
\end{aligned}
$$

so $\{(u_n, v_n)\}$ is bounded in W.

Now, we establish the existence of local minimum for $\ell_{\lambda,\mu}$ on $N^+_{\lambda,\mu}$ and $N^-_{\lambda,\mu}$. For simplicity, let $\Lambda^* = \{\Lambda_0 \cap \Lambda_1\}$ and $\Lambda^{**} = \Lambda_1$ where Λ_0 and Λ_1 are given in the previous section.

First, we establish the existence of local minimum for $\ell_{\lambda,\mu}$ on $N^+_{\lambda,\mu}$ and $N^-_{\lambda,\mu}$.

Proposition 3.3 If $G(u, v) > 0$, then for $(\lambda, \mu) \in \Lambda^*$ we have
(i) there exists a minimizer of $\ell_{\lambda,\mu}$ on $\mathcal{N}^+_{\lambda,\mu}(\Omega)$,
(ii) if $q \leqslant r$ and $J(u, v) \leqslant 0$, then there exists a minimizer of $\ell_{\lambda,\mu}$ on $\mathcal{N}^-_{\lambda,\mu}(\Omega)$.

Proof. (i) By arguing as in Lemma 1.1, $\ell_{\lambda,\mu}$ is bounded from below on $\mathcal{N}_{\lambda,\mu}(\Omega)$ and so on $\mathcal{N}^+_{\lambda,\mu}(\Omega)$. Let $\{(u_n, v_n)\}$ be a minimizing sequence for $\ell_{\lambda,\mu}$ on $\mathcal{N}^+_{\lambda,\mu}(\Omega)$, i.e.

$$\lim_{n\to\infty} \ell_{\lambda,\mu}(u_n, v_n) = \inf_{(u,v)\in\mathcal{N}^+_{\lambda,\mu}} \ell_{\lambda,\mu}(u, v) = c.$$

From Ekeland's variational principle [10] we have

$$\langle \ell'_{\lambda,\mu}(u_n, v_n), (u_n, v_n)\rangle \to 0,$$

combining the compact embedding Theorem [5] and Lemma 3.2, we obtain that there exists a subsequence $\{(u_n, v_n)\}$ and (u_1, v_1) in W such that

$$\begin{cases} u_n \rightharpoonup u_1, & v_n \rightharpoonup v_1 \text{ weakly in } \quad W^{1,2}(\Omega), \\ u_n \to u_1, & v_n \to v_1 \quad \text{strongly in } L^{\alpha}(\Omega) \text{ and } L^{\alpha}(\partial\Omega) \text{ for} 1 \leqslant \alpha < 2^*, \end{cases} \tag{20}$$

and $(u_n(x), v_n(x)) \to (u_1(x), v_1(x))$, a.e.

By Corollary 2.6(i) for $(u_1, v_1) \in W \setminus \{(0,0)\}$, there exists t_1 such that $(t_1u_1, t_1v_1) \in N^+_{\lambda,\mu}$ and so $\phi'_{u_1,v_1}(t_1) = 0$. Now we show that $(u_n, v_n) \to (u_1, v_1)$ in W. Suppose this is false, then

$$M(u_1, v_1) < \liminf_{n\to\infty} M(u_n, v_n), \tag{21}$$

also we have

$$\begin{aligned} \phi'_{u_n,v_n}(t) &= tM(u_n, v_n) - \lambda t^{p-1}F(u_n, v_n) \\ &\quad - \mu t^{q-1}H(u_n, v_n) - G(u_n, v_n) - t^{r-1}J(u_n, v_n), \end{aligned} \tag{22}$$

and

$$\begin{aligned} \phi'_{u_1,v_1}(t) &= tM(u_1, v_1) - \lambda t^{p-1}F(u_1, v_1) \\ &\quad - \mu t^{q-1}H(u_1, v_1) - G(u_1, v_1) - t^{r-1}J(u_1, v_1), \end{aligned} \tag{23}$$

so from (20)–(23) and Remark *1*, $\phi'_{u_n,v_n}(t_1) > \phi'_{u_1,v_1}(t_1) = 0$ for n sufficiently large. Since $\{(u_n, v_n)\} \subseteq N^+_{\lambda,\mu}(\Omega)$, by considering the possible fibering maps it is easy to see that, $\phi'_{u_n,v_n}(t) < 0$ for $0 < t < 1$ and $\phi'_{u_n,v_n}(1) = 0$ for all n. Hence, we must have $t_1 > 1$, but $(t_1u_1, t_1v_1) \in N^+_{\lambda,\mu}$ and so

$$\begin{aligned} \ell_{\lambda,\mu}(t_1u_1, t_1v_1) &= \phi_{u_1,v_1}(t_1) < \phi_{u_1,v_1}(1) \\ &< \lim_{n\to\infty} \phi_{u_n,v_n}(1) = \lim_{n\to\infty} \ell_{\lambda,\mu}(u_n, v_n) = \inf_{(u,v)\in\mathcal{N}^+_{\lambda,\mu}} \ell_{\lambda,\mu}(u, v), \end{aligned}$$

which is a contradiction. Therefore, $(u_nv_n) \to (u_1, v_1)$ in W and so

$$\ell_{\lambda,\mu}(u_1, v_1) = \lim_{n\to\infty} \ell_{\lambda,\mu}(u_n, v_n) = \inf_{(u,v)\in\mathcal{N}^+_{\lambda,\mu}} \ell_{\lambda,\mu}(u, v).$$

Thus, (u_1, v_1) is a minimizer for $\ell_{\lambda,\mu}$ on $\mathcal{N}^+_{\lambda,\mu}(\Omega)$.
Proof. (ii) By Corollary 2.4, we have $\ell_{\lambda,\mu}(u, v) > \epsilon > 0$ for all $(u, v) \in \mathcal{N}^-_{\lambda,\mu}$, i.e.

$$\inf_{(u,v)\in\mathcal{N}^-_{\lambda,\mu}} \ell_{\lambda,\mu}(u, v) > 0,$$

hence, there exists a minimizing sequence $\{(u_n, v_n)\} \subseteq \mathcal{N}^-_{\lambda,\mu}(\Omega)$ such that

$$\lim_{n\to\infty} \ell_{\lambda,\mu}(u_n, v_n) = \inf_{(u,v)\in\mathcal{N}^-_{\lambda,\mu}} \ell_{\lambda,\mu}(u, v) > 0. \tag{24}$$

Similar to the argument in the proof of (i), we find that, $\{(u_n, v_n)\}$ is bounded in W and also the results obtained in (20) are satisfied for $\{(u_n, v_n)\}$ and $\{(u_2, v_2)\}$. Since $(u_n, v_n) \in \mathcal{N}^-_{\lambda,\mu}(\Omega)$, so by (8) $\phi'_{u_n,v_n}(1) = 0, \phi''_{u_n,v_n}(1) < 0$ and by Lemma 2.2, $F(u_n, v_n) > 0$. Letting $n \to \infty$, we see that $\phi'_{u_2,v_2}(1) = 0, \phi''_{u_2,v_2}(1) \leqslant 0$ and $F(u_2, v_2) \geqslant 0$. If $F(u_2, v_2) = 0$, then by (7) and (8) we have

$$(2-q)M(u,v) + (q-1)G(u,v) + (q-r)J(u,v) \leqslant 0,$$

which is a contradiction with our assumptions. So $F(u_2, v_2) > 0$ and by Corollary 2.6(ii) there exists $t_2 > 0$ such that $(t_2u_2, t_2v_2) \in \mathcal{N}^-_{\lambda,\mu}(\Omega)$. We claim that $(u_n, v_n) \to (u_2, v_2)$ in W, suppose that this is false, so

$$M(u_2, v_2) < \liminf_{n\to\infty} M(u_n, v_n), \tag{25}$$

but $(u_n, v_n) \in \mathcal{N}^-_{\lambda,\mu}$ and so $\ell_{\lambda,\mu}(u_n, v_n) \geqslant \ell_{\lambda,\mu}(tu_n, tv_n)$ for all $t \geqslant 0$. Therefore, by considering (7), (24) and (25) and Remark *1*, we can write

$$\begin{aligned} \ell_{\lambda,\mu}(t_2u_2, t_2v_2) &= \phi_{u_2,v_2}(t_2) < \lim_{n\to\infty} \phi_{u_2,v_2}(t_2) \\ &= \lim_{n\to\infty} \ell_{\lambda,\mu}(t_2u_n, t_2v_n) \leqslant \lim_{n\to\infty} \ell_{\lambda,\mu}(u_n, v_n) = \inf_{(u,v)\in\mathcal{N}^-_{\lambda,\mu}} \ell_{\lambda,\mu}(u,v), \end{aligned}$$

which is a contradiction. Therefore, $(u_n, v_n) \to (u_2, v_2)$ in W and so the proof is complete.

Lemma 3.4 If $G(u,v) > 0$, then for $(\lambda, \mu) \in \Lambda^*$, the functional $\ell_{\lambda,\mu}(u,v)$ satisfies (PS) condition on W.

Proof. If $\ell_{\lambda,\mu}(u_n, v_n)$ is bounded and $\ell_{\lambda,\mu}(u_n, v_n) \to 0$, then using Lemma 3.2, (u_n, v_n) is bounded in W. Also, similar to the argument in the proof of Proposition 3.3(i) we find that, the sequence (u_n, v_n), has a convergent subsequence and this completes the proof.

Proof of Theorem 1.3. By Proposition 3.3 there exist $(u_1, v_1) \in N^+_{\lambda,\mu}(\Omega)$ and $(u_2, v_2) \in N^-_{\lambda,\mu}(\Omega)$ such that $\ell_{\lambda,\mu}(u_1, v_1) = \inf_{(u,v)\in N^+_{\lambda,\mu}} \ell_{\lambda,\mu}(u,v)$ and $\ell_{\lambda,\mu}(u_2, v_2) = \inf_{(u,v)\in N^-_{\lambda,mu}} \ell_{\lambda,\mu}(u,v)$ and by Lemmas 1.2 and 2.1, (u_1, v_1) and (u_2, v_2) are critical points of $\ell_{\lambda,\mu}$ on W and hence are weak solutions of problem (1). On the other hand, $\ell_{\lambda,\mu}(u,v) = \ell_{\lambda,\mu}(|u|, |v|)$, so we may assume that (u_1, v_1) and (u_2, v_2) are positive solutions. Also, since $N^+_{\lambda,\mu} \cap N^-_{\lambda,\mu} = \emptyset$, this implies that (u_1, v_1) and (u_2, v_2) are distinct and the proof is complete.

For the proof of Theorem 1.4. we need the Mountain Pass Lemma.

Mountain Pass Lemma. (see [16]) Let X be a real Banach space with the norm $\|.\|$ and $J \in C^1(X, \mathbb{R})$, $J(0) = 0$. Assume
(i) the function $J(u)$ on X satisfies the (PS) condition;
(ii) there are β, $\rho > 0$ such that $J(u) \geqslant \beta$, $\|u\| = \rho$;
(iii) there is $e \in X$, $\|e\| \geqslant \rho$ such that $J(e) \leqslant 0$.
then $c_0 = \inf_{\phi\in\Psi} \max_{t\in[0,1]} J(\phi(t))$ is a critical value of $J(u)$ with $0 < \beta \leqslant c_0 < \infty$, where $\Psi = \{\phi \in (C[0,1], X), \phi(0) = 0, \phi(1) = e\}$.

Proof of Theorem 1.4. From (7), it is clear that $\ell_{\lambda,\mu}(u,v) \in C^1(W, R)$ and

$\ell_{\lambda,\mu}(0,0) = 0$. Since $F(u,v) > 0$, then $\lim_{t\to\infty} \ell(tu,tv) = -\infty$, this means that there exists $t_0 > 0$ such that $\ell_{\lambda,\mu}(t_0 u, t_0 v) < 0$. Also by using corollary 2.4. we know that $\ell_{\lambda,\mu}(u,v) > \epsilon > 0$ and by Lemma 3.4. $\ell_{\lambda,\mu}(u,v)$ satisfies the (PS) condition on W. Now application of the Mountain Pass Lemma gives Theorem 1.4.

References

[1] A. Aghajani and J. Shamshiri and F. M. Yaghoobi, *Existence and multiplicity of positive solutions for a class of nonlinear elliptic problems*, Turk. J. Math (2012) doi:10.3906/mat-1107-23.

[2] A. Aghajani and J. Shamshiri, *Multilicity of positive solutions for quasilinear elliptic p-Laplacian systems*, E. J. D. E. 111 (2012) 1-16.

[3] A. Aghajani and F.M. Yaghoobi and J. Shamshiri, *Multiplicity of positive solutions for a class of quasilinear elliptic p-Laplacian problems with nonlinear boundary conditions*, Journal of Information and computing Science 8 (2013) 173-182.

[4] Y. Bozhkov and E. Mitidieri, *Existence of multiple solutions for quasilinear systems via fibering method*, Journal of Differential Equations 190 (2003) 239-267.

[5] H. Brezis, *Functional Analysis, Sobolev Spaces and Partial Differential Equations*, Springer, New York, 2010.

[6] K. J. Brown and T.-F.Wu, *A semilinear elliptic system involving nonlinear boundary condition and sign-changing weight function*, J. Math.Anal. Appl. 337 (2008) 1326-1336.

[7] K. J. Brown and Y. Zhang, *The Nehari manifold for a semilinear elliptic problem with a sign changing weight function*, J. Differential Equations 193 (2003) 481-499.

[8] C. M. Chu and C. L. Tang, *Existence and multiplicity of positive solutions for semilinear elliptic systems with Sobolev critical exponents*, Nonlinear Anal 71 (2009) 5118-5130.

[9] P. Drabek and S. I. Pohozaev, *Positive solutions for the p-Laplacian: application of the fibering method*, Proc. Royal Soc. Edinburgh Sect. A 127 (1997) 703-726.

[10] I. Ekeland, *On the variational principle*, J. Math. Anal. Appl. 47 (1974) 324-353.

[11] H. Fan, *Multiple positive solutions for a critical elliptic system with concave and convex nonlinearities*, Nonlinear Anal. Real World Appl. 18 (2014) 1422.

[12] M. F. Furtado a and J. P. P. da Silva, *Multiplicity of solutions for homogeneous elliptic systems with critical growth*, J.Math. Anal. Appl. 385 (2012) 770-785.

[13] T. -s. Hsu, *Multiple positive solutions for a quasilinear elliptic system involving concave-convex nonlinearities and sign changing weight functions*, Internat. J. Math. Math. Sci. (2012) doi:10.1155 (2012) 109-214.

[14] Feng-Yun Lu, *The Nehari manifold and application to a semilinear elliptic system*, Nonlinear Analysis 71 (2009) 3425-3433

[15] Y. Shen and J. Zhang, *Multiplicity of positive solutions for a semilinear p-Laplacian system with Sobolev critical exponent*, Nonlinear Analysis 74 (2011) 1019-1030.

[16] M. Struwe, *Variational methods*, Springer, Berlin, 1990.

[17] T. F. Wu, *The Nehari manifold for a semilinear elliptic system involving sign-changing weight functions*, Nonlin. Analysis 68(6)(2008), 1733-1745.

16

Probability of having n^{th}-roots and n-centrality of two classes of groups

M. Hashemi[a*], M. Polkouei[a]

[a]*Faculty of Mathematical Sciences, University of Guilan, P.O.Box 41335-19141, Rasht, Iran.*

Abstract. In this paper, we consider the finitely 2-generated groups $K(s,l)$ and G_m as follows;

$$K(s,l) = \langle a, b | ab^s = b^l a,\ ba^s = a^l b \rangle,$$
$$G_m = \langle a, b | a^m = b^m = 1,\ [a,b]^a = [a,b],\ [a,b]^b = [a,b] \rangle$$

and find the explicit formulas for the probability of having n^{th}-roots for them. Also we investigate integers n for which, these groups are n-central.

Keywords: Nilpotent groups, n^{th}-roots, n-central groups

1. Introduction

Let $n > 1$ be an integer. An element a of group G is said to have an nth-root b in G, if $a = b^n$. The probability that a randomly chosen element in G has an nth-root, is given by

$$P_n(G) = \frac{|G^n|}{|G|}$$

---

*Corresponding author.
E-mail address: m_hashemi@guilan.ac.ir (M. Hashemi).

where $G^n = \{a \in G | a = b^n, for \quad some \quad b \in G\} = \{x^n | x \in G\}$. In [5], the probability $P_n(G)$ for Dihedral groups D_{2m} and Quaternion groups Q_{2^m} for every integer $m \geqslant 3$ have been computed. Also, in [4] the probability that Hamiltonian groups may have n^{th}-roots have been calculated. For $n > 1$, a group G is said to be n-central if $[x^n, y] = 1$ for all $x, y \in G$. In [6], some aspects of n-central groups have been investigated.

First, we state the following Lemma without proof.

Lemma 1.1 If G is a group and $G' \subseteq Z(G)$, then the following hold for every integer k and $u, v, w \in G$:

(i) $[uv, w] = [u, w][v, w]$ and $[u, vw] = [u, v][u, w]$;
(ii) $[u^k, v] = [u, v^k] = [u, v]^k$;
(iii) $(uv)^k = u^k v^k [v, u]^{k(k-1)/2}$.

Now, we state some lemmas which can be found in [1, 2].

Lemma 1.2 The groups $K(s, l) = \langle a, b | ab^s = b^l a, \ ba^s = a^l b \rangle$ where $(s, l) = 1$, have the following properties:

(i) $|K(s, l)| = |l - s|^3$, if $(s, l) = 1$ and is infinite otherwise;
(ii) if $(s, l) = 1$ then $|a| = |b| = (l - s)^2$;
(iii) if $(s, l) = 1$, then $a^{l-s} = b^{s-l}$.

Lemma 1.3 (i) For every $l \geqslant 3$, $K(s, l) \cong K(1, 2 - l)$.
(ii) For every $i \geqslant 2$ and $(s, i) = 1$, $K(s, s + i) \cong K(1, i + 1)$.

Note that if $(s, l) = 1$, then $K(s, l) \cong K(1, l - s + 1)$ which we can write as K_m where $m = l - s + 1$.

Lemma 1.4 Every element of K_m can be uniquely presented by $x = a^\beta b^\gamma a^{(m-1)\delta}$, where $1 \leqslant \beta, \gamma, \delta \leqslant m - 1$.

Lemma 1.5 In K_m, $[a, b] = b^{m-1} \in Z(K_m)$.

The following lemma can be seen in [3].

Lemma 1.6 Let $G_m = \langle a, b | a^m = b^m = 1, \ [a, b]^a = [a, b], \ [a, b]^b = [a, b] \rangle$ where $m \geqslant 2$, then we have

(i) every element of G_m can be uniquely presented by $a^i b^j [a, b]^t$, where $1 \leqslant i, j, t \leqslant m$.
(ii) $|G_m| = m^3$.

In this paper, we consider the groups K_m and G_m which are nilpotent groups of nilpotency class two. In section 2, we compute the probability of having n^{th}-root of K_m and G_m. Section 3 is devoted to finding integers n for which, K_m and G_m are n-central.

2. The probability of having n^{th}-roots

In this section we consider groups K_m and G_m and find the probability of having n^{th}-roots. Here for $m \in \mathbb{Z}$, by m^* we mean the arithmetic inverse of m.

Proposition 2.1 For integers $m, n \geqslant 2$;

(1) If $G = K_m$ and $x \in G$, then we have

$$x^n = a^{n\beta} b^{n\gamma} a^{(m-1)(n\delta + \frac{n(n-1)}{2}\beta\gamma)};$$

(2) If $G = G_m$ and $x \in G$, then we have

$$x^n = a^{ni}b^{nj}[a,b]^{nt-\frac{n(n-1)}{2}ij}.$$

Proof. We use an induction method on n. By Lemma 1.4, the assertion holds for $n = 1$. Now, let

$$x^n = a^{n\beta}b^{n\gamma}a^{(m-1)(n\delta+\frac{n(n-1)}{2}\beta\gamma)}.$$

Then

$$x^{n+1} = a^{\beta}b^{\gamma}a^{(m-1)\delta}a^{n\beta}b^{n\gamma}a^{(m-1)(n\delta+\frac{n(n-1)}{2}\beta\gamma)}$$

By Lemma 1.2, $a^{(m-1)\delta} = b^{(1-m)\delta}$. So

$$\begin{aligned} x^{n+1} &= a^{\beta}b^{\gamma}a^{n\beta}b^{n\gamma}a^{(m-1)((n+1)\delta+\frac{n(n-1)}{2}\beta\gamma)} \\ &= a^{(n+1)\beta}[b,a]^{n\beta\gamma}b^{(n+1)\gamma}a^{(m-1)((n+1)\delta+\frac{n(n-1)}{2}\beta\gamma)}. \end{aligned}$$

Since K_m is a group of nilpotency class two, $G' \subseteq Z(G)$. Hence by Lemma 1.1 we have

$$x^{n+1} = a^{(n+1)\beta}b^{(n+1)\gamma}a^{(m-1)((n+1)\delta+\frac{n(n+1)}{2}\beta\gamma)}.$$

The second part can be proved similarly. ■

Theorem 2.2 Let $G = K_m$, where $m \geqslant 2$. Then

$$P_n(G) = \begin{cases} \frac{2}{d^3} & if\ n\ be\ even,\ (\frac{n}{2}, m-1) = \frac{d}{2}\ and\ \frac{m-1}{d}\ be\ odd; \\ \frac{1}{d^3} & otherwise, \end{cases}$$

where $(n, m-1) = d$.

Proof. Let $a^{\beta}b^{\gamma}a^{(m-1)\delta}$ be an element of G^n where $1 \leqslant \beta, \gamma, \delta \leqslant m-1$. If $x = (x_1)^n$ when $a^{\beta_1}b^{\gamma_1}a^{(m-1)\delta_1} \in G$, $1 \leqslant \beta_1, \gamma_1, \delta_1 \leqslant m-1$, then by Proposition 2.1 we have

$$\begin{aligned} a^{\beta}b^{\gamma}a^{(m-1)\delta} &= (a^{\beta_1}b^{\gamma_1}a^{(m-1)\delta_1})^n \\ &= a^{n\beta_1}b^{n\gamma_1}a^{(m-1)(n\delta_1+\frac{n(n-1)}{2}\beta_1\gamma_1)}. \end{aligned}$$

By uniqueness of presentation of G, we obtain

$$\begin{cases} n\beta_1 \equiv \beta \pmod{m-1} \\ n\gamma_1 \equiv \gamma \pmod{m-1} \\ n\delta_1 + \frac{n(n-1)}{2}\beta_1\gamma_1 \equiv \delta \pmod{m-1}. \end{cases} \qquad (1)$$

Now let $(n, m-1) = d$. The first congruence of the system (1) has the solution

$$\beta_1 \equiv (\frac{n}{d})^*(\frac{\beta}{d}) \quad (mod\ \frac{m-1}{d})$$

if and only if $d \mid \beta$. Then

$$\beta \in \{d, 2d, \ \ldots, \frac{m-1}{d} \times d\}.$$

This means that β has $\frac{m-1}{d}$ choices. Similarly, by second equation of System (1) we get

$$\gamma \in \{d, 2d, \ \ldots, \frac{m-1}{d} \times d\}.$$

So γ admits $\frac{m-1}{d}$ values.
Now for finding the number of values of δ, we consider two cases, where n is odd or even.

First let n be an odd integers. Then

$$n(\delta_1 + \frac{n(n-1)}{2}\beta_1\gamma_1) \equiv \delta \quad (mod\ m-1).$$

Since $(n, m-1) = d$, we get

$$\delta_1 \equiv (\frac{n}{d})^* \frac{\delta}{d} - \frac{n(n-1)}{2}\beta_1\gamma_1 \quad (mod\ \frac{m-1}{d})$$

provided that $d \mid \delta$. So

$$\delta \in \{d, 2d, \ \ldots, \frac{m-1}{d} \times d\}.$$

Therefore in this case we have $\frac{m-1}{d}$ choices for δ. By the above facts, we have

$$\begin{aligned} \mid G^n \mid &= \mid \{a^\beta b^\gamma a^{(m-1)\delta} \mid \beta \in \{d, \ldots, \frac{m-1}{d}d\}, \gamma \in \{d, \ldots, \frac{m-1}{d}d\}, \delta \in \{d, \ldots, \frac{m-1}{d}d\}\} \mid \\ &= \mid \{(\beta, \gamma, \delta) \mid \{\beta \in \{d, \ldots, \frac{m-1}{d}d\}, \gamma \in \{d, \ldots, \frac{m-1}{d}d\}, \delta \in \{d, \ldots, \frac{m-1}{d}d\}\} \mid \\ &= \frac{m-1}{d} \times \frac{m-1}{d} \times \frac{m-1}{d} = (\frac{m-1}{d})^3. \end{aligned}$$

So

$$P_n(G) = \frac{\mid G^n \mid}{\mid G \mid} = \frac{(m-1/d)^3}{(m-1)^3} = \frac{1}{d^3}.$$

Now suppose n be an even integer. Then $(\frac{n}{2}, m-1) = d$ or $(\frac{n}{2}, m-1) = \frac{d}{2}$.
Case 1. Let $(\frac{n}{2}, m-1) = d$. Then

$$\frac{n}{2}(2\delta_1 + (n-1)\beta_1\gamma_1) \equiv \delta \quad (mod\ m-1).$$

So

$$2\delta_1 \equiv (\frac{n}{2d})^* \frac{\delta}{d} - (n-1)\beta_1\gamma_1 \quad (mod\ \frac{m-1}{d}).$$

Since $(\frac{n}{2}, m-1) = d$, $(\frac{m-1}{d}, 2) = 1$. Hence, the above congruence holds if and only if $d \mid \delta$. Therefore

$$\delta \in \{d, 2d, \ \ldots\ , \frac{m-1}{d} \times d\}.$$

So

$$| G^n | = | \{(\beta, \gamma, \delta) \mid \{\beta \in \{d, \ldots, \frac{m-1}{d}d\}, \gamma \in \{d, \ldots, \frac{m-1}{d}d\}, \delta \in \{d, \ldots, \frac{m-1}{d}d\}\} |$$
$$= (\frac{m-1}{d})^3$$

and consequently

$$P_n(G) = \frac{1}{d^3}.$$

Case 2. Let $(\frac{n}{2}, m-1) = \frac{d}{2}$. Then

$$\frac{n}{d}(2\delta_1 + (n-1)\beta_1\gamma_1) \equiv \frac{2\delta}{d} \quad (mod\ \frac{2(m-1)}{d}).$$

Hence

$$2\delta_1 \equiv (\frac{n}{d})^* \frac{2\delta}{d} - (n-1)\beta_1\gamma_1 \quad (mod\ \frac{2(m-1)}{d}). \quad (2)$$

So, we must have $2 \mid \beta_1\gamma_1$. Suppose $2 \mid \gamma_1$. Now by congruence

$$\gamma_1 \equiv (\frac{n}{d})^* \frac{\gamma}{d} \quad (mod\ \frac{m-1}{d}) \quad (3)$$

we consider two subcases:

Subcase 2.a. Let $\frac{(m-1)}{d}$ be an even integer. Now since

$$\frac{n}{d}(\frac{n}{d})^* \equiv 1 \quad (mod\ \frac{m-1}{d}),$$

both $\frac{n}{d}$ and $(\frac{n}{d})^*$ are odd. Since $2 \mid \gamma_1$, By congruence (3) we get $2 \mid \frac{\gamma}{d}$. It means that

$$\gamma \in \{2d, 4d, \ldots, \frac{m-1}{2d} \times 2d\}.$$

Hence the number of values of γ is $\frac{m-1}{2d}$. On the other hand according to congruence (2), $\frac{d}{2} \mid \delta$. Therefore

$$\delta \in \{\frac{d}{2}, d, \ldots, \frac{2(m-1)}{d} \times \frac{d}{2}\}.$$

So δ admits $\frac{2(m-1)}{d}$ values. Consequently

$$| G^n |= \frac{m-1}{d} \times \frac{m-1}{2d} \times \frac{2(m-1)}{d} = (\frac{m-1}{d})^3$$

and

$$P_n(G) = \frac{1}{d^3}.$$

Case 2.b. Let $\frac{(m-1)}{d}$ be an odd integer and $\gamma \in \{d, 2d, \ldots, \frac{m-1}{d}d\}$. If

$$\gamma_1 \equiv \frac{n}{d}(\frac{n}{d})^* \ (mod\ \frac{m-1}{d})$$

and γ_1 be an even integer, then we get the desired result. Otherwise, instead of γ_1, we put $\gamma_1 + \frac{m-1}{d}$. So for each

$$\gamma \in \{d, 2d, \ldots, \frac{m-1}{d} \times d\},$$

the congruence holds. It means that the number of choices for γ is equal to $\frac{m-1}{d}$. Finally, we get

$$| G^n |= \frac{m-1}{d} \times \frac{m-1}{d} \times \frac{2(m-1)}{d} = 2(\frac{m-1}{d})^3$$

and

$$P_n(G) = \frac{2}{d^3}.$$

■

Theorem 2.3 Let $G = G_m$, where $m \geqslant 2$. Then

$$P_n(G) = \begin{cases} \frac{2}{d^3} \ if\ n\ be\ even,\ (\frac{n}{2}, m) = \frac{d}{2}\ and\ \frac{m}{d}\ be\ odd; \\ \frac{1}{d^3}\ otherwise, \end{cases}$$

where $(n, m) = d$.

Proof. Let $a^i b^j [a, b]^t$ be an element of G^n where $1 \leqslant i, j, t \leqslant m$. If $x = (x_1)^n$ when $a^{i_1} b^{j} [a, b]^{t_1} \in G$, $1 \leqslant i_1, j_1, t_1 \leqslant m$, then by Proposition 2.1 we have

$$\begin{aligned} a^i b^j [a, b]^t &= (a^{i_1} b^{j_1} [a, b]^{t_1})^n \\ &= a^{ni_1} b^{nj_1} [a, b]^{nt_1 - \frac{n(n-1)}{2} i_1 j_1}. \end{aligned}$$

By uniqueness of presentation of G, we obtain

$$\begin{cases} ni_1 \equiv i \quad (mod\ m) \\ nj_1 \equiv j \quad (mod\ m) \\ nt_1 - \frac{n(n-1)}{2} i_1 j_1 \equiv t \quad (mod\ m). \end{cases}$$

The obtained congruence system is exactly similar to System (1). So it can be solve, similarly. ■

3. n-centrality

In this section, we again consider groups K_m, G_m and investigate n-centrality for them.

Theorem 3.1 Let $G = K_m$, where $m \geqslant 2$. Then for $n > 1$, the group G is n-central if and only if $m - 1 \mid n$.

Proof. By Proposition 2.1 and Lemma 1.1, we get

$$x^n y = a^{n\beta_1+\beta_2} b^{n\gamma_1+\gamma_2} a^{(m-1)(n\delta_1+\delta_2+\frac{n(n-1)}{2}\beta_1\gamma_1+n\beta_2\gamma_1)}.$$

Also we obtain

$$yx^n = a^{n\beta_1+\beta_2} b^{n\gamma_1+\gamma_2} a^{(m-1)(n\delta_1+\delta_2+\frac{n(n-1)}{2}\beta_1\gamma_1+n\beta_1\gamma_2)}.$$

We know that G is n-central if and only if $x^n y = yx^n$, for all $x, y \in G$. Furthermore by uniqueness of presentation of $x^n y$ and yx^n, we see that $x^n y = yx^n$ if and only if

$$n\delta_1 + \delta_2 + \frac{n(n-1)}{2}\beta_1\gamma_1 + n\beta_2\gamma_1 \equiv n\delta_1 + \delta_2 + \frac{n(n-1)}{2}\beta_1\gamma_1 + n\beta_1\gamma_2 \pmod{m-1}.$$

This is equivalent to

$$n(\beta_1\gamma_2 - \beta_2\gamma_1) \equiv 0 \pmod{m-1}.$$

Now since this holds for all $x, y \in G$, $m - 1 \mid n$. ■

Theorem 3.2 Let $G = G_m$, where $m \geqslant 2$. Then for $n > 1$, the group G is n-central if and only if $m \mid n$.

Proof. By Proposition 2.1 and Lemma 1.1, we get

$$x^n y = a^{ni_1+i_2} b^{nj_1+j_2} [a, b]^{nt_1+t_2-\frac{n(n-1)}{2}i_1j_1-ni_2j_1}.$$

Also we obtain

$$yx^n = a^{ni_1+i_2} b^{nj_1+j_2} [a, b]^{nt_1+t_2-\frac{n(n-1)}{2}i_1j_1-ni_1j_2}.$$

We know that G is n-central if and only if $x^n y = yx^n$, for all $x, y \in G$. Furthermore by uniqueness of presentation of $x^n y$ and yx^n, we see that $x^n y = yx^n$ if and only if

$$nt_1 + t_2 - \frac{n(n-1)}{2}i_1j_1 - ni_2j_1 \equiv nt_1 + t_2 - \frac{n(n-1)}{2}i_1j_1 - ni_1j_2 \pmod{m}.$$

This is equivalent to

$$n(i_1j_2 - i_2j_1) \equiv 0 \pmod{m}.$$

Now since this holds for all $x, y \in G$, $m \mid n$. ■

References

[1] C. M. Campbell, P. P. Campel, H. Doostie and E. F. Robertson, Fibonacci length for metacyclian groups. Algebra Colloq. 11 (2004), 215-222.

[2] C. M. Campbell, E. F. Robertson, On a group presentation due to Fox. Canada. Math. Bull. 19 (1967), 247-248.

[3] H. Doostie, M. Hashemi, Fibonacci lengths involving the Wall number $K(n)$. J. Appl. Math. Computing. 20 (2006), 171-180.

[4] A. Sadeghieh, H. Doostie And M. Azadi, Certain numerical results on the Fibonacci length and n^{th}-roots of Hamiltonian groups. International Mathematical Forum. 39 (2009), 1923-1938.

[5] A. Sadeghieh, H. Doostie, The n-th roots of elements in finite groups. Mathematical Sciences. 4 (2008), 347-356.

[6] C. Delizia, A. Tortora and A. Abdollahi, Some special classes of n-abelian groups. International journal of Group Theory. 1 (2012), 19-24.

17

On categories of merotopic, nearness, and filter algebras

Vijaya L. Gompa*

Department Head and Professor of Mathematics, Jacksonville State University, Jacksonville, AL 36265.

Abstract. We study algebraic properties of categories of Merotopic, Nearness, and Filter Algebras. We show that the category of filter torsion free abelian groups is an epireflective subcategory of the category of filter abelian groups. The forgetful functor from the category of filter rings to filter monoids is essentially algebraic and the forgetful functor from the category of filter groups to the category of filters has a left adjoint.

Keywords: Universal algebra, topological algebra, nearness spaces, merotopic spaces, filter spaces.

1. Introduction

We first describe three categories which contain ***Top***, the category of topological spaces (sometimes with a separation axiom). They are ***Mer***, the category of the merotopic spaces of Katetov [8], ***Near***, the category of nearness spaces of Herrlich [4], and ***Fil***, the category of filter spaces of Katetov [8].

Let X be a set and $\mathbf{P^2(X)}$ be the set whose members are all collections of subsets of X. For any member $\mathcal{A}$ of $\mathbf{P^2(X)}$, we write

$$\sec \mathcal{A} := \{B \subseteq X : A \cap B \neq \phi \text{ for all } A \in \mathcal{A}\}$$

*Corresponding author.
E-mail address: vgompa@jsu.edu (Vijaya L. Gompa).

and

$$\text{stack}_X \mathcal{A} := \{E \subseteq X : A \subseteq E \text{ for some member A of } \mathcal{A}\}.$$

A member $\mathcal{A}$ of $\mathbf{P^2(X)}$ is called a **filter** iff $X \in \mathcal{A}$, $\phi \notin \mathcal{A}$, intersection of any two sets in $\mathcal{A}$ is in $\mathcal{A}$, and all supersets of members of $\mathcal{A}$ are in $\mathcal{A}$.

For members $\mathcal{A}$ and $\mathcal{B}$ of $\mathbf{P^2(X)}$, the **join** of $\mathcal{A}$ and $\mathcal{B}$, denoted by the symbol $\mathcal{A} \vee \mathcal{B}$, is the collection of all subsets of X of the form $A \cup B$, where $A \in \mathcal{A}$ and $B \in \mathcal{B}$, and the **meet** of $\mathcal{A}$ and $\mathcal{B}$, denoted by $\mathcal{A} \wedge \mathcal{B}$, is the collection of all subsets of the form $A \cap B$ with A and B belonging to $\mathcal{A}$ and $\mathcal{B}$ respectively. We say $\mathcal{A}$ **refines** $\mathcal{B}$ iff for each member A of $\mathcal{A}$ there exists a member B of $\mathcal{B}$ containing A. $\mathcal{A}$ **corefines** $\mathcal{B}$ iff for each $A \in \mathcal{A}$ corresponds a set $B \in \mathcal{B}$ contained in A.

A pair (X, ξ) of a set X and a subset ξ of $\mathbf{P^2(X)}$ is said to be a **prenearness space** iff ξ is a nonempty proper subset of $\mathbf{P^2(X)}$ containing all members of $\mathbf{P^2(X)}$ with nonempty intersection and all corefinements of each of its members.

The prenearness space (X, ξ) is called a **merotopic space** iff ξ containing the join of two members of $\mathbf{P^2(X)}$ means one of them belongs to ξ, in other words, for any $\mathcal{A}$ and $\mathcal{B}$ in $\mathbf{P^2(X)}$ whose join $\mathcal{A} \vee \mathcal{B}$ is an element of ξ, either $\mathcal{A}$ or $\mathcal{B}$ is already in ξ.

The merotopic space (X, ξ) is called a **nearness space** iff ξ contains all members $\mathcal{A}$ in $\mathbf{P^2(X)}$ with the property that the associated closure collection

$$cl_\xi \mathcal{A} := \{cl_\xi A : A \in \mathcal{A}\}$$

belongs to ξ, where

$$cl_\xi A := \{x \in X : \{A, \{x\}\} \in \xi\}.$$

If (X, ξ) and (Y, η) are two merotopic, prenearness, or nearness spaces, then a mapping $f : X \longrightarrow Y$ is called **uniformly continuous** iff $f[\mathcal{A}] := \{f(A) : A \in \mathcal{A}\}$ is a member of η for each $\mathcal{A} \in \xi$.

P − Near, ***Mer***, and ***Near*** are the categories with objects which are prenearness, merotopic, and nearness spaces respectively with uniformly continuous mappings as morphisms. ***Mer*** is a bicoreflective full subcategory of ***P − Near*** and ***Near*** is a bireflective full subcategory of ***Mer*** (see [5]).

If $\boldsymbol{X}$ denotes any one of these categories, any $\boldsymbol{X}$-object is simply denoted by X if the $\boldsymbol{X}$-structure ξ on X is understood. Moreover, a collection $\mathcal{A}$ of subsets of X is said to be

(1) **near** in X provided $\mathcal{A}$ is a member of ξ,

(2) **micromeric** in X provided the collection sec $\mathcal{A}$ is near in X,

(3) **far** in X provided $\mathcal{A}$ is not near in X,

(4) **uniform cover** of X provided the collection $\{X \backslash A : A \in \mathcal{A}\}$ is far in X.

$\mathcal{A}$ is called a **stack** on X iff $\mathcal{A} = \text{stack}_X \mathcal{A}$. The structure of a merotopic space is determined by the set of merotopic stacks because a collection is micromeric iff its stack in X is micromeric.

A filter on X is said to be a **Cauchy filter** iff it is micromeric on X. X is called a **filter-merotopic** (or just a **filter**) **space** iff every micromeric stack contains a Cauchy filter. The full subcategory of ***Mer*** whose objects are all filter spaces will be denoted by ***Fil***. This category ***Fil*** is cartesian closed (see [9]) and is bicoreflective and hereditary in ***Mer***.

A family $\Omega = (n_j)_{j\in J}$ of natural numbers indexed by some set J is called a **type**. The index set J is called the **order** of Ω. In the following, we let a type $\Omega = (n_j)_{j\in J}$ be fixed. A pair $(|A|, (\omega_j)_{j\in J})$ of a set $|A|$ and a family $\omega_j : |A|^{n_j} \longrightarrow |A|$ $(j \in J)$ of mappings is called an **Ω-algebra** (see, for example, [2]). For the sake of simplicity, we write A instead of the pair $(|A|, (\omega_j)_{j\in J})$ and $\omega_{j,A}$ for the **n_j-ary operation** ω_j on A. If the Ω-algebra A is clear from the context, we drop the suffix A in denoting its n_j-ary $(j \in J)$ operation. If A and B are Ω-algebras, then a mapping $f : |A| \longrightarrow |B|$ is said to be an **Ω-morphism** $f : A \longrightarrow B$ iff for each $j \in J$, $f \circ \omega_{j,A} = \omega_{j,B} \circ f^n$ where $n = n_j$ and $f^n : |A|^n \longrightarrow |B|^n$ is the mapping with the obvious definition $(a_1, \ldots, a_n) \longrightarrow (fa_1, \ldots, fa_n)$.

The symbol $\boldsymbol{Alg(\Omega)}$ denotes the category whose objects are Ω-algebras and whose morphisms are Ω-morphisms.

Let $\boldsymbol{X}$ be a construct with finite concrete powers and $\boldsymbol{A}$ be a subcategory of $\boldsymbol{Alg(\Omega)}$. By a **paired object** (from $\boldsymbol{X}$ and $\boldsymbol{A}$) is meant an ordered pair (X, A) where X and A are objects in $\boldsymbol{X}$ and $\boldsymbol{A}$ respectively with the same underlying set such that, for each $j \in J$, the $n(= n_j)$-ary operation $\omega_{j,A} : |A|^n \longrightarrow |A|$ on A is an $\boldsymbol{X}$-morphism

$$\omega_{j,A} : X^n \longrightarrow X.$$

In this case, we write $\omega_{j,X}$ for the $\boldsymbol{X}$-morphism from X^n to X whose underlying function is $\omega_{j,A}$. If (X, A) and (X', A') are two paired objects (from $\boldsymbol{X}$ and $\boldsymbol{A}$), then an $\boldsymbol{X}$-morphism $f : X \longrightarrow X'$ that is also an $\boldsymbol{A}$-morphism $f : A \longrightarrow A'$ is called a **paired morphism** (from $\boldsymbol{X}$ and $\boldsymbol{A}$) and is denoted by $f : (X, A) \longrightarrow (X', A')$. The category of all paired objects (from $\boldsymbol{X}$ and $\boldsymbol{A}$) together with paired morphisms (from $\boldsymbol{X}$ and $\boldsymbol{A}$) is called the **paired category** (from $\boldsymbol{X}$ and $\boldsymbol{A}$). We denote this category by $\boldsymbol{X\diamond A}$.

In this work, we assume that all subcategories are isomorphism closed. The fact that the most of the natural subcategories fall into this class justifies our assumption. Unless otherwise stated, $\boldsymbol{X}$ and $\boldsymbol{Y}$ denote arbitrary constructs with finite concrete powers, and $\boldsymbol{A}$ represents any subcategory of $\boldsymbol{Alg(\Omega)}$. We write $|X|$ for the underlying set of an object X in a construct. For the sake of simplicity, we will denote an object (X, A) in the paired category $\boldsymbol{X\diamond A}$ (from $\boldsymbol{X}$ and $\boldsymbol{A}$) either by X or by A. We will use a similar identification for morphisms in the paired category.

2. Essentially algebraic and algebraic subcategories

We explore algebraic properties of paired categories with the following lemma whose proof can be found in [3].

Lemma 2.1 Suppose that $\boldsymbol{X}$ is monotopological, $\boldsymbol{A}$ is a subcategory of $\boldsymbol{Alg}(\Omega')$, $\boldsymbol{B}$ is an essentially algebraic subcategory of $\boldsymbol{Alg}(\Omega)$, and $H : \boldsymbol{B} \longrightarrow \boldsymbol{A}$ is a concrete functor such that the association

$$\tilde{H}(X, B) := (X, HB)$$

for any $\boldsymbol{X\diamond B}$-object (X, B) and $\tilde{H}f := (f, Hf)$ for any paired morphism

$$f : (X, B) \longrightarrow (X', B'),$$

is a concrete functor $\tilde{H} : \boldsymbol{X\diamond B} \longrightarrow \boldsymbol{X\diamond A}$ in the commutative square diagram

$$\begin{array}{ccc} \boldsymbol{X}\diamond\boldsymbol{B} & \xrightarrow{\tilde{H}} & \boldsymbol{X}\diamond\boldsymbol{A} \\ T\downarrow & & \downarrow T' \\ \boldsymbol{B} & \xrightarrow[H]{} & \boldsymbol{A} \end{array},$$

where T and T' are forgetful functors.

Then the following hold.

(a) If H has a left adjoint, then $\tilde{H}$ also has a left adjoint and $\tilde{H}$ is (generating, monosource) - factorizable.

(b) If H reflects isomorphisms, then $\tilde{H}$ reflects isomorphisms.

(c) If H is essentially algebraic, then $\tilde{H}$ is essentially algebraic.

Proposition 2.2 If $\boldsymbol{X}$ is any one of the categories $\boldsymbol{Mer}$, $\boldsymbol{P-Near}$, $\boldsymbol{Near}$, or $\boldsymbol{Fil}$, then the forgetful functors $\boldsymbol{X}\diamond\boldsymbol{Rng}\rightarrow \boldsymbol{X}\diamond\boldsymbol{Ab}$ and $\boldsymbol{X}\diamond\boldsymbol{Rng}\rightarrow \boldsymbol{X}\diamond\boldsymbol{Mon}$ ($\boldsymbol{Rng}$, $\boldsymbol{Ab}$, and $\boldsymbol{Mon}$ are the categories of rings, abelian groups, or monoids repectively) are essentially algebraic.

Proof. Since the forgetful functors $\boldsymbol{Rng} \rightarrow \boldsymbol{Ab}$ and $\boldsymbol{Rng} \rightarrow \boldsymbol{Mon}$ are essentially algebraic (see [1, 23.18]), the associated forgetful functors are essentially algebraic by Lemma 2.1 (part (c)). ■

Lemma 2.3 Suppose that $\boldsymbol{X}$ is monotopological, $\boldsymbol{A}$ is a subcategory of $\boldsymbol{Alg}(\Omega')$, and $\boldsymbol{B}$ is an essentially algebraic subcategory of $\boldsymbol{Alg(\Omega)}$.

(1) If $\boldsymbol{B}$ is also a reflective subcategory of $\boldsymbol{A}$, then $\boldsymbol{X}\diamond\boldsymbol{B}$ is a reflective subcategory of $\boldsymbol{X}\diamond\boldsymbol{A}$.

(2) If $\boldsymbol{B}$ is also an epireflective subcategory of $\boldsymbol{A}$, then $\boldsymbol{X}\diamond\boldsymbol{B}$ is an epireflective subcategory of $\boldsymbol{X}\diamond\boldsymbol{A}$.

Proof. The inclusion map $H : \boldsymbol{B} \rightarrow \boldsymbol{A}$ is essentially algebraic, and hence $\tilde{H}$, defined as in Lemma 2.1, is essentially algebraic (by Lemma 2.1). Thus $\boldsymbol{X}\diamond\boldsymbol{B}$ is a reflective subcategory of $\boldsymbol{X}\diamond\boldsymbol{A}$.

For the second part, if $\boldsymbol{B}$ is also an epireflective subcategory of $\boldsymbol{A}$, then $H : \boldsymbol{B} \rightarrow \boldsymbol{A}$ has a left adjoint, and hence $\tilde{H}$ has left adjoint (by Lemma 2.1), which in turn implies that $\boldsymbol{X}\diamond\boldsymbol{B}$ is an epireflective subcategory of $\boldsymbol{X}\diamond\boldsymbol{A}$. ■

Proposition 2.4 Suppose that $\boldsymbol{X}$ is any one of the categories $\boldsymbol{Mer}$, $\boldsymbol{P-Near}$, $\boldsymbol{Near}$, or $\boldsymbol{Fil}$. Then

(a) $\boldsymbol{X}\diamond\boldsymbol{Ab}$ is an epireflective subcategory of $\boldsymbol{X}\diamond\boldsymbol{Grp}$, where $\boldsymbol{Grp}$ is the category of groups.

(b) If $\boldsymbol{B}$ is the category $\boldsymbol{TfAb}$ of Torsion free abelian groups or the category $\boldsymbol{Ab}_n$ of abelian groups annihilated by a fixed integer n, then $\boldsymbol{X}\diamond\boldsymbol{B}$ is an epireflective subcategory of $\boldsymbol{X}\diamond\boldsymbol{Ab}$.

and

(c) $\boldsymbol{X}\diamond\boldsymbol{B}$ is an epireflective subcategory of $\boldsymbol{X}\diamond\boldsymbol{Alg(\Omega)}$ where $\boldsymbol{B}$ is any SP-class of Ω-algebras.

Proof. This proposition is a consequence of the second part of Lemma 2.3 because $\boldsymbol{Ab}$ is an epireflective subcategory of $\boldsymbol{Grp}$; $\boldsymbol{TfAb}$ and $\boldsymbol{Ab}_n$ are epireflective subcategories of

$\boldsymbol{Ab}$ (see [6, 26.2 (2) (b)]); and any SP-class of Ω-algebras is an epireflective subcategory of $\boldsymbol{Alg(\Omega)}$ ([7]). ■

The following lemma, whose proof can be found in [3], shows that algebraic subcategories of $\boldsymbol{Alg(\Omega)}$ can be paired with well-fibred topological categories to yield algebraic subcategories of paired categories.

Lemma 2.5 Suppose that $\boldsymbol{X}$ is a well-fibred topological category with finitely productive quotients, $\boldsymbol{A}$ is a subcategory of $\boldsymbol{Alg}(\Omega')$, $\boldsymbol{B}$ is an algebraic subcategory of $\boldsymbol{Alg}(\Omega)$, and $H : \boldsymbol{B} \longrightarrow \boldsymbol{A}$ is a concretely algebraic functor that induces a concrete functor $\tilde{H} : \boldsymbol{X \diamond B} \longrightarrow \boldsymbol{X \diamond A}$ as defined in the commutative square diagram in Lemma 2.1. Then $\tilde{H}$ is algebraic.

Proposition 2.6 The functors $\boldsymbol{Mer \diamond Rng} \longrightarrow \boldsymbol{Mer \diamond Ab}$ and $\boldsymbol{Mer \diamond Vec} \longrightarrow \boldsymbol{Mer \diamond Ab}$ are algebraic, where $\boldsymbol{Vec}$ is the category of vector spaces.

Proof. Since the forgetful functors $\boldsymbol{Rng} \longrightarrow \boldsymbol{Ab}$ and $\boldsymbol{Vec} \longrightarrow \boldsymbol{Ab}$ are algebraic [6, 32.20], it follows from Lemma 2.5 that $\tilde{H}$, described in the commutative diagram in Lemma 2.1, has left adjoint, where $\boldsymbol{B}$ is $\boldsymbol{Rng}$ or $\boldsymbol{Vec}$. ■

Lemma 2.7 If $\boldsymbol{Y}$ is a coreflective subcategory of $\boldsymbol{X}$ and the pair (X, A) is an object in $\boldsymbol{X \diamond A}$ such that X is an object in $\boldsymbol{Y}$ then (X, A) is also an object in $\boldsymbol{Y \diamond A}$.

Proof. We need to prove that each algebraic operation on A is a $\boldsymbol{Y}$-morphism. Let $j \in J$ and $n = n_j$. To avoid ambiguity, let us use the symbol Y to indicate X regarded as a $\boldsymbol{Y}$-object and write Y^n and X^n for the products of X to itself n times in the categories $\boldsymbol{Y}$ and $\boldsymbol{X}$ respectively. Since $\boldsymbol{Y}$ is coreflective in $\boldsymbol{X}$, the $\boldsymbol{Y}$-product Y^n is the $\boldsymbol{Y}$-coreflection of the $\boldsymbol{X}$-product X^n. Therefore, any $\boldsymbol{X}$-morphism $X^n \longrightarrow X$ is also a $\boldsymbol{Y}$-morphism

$$Y^n \longrightarrow Y.$$

In particular, the n_j-ary operation on A being an $\boldsymbol{X}$-morphism $\omega_{j,X} : X^n \longrightarrow X$ is indeed a $\boldsymbol{Y}$-morphism $\omega_{j,Y} : Y^n \longrightarrow Y$. ■

Proposition 2.8 Let $\boldsymbol{A}$ be a subcategory of $\boldsymbol{Alg(\Omega)}$. A merotopic space which is an object of $\boldsymbol{P - Near \diamond A}$ is also an object of $\boldsymbol{Mer \diamond A}$.

Proof. The result follows from Lemma 2.7, because $\boldsymbol{Mer}$ is a bicoreflective subcategory of $\boldsymbol{P - Near}$. ■

Lemma 2.9 If $\boldsymbol{Y}$ is a subcategory of $\boldsymbol{X}$ such that concrete powers in $\boldsymbol{Y}$ agree with concrete powers in $\boldsymbol{X}$ then $\boldsymbol{Y \diamond A}$ is a subcategory of $\boldsymbol{X \diamond A}$.

In particular, if $\boldsymbol{Y}$ is an epireflective subcategory of $\boldsymbol{X}$, then $\boldsymbol{Y \diamond A}$ is a subcategory of $\boldsymbol{X \diamond A}$.

Proof. Let (Y, A) be any object in $\boldsymbol{Y \diamond A}$. For each $j \in J$, the n_j-th product Y^{n_j} of Y in the category $\boldsymbol{Y}$ is the same as the n_j-th product of Y in the category $\boldsymbol{X}$ and the n_j-ary operation $\omega_{j,Y} : Y^{n_j} \longrightarrow Y$, being a morphism in $\boldsymbol{Y}$, must be a morphism in $\boldsymbol{X}$. Thus (Y, A) is also an $\boldsymbol{X \diamond A}$-object. Obviously $\boldsymbol{Y \diamond A}$-morphisms are also morphisms in $\boldsymbol{X \diamond A}$.

If $\boldsymbol{Y}$ is an epireflective subcategory of $\boldsymbol{X}$, then the products in $\boldsymbol{Y}$ do agree with those in $\boldsymbol{X}$ so that $\boldsymbol{Y \diamond A}$ is a subcategory of $\boldsymbol{X \diamond A}$ by what was proved above. ■

Proposition 2.10 Let $\boldsymbol{A}$ be a subcategory of $\boldsymbol{Alg(\Omega)}$. A nearness space is an object of $\boldsymbol{Near \diamond A}$ iff it is an object of $\boldsymbol{Mer \diamond A}$.

In particular, $\boldsymbol{Near \diamond A}$ is a subcategory of $\boldsymbol{Mer \diamond A}$.

Proof. A near Ω-algebra is a merotopic Ω-algebra by Lemma 2.9 because ***Near*** is a bireflective subcategory of ***Mer***. The converse is trivial since the products in ***Near*** agree with the products in ***Mer***. ■

Let $\boldsymbol{A}$ be a subcategory of $\boldsymbol{Alg(\Omega)}$ and $\boldsymbol{X}$ be any one of the categories $\boldsymbol{P-Near}$, ***Mer***, or ***Near***. If X is an $\boldsymbol{A}$-object, $(X_i)_{i\in I}$ is a family of $\boldsymbol{X\diamond A}$-objects and

$$f_i : X \longrightarrow X_i$$

is an $\boldsymbol{A}$-homomorphism for each $i \in I$, then the initial structure with respect to $(f_i)_{i\in I}$ in $\boldsymbol{X}$ makes X an $\boldsymbol{X\diamond A}$-object. A subspace of an $\boldsymbol{X\diamond A}$-object which is also a sub-$\boldsymbol{A}$-object is itself an $\boldsymbol{X\diamond A}$-object. Thus we can say that a sub-$\boldsymbol{X\diamond A}$-object of an $\boldsymbol{X\diamond A}$-object is a subspace of an $\boldsymbol{X}$-space which is also a sub-$\boldsymbol{A}$-object. For instance, a submerotopic group Y of a merotopic group X is a merotopic subspace of X which is a subgroup of X. If concrete products exist in $\boldsymbol{A}$ and if $(X_i)_{i\in I}$ is a family of $\boldsymbol{X\diamond A}$-objects, then the cartesian product ΠX_i is also an $\boldsymbol{X\diamond A}$-object with the initial $\boldsymbol{X}$-structure with respect to the natural projections.

The initial $\boldsymbol{X}$-structures in $\boldsymbol{P-Near}$ and ***Mer*** are different, while those in ***Near*** coincide with the construction in ***Mer***.

Let X be a set, $((X_i, \xi_i))_{i\in I}$ be a family of $\boldsymbol{X}$-spaces and $f_i : X \longrightarrow X_i$ be a map for each $i \in I$. Define

$$\xi := \{\mathcal{A} \in \boldsymbol{P}^2(X) : f_i[\mathcal{A}] \in \xi_i \text{ for all } i \in I\}.$$

Then ξ is a prenearness structure on X, initial with respect to the family $(f_i : X \longrightarrow X_i)_{i\in I}$. If each X_i is a merotopic (or nearness) space, then the merotopic reflection of ξ, defined as the collection of all members A of $\mathbf{P^2}(\boldsymbol{\Pi}X_i)$ such that there is no finite join of elements in $\mathbf{P^2}(\boldsymbol{\Pi}X_i)\backslash\xi$ which corefines $\mathcal{A}$, is the merotopic (or nearness, respectively) structure on X, that is initial with respect to the family $(f_i : X \longrightarrow X_i)_{i\in I}$.

We say that **final epi sinks are finitely productive** in $\boldsymbol{X}$ iff the product

$$(f_i \times g_k : X_i \times Y_k \longrightarrow X \times Y, X \times Y)_{i\in I, k\in K}$$

of any two final epi sinks $(f_i : X_i \longrightarrow X, X)_{i\in I}$ and $(g_k : Y_k \longrightarrow Y, Y)_{k\in K}$ in $\boldsymbol{X}$ is final in $\boldsymbol{X}$.

Lemma 2.11 Suppose that final epi sinks are finitely productive in $\boldsymbol{X}$, A is an $\boldsymbol{A}$-object, $(X_i)_{i\in I}$ is a family of $\boldsymbol{X}$-objects, and

$$(f_i :\mid X_i \mid\longrightarrow \mid A \mid)_{i\in I}$$

is a class of functions Ω-admissible to A. If X is an $\boldsymbol{X}$-object with the same underlying set as A such that $(f_i : X_i \longrightarrow X, X)_{i\in I}$ is a final epi sink in $\boldsymbol{X}$, then (X, A) is an $\boldsymbol{X\diamond A}$-object.

Proof. Since X has the final structure with respect to $f_i : X_i \longrightarrow X$, for any positive integer n, X^n has the final structure with respect to $f_{i_1} \times \ldots \times f_{i_n}, i_1 \in I, .., i_n \in I$, by hypothesis. Let $j \in J$ and $n = n_j$. We have to show that $\omega_{j,A}$ is an $\boldsymbol{X}$-morphism. However, $\omega_{j,A} \circ (f_{i_1} \times \ldots \times f_{i_n}), i_1 \in I, \ldots, i_n \in I$, being one of the $f_i's$ as (f_i) is Ω-admissible to A, is an $\boldsymbol{X}$-morphism. Consequently, the n_j-ary operation $\omega_{j,A}$ on A is an

X-morphism

$$\omega_{j,X} : X^n \longrightarrow X.$$

This being true for each $j \in J$, (X, A) is an ***X◇A***-object. ■

Proposition 2.12 Let ***A*** be a subcategory of ***Alg*(Ω)**. Suppose that $(X_i)_{i \in I}$ is a family of ***Fil◇A***-objects, X is an ***A***-object, and $f_i : X_i \longrightarrow X$ is a function for each i ∈I. If (f_i) is Ω-admissible to X (that is, for each $j \in J, n = n_j, i_1, \ldots, i_n \in I$ there exists $j \in I$ such that $\omega_{j,A} \circ (f_{i_1} \times \ldots \times f_{i_n}) = f_j$), then X becomes a ***Fil◇A***-object with the final filter structure with respect to $(f_i)_{i \in I}$.

The quotient of a ***Fil◇A***-object is nothing but the quotient of the corresponding filter space.

Proof. Note that final epi sinks in the category ***Fil*** are finitely productive (see [8]). Since X has the final structure with respect to $f_i : X_i \longrightarrow X$, for any positive integer n, X^n has the final structure with respect to $f_{i_1} \times \ldots \times f_{i_n}, i_1 \in I, .., i_n \in I$, by hypothesis. Let $j \in J$ and $n = n_j$. We have to show that $\omega_{j,A}$ is a ***Fil***-morphism. However,

$$\omega_{j,A} \circ (f_{i_1} \times \ldots \times f_{i_n}), \quad i_1 \in I, \ldots, i_n \in I,$$

being one of the $f_i's$ as (f_i) is Ω-admissible to A, is a ***Fil***-morphism. Consequently, the n_j-ary operation $\omega_{j,A}$ on A is a ***Fil***-morphism $\omega_{j,X} : X^n \longrightarrow X$. This being true for each $j \in J$, (X, A) is a ***Fil◇A***-object.

To prove the second part, assume that (X, A) is a ***Fil◇A***-object, $f : A \longrightarrow A'$ is an ***A***-morphism, and $f : X \longrightarrow X'$ is a quotient map in ***Fil***. We show that (X', A') is a ***Fil◇A***-object.

Let $j \in J$, $n = n_j$ and $\omega = \omega_{j,A}$, $\omega' = \omega_{j,A'}$ be n-ary operations on A and A' respectively. Since ***Fil*** has finitely productive quotients, X'^n is a quotient of X^n with respect to $f^n : X^n \longrightarrow X'^n$. Thus ω' is a ***Fil***-morphism iff $\omega' \circ f^n$ is a ***Fil***-morphism. However, because f is an Ω-homomorphism, we have the commutative diagram,

$$\begin{array}{ccc} |X^n| & \xrightarrow{\ \omega\ } & |X| \\ {\scriptstyle f^n}\downarrow & & \downarrow{\scriptstyle \omega} \\ |X'^n| & \xrightarrow[f]{} & |X'|, \end{array}$$

which shows that $\omega' \circ f^n$, being equal to $f \circ \omega$, is a ***Fil***-morphism. This being true for each $j \in J$, (X', A') is a ***Fil◇A***-object.

It remains to show that $f : (X, A) \longrightarrow (X', A')$ is a quotient map in ***Fil◇A***. Let (X'', A'') be any object in ***Fil◇A*** and $g : |X'| \longrightarrow |X''|$ be a function between the two sets such that $g \circ f$ is a ***Fil◇A***-morphism. Then g is a ***Fil***-morphism between X' and X'' because $g \circ f$ is one such and $f : X \longrightarrow X'$ is a quotient map in ***Fil***. Similarly g is also an Ω-homomorphism. Thus $g : (X', A') \longrightarrow (X'', A'')$ is a ***Fil◇A***-morphism. ■

In [3], it is proved that, the forgetful functor U : ***X◇A*** $\longrightarrow$ ***X*** has a left adjoint whenever ***X*** is cartesian closed topological category and ***A*** is algebraic. Therefore, The forgetful functor U : ***Fil◇Grp*** $\longrightarrow$ ***Fil*** is algebraic. Here we give a constructive proof to show that U has a left adjoint.

Proposition 2.13 The forgetful functor $U : \boldsymbol{Fil\text{-}Grp} \longrightarrow \boldsymbol{Fil}$ has a left adjoint.

Proof. Suppose that X is an arbitrary filter space. Let A be the free group generated by X and $u : X \hookrightarrow A$ be the inclusion map. For each $n \in \mathbf{N}$ and for any subset L of $\mathbf{N}_n := \{1, 2, \ldots, n\}$, define $h_{n,L} : A^n \longrightarrow A$ by

$$h_{n,L}(a_1, \ldots, a_n) := \beta_{\chi_L(1)}(a_1) \ldots \beta_{\chi_L(n)}(a_n),$$

where χ_L is the characteristic function of L (i.e., $\chi_L(r) = 1$ if $r \in L$ and $\chi_L(r) = 0$ if $r \notin L$), $\beta_0 :=\mathrm{id}_A$, and β_1 is the inversion on A. Equip A with the final structure in the category $\boldsymbol{Fil}$ with respect to $(h_{n,L} \circ u^n : X^n \longrightarrow A)_{n\in\mathbf{N},L\subseteq\mathbf{N}_n}$.

We now show that A is a filter group. Noting that

$$\alpha \circ (h_{n,L} \circ u^n \times h_{m,K} \circ u^m) = h_{n+m,L\cup(n+K)} \circ u^{n+m}$$

is uniformly continuous and that $A \times A$ has the final structure (because $h_{n,L} \circ u^n$'s form an epi sink in $\boldsymbol{Fil}$ and $\boldsymbol{Fil}$ is cartesian closed) with respect to $h_{n,L} \circ u^n \times h_{m,K} \circ u^m$ ($n \in \mathbf{N}$, $m \in \mathbf{N}$, $L \subseteq \mathbf{N}_n$, $K \subseteq \mathbf{N}_m$), we conclude that the multiplication α on A is uniformly continuous. Since $\beta_1 \circ (h_{n,L} \circ u^n) = (h_{n,L'} \circ u^n) \circ r_n$, where r_n is the map from X^n to X^n which assigns $(x_n, \ldots, x_1)$ to $(x_1, \ldots, x_n)$ and L' is the complement of L in $\mathbf{N}_n, \beta_1$ is uniformly continuous. Hence A is a filter group. Finally we show that (u, A) is a universal map for X. Clearly u is uniformly continuous since $u = h_{1,\phi}\circ$u. Let A' be any filter group and $f : X \longrightarrow A'$ be uniformly continuous. Since A is a free group on X, there exists a unique group homomorphism $\bar{f} : A \longrightarrow A'$ which extends f, i.e., $\bar{f} \circ u =$f. It remains to show that $\bar{f}$ is uniformly continuous. It is enough to show, for each $n \in \mathbf{N}$ and for each subset L of $\mathbf{N}_n$, that the composition $\bar{f} \circ (h_{n,L} \circ u^n)$ is uniformly continuous. But $\bar{f} \circ (h_{n,L} \circ u^n) = h_{n,L} \circ f^n$ and $h_{n,L} \circ f^n$ is obviously uniformly continuous. This shows U has a left adjoint. ■

Acknowledgments

The author wishes to acknowledge thanks to Prof. H. Lamar Bentley for his helpful suggestions.

References

[1] J. Adámek, H. Herrlich and G. E. Strecker, Abstract and Concrete Categories. John Wiley & Sons, Inc., New York, 1990.

[2] P. M. Cohn, Universal Algebra. Harper and Row, Publishers, New York, 1965.

[3] V. L. Gompa, Essentially algebraic functors and topological algebra. Indian Journal of Mathematics, 35, (1993), 189-195.

[4] H. Herrlich, A concept of nearness. General Topology and its Applications 4 (1974), 191-212.

[5] H. Herrlich, Topological structures. In: Topological structures. Math. Centre Tracts 52 (1974), 59-122.

[6] H. Herrlich, G. E. Strecker, Category Theory. Allyn and Bacon, Boston, 1973.

[7] Y. H. Hong, Studies on categories of universal topological algebras. Doctoral Dissertation, McMaster University, 1974.

[8] M. Katetov, On continuity structures and spaces of mappings. Comment. Math. Univ. Carol. 6 (1965), 257 - 278.

[9] M. Katetov, Convergence structures. General Topology and its Applications II, Academic Press, New York (1967), 207-216.

[10] O. Wyler, On the categories of general topology and topological algebras. Arc. Math. (Basel) 22 (1971), 7-17.

18

Recognition by prime graph of the almost simple group PGL(2, 25)

A. Mahmoudifar*

Department of Mathematics, Tehran North Branch, Islamic Azad University, Tehran, Iran.

Abstract. Throughout this paper, every groups are finite. The prime graph of a group G is denoted by $\Gamma(G)$. Also G is called recognizable by prime graph if for every finite group H with $\Gamma(H) = \Gamma(G)$, we conclude that $G \cong H$. Until now, it is proved that if k is an odd number and p is an odd prime number, then $\mathrm{PGL}(2, p^k)$ is recognizable by prime graph. So if k is even, the recognition by prime graph of $\mathrm{PGL}(2, p^k)$, where p is an odd prime number, is an open problem. In this paper, we generalize this result and we prove that the almost simple group $\mathrm{PGL}(2, 25)$ is recognizable by prime graph.

Keywords: Linear group, almost simple group, prime graph, element order, Frobenius group.

1. Introduction

Let $\mathbb{N}$ denotes the set of natural numbers. If $n \in \mathbb{N}$, then we denote by $\pi(n)$, the set of all prime divisors of n. Let G be a finite group. The set $\pi(|G|)$ is denoted by $\pi(G)$. Also the set of element orders of G is denoted by $\pi_e(G)$. We denote by $\mu(S)$, the maximal numbers of $\pi_e(G)$ under the divisibility relation. The *prime graph* of G is a graph whose vertex set is $\pi(G)$ and two distinct primes p and q are joined by an edge (and we write $p \sim q$), whenever G contains an element of order pq. The prime graph of G is denoted by $\Gamma(G)$. A finite group G is called *recognizable by prime graph* if for every finite group H such that $\Gamma(H) = \Gamma(G)$, then we have $G \cong H$.

*Corresponding author.
E-mail address: alimahmoudifar@gmail.com (A. Mahmoudifar).

In [10], it is proved that if p is a prime number which is not a Mersenne or Fermat prime and $p \neq 11$, 19 and $\Gamma(G) = \Gamma(\mathrm{PGL}(2,p))$, then G has a unique nonabelian composition factor which is isomorphic to $\mathrm{PSL}(2,p)$ and if $p = 13$, then G has a unique nonabelian composition factor which is isomorphic to $\mathrm{PSL}(2,13)$ or $\mathrm{PSL}(2,27)$. In [3], it is proved that if $q = p^{\alpha}$, where p is a prime and $\alpha > 1$, then $\mathrm{PGL}(2,q)$ is uniquely determined by its element orders. Also in [1], it is proved that if $q = p^{\alpha}$, where p is an odd prime and α is an odd natural number, then $\mathrm{PGL}(2,q)$ is uniquely determined by its prime graph. In this paper as the main result we consider the recognition by prime graph of almost simple group $\mathrm{PGL}(2,25)$.

2. Preliminary Results

Lemma 2.1 ([8]) Let G be a finite group and $|\pi(G)| \geqslant 3$. If there exist prime numbers r, s, $t \in \pi(G)$, such that $\{tr, ts, rs\} \cap \pi_e(G) = \emptyset$, then G is non-solvable.

Lemma 2.2 (see [20]) Let G be a Frobenius group with kernel F and complement C. Then every subgroup of C of order pq, with p, q (not necessarily distinct) primes, is cyclic. In particular, every Sylow subgroup of C of odd order is cyclic and a Sylow 2-subgroup of C is either cyclic or generalized quaternion group. If C is a non-solvable group, then C has a subgroup of index at most 2 isomorphic to $SL(2,5) \times M$, where M has cyclic Sylow p-subgroups and $(|M|, 30) = 1$.

Using [14, Theorem A], we have the following result:

Lemma 2.3 Let G be a finite group with $t(G) \geqslant 2$. Then one of the following holds:
(a) G is a Frobenius or 2-Frobenius group;
(b) there exists a nonabelian simple group S such that $S \leqslant \overline{G} := G/N \leqslant \mathrm{Aut}(S)$ for some nilpotent normal subgroup N of G.

Lemma 2.4 ([21]) Let $G = L_n^{\varepsilon}(q)$, $q = p^m$, be a simple group which acts absolutely irreducibly on a vector space W over a field of characteristic p. Denote $H = W \rtimes G$. If $n = 2$ and q is odd then $2p \in \pi_e(H)$.

3. Main Results

Theorem 3.1 The almost simple group $PGL(2,25)$ is recognizable by prime graph.

Proof. Throughout this proof, we suppose that G is a finite group such that $\Gamma(G) = \Gamma(\mathrm{PGL}(2,25))$.

First of all, we remark that by [19, Lemma 7], we have $\mu(\mathrm{PGL}(2,25)) = \{5, 24, 26\}$. Therefore, the prime graph of $\mathrm{PGL}(2,25)$ has two connected components which are $\{5\}$ and $\pi(5^4-1)$. Also we conclude that the subsets $\{2,5\}$ and $\{3,5,13\}$ are two independent subsets of $\Gamma(G)$. In the sequel, we prove that G is neither a Frobenius nor a 2-Frobenius group.

Let $G = K : C$ be a Frobenius group with kernel K and complement C. By Lemma 2.2, we know that K is nilpotent and $\pi(C)$ is a connected component of the prime graph of G. Hence we conclude that $\pi(K) = \{5\}$ and $\pi(C) = \{2,3,13\}$, since 5 is an isolated vertex in $\Gamma(G)$.

If C is non-solvable, then by Lemma 2.2, C consists a subgroup isomorphic to $SL(2,5)$. This implies that $5 \in \pi(SL(2,5)) \subseteq \pi(C)$, which is a contradiction since $\pi(C) = \{2,3,13\}$. Therefore, C is solvable and so it contains a $\{3,13\}$-Hall subgroup, say H. Since K is a normal subgroup of G, KH is a subgroup of G. Also we have $\pi(KH) = \{3, 5, 13\}$. Thus KH is a subgroup of odd order and so it is a solvable subgroup of G. On the other hand, in the prime graph of G, the subset $\{3, 5, 13\}$ is independent. Hence KH is a solvable subgroup of G such that its prime graph contains no edge, which contradicts to Lemma 2.1. Therefore, we get that G is a Frobenius group.

Let G be a 2-Frobenius group with the normal series $1 \lhd H \lhd K \lhd G$, where K is a Frobenius group with kernel H and G/H is a Frobenius group with kernel K/H. We know that G is a solvable group. This implies that G contains a $\{3, 5, 13\}$-Hall subgroup, say T. Again similar to the previous discussion, we get a contraction.

By the above argument, the finite group G is neither Frobenius nor 2-Frobenius. So by Lemma 2.3, we conclude that there exists a nonabelian simple group S such that:

$$S \leqslant \overline{G} := \frac{G}{K} \leqslant \mathrm{Aut}(S)$$

in which K is the Fitting subgroup of G. We know that $\pi(S) \subseteq \pi(G)$. Since $\pi(G) = \{2, 3, 5, 13\}$, so by [13, Table 8], we get that S is isomorphic to one of the simple group A_5, A_6, $\mathrm{PSU}_4(2)$, ${}^2F_4(2)'$, $\mathrm{PSU}_3(4)$, $\mathrm{PSL}_3(3)$, $S_4(5)$, or $\mathrm{PSL}_2(25)$. Now we consider each possibility for the simple group S, step by step.

Step 1. Let S be isomorphic to the alternating group A_5 or A_6. Since $\pi(S) \cup \pi(\mathrm{Out}(S)) = \{2, 3, 5\}$, we conclude that $13 \in \pi(K)$. We know that the alternating groups A_5 and A_6 consist a Frobenius subgroup $2^2 : 3$. Hence since $13 \in \pi(K)$, by [17, Lemma 3.1], we deduce that $13 \sim 3$, which is a contradiction.

Step 3.2. Let S be isomorphic to the simple group $\mathrm{PSU}_4(2)$. By [5], the finite group S contains a Frobenius group $2^2 : 3$, so similar to the above argument we get a contradiction.

Step 3.3. Let S be isomorphic to the simple group ${}^2F_4(2)'$. By [4], in the prime graph of the simple group S, 5 is not an isolated vertex.

Step 3.4. Let S be isomorphic to the simple group $\mathrm{PSU}_3(4)$. Again by [4], in the prime graph of the simple group S, 5 is not an isolated vertex.

Step 3.5. Let S be isomorphic to the simple group $\mathrm{PSL}_3(3)$. Since $\pi(\mathrm{PSL}_3(3)) = \{2, 3, 13\}$, we get that $5 \in \pi(K)$. On the other hand Sylow 3-subgroups of $\mathrm{PSL}_3(3)$ are not cyclic. Hence $5 \notin \pi(K)$, since 5 and 3 are nonadjacent in $\Gamma(G)$.

Step 3.5. Let S be isomorphic to the simple group $S_4(5)$. Again by [4], in the prime graph of the simple group S, 5 is not an isolated vertex.

Step 3.4. Let S be isomorphic to $\mathrm{PSL}_2(25)$. Hence $\mathrm{PSL}_2(25) \leqslant \bar{G} \leqslant \mathrm{Aut}(\mathrm{PSL}_2(25))$.

Let $\pi(K)$ contains a prime r such that $r \neq 5$. Since K is nilpotent, we may assume that K is a vector space over a field with r elements. Hence the prime graph of the semidirect product $K \rtimes \mathrm{PSL}_2(25)$ is a subgraph of $\Gamma(G)$. Let B be a Sylow 5-subgroup of $\mathrm{PSL}_2(25)$. We know that B is not cyclic. On the other hand $K \rtimes B$ is a Frobenius group such that $\pi(K \rtimes B) = \{r, 5\}$. Hence B should be cyclic which is a contradiction.

Let $\pi(K) = \{5\}$. In this case, by Lemma 2.4, we get that there is an edge between 2 and 5 in the prime graph of G which is a contradiction. Therefore, by the above discussion, we deduce that $K = 1$. Also this implies that $\mathrm{PSL}_2(25) \leqslant G \leqslant \mathrm{Aut}(\mathrm{PSL}_2(25))$.

We know that $\mathrm{Aut}(\mathrm{PSL}_2(25)) \cong Z_2 \times Z_2$. Since in the prime graph of $\mathrm{PSL}_2(25)$ there is not any edge between 13 and 2, we get that $G \ncong \mathrm{PSL}_2(25)$. Also if $G = \mathrm{PSL}_2(25) : \langle \theta \rangle$, where θ is a field automorphism, then we get that 2 and 5 are adjacent in G, which is a contradiction. If $G = \mathrm{PSL}_2(25) : \langle \gamma \rangle$, where γ is a diagonal-field automorphism, then we get that G does not contain any element with order $2 \cdot 13$ (see [3, Lemm 12]), which is contradiction, since in $\Gamma(G)$, $2 \sim 13$. This argument shows that $G \cong \mathrm{PGL}_2(25)$, which completes the proof. ■

References

[1] Z. Akhlaghi, M. Khatami and B. Khosravi, Characterization by prime graph of $\mathrm{PGL}(2, p^k)$ where p and k are odd, Int. J. Algebra Comp. 20 (7) (2010), 847-873.
[2] A. A. Buturlakin, Spectra of Finite Symplectic and Orthogonal Groups, Siberian Advances in Mathematics, 21 (3) (2011), 176-210.
[3] G. Y. Chen, V. D. Mazurov, W. J. Shi, A. V. Vasil'ev and A. Kh. Zhurtov, Recognition of the finite almost simple groups $PGL_2(q)$ by their spectrum, J. Group Theory 10(1) (2007), 71-85.
[4] J. H. Conway, R. T. Curtis, S. P. Norton, R. A. Parker and R. A. Wilson, Atlas of Finite Groups, Oxford University Press, Oxford, 1985.
[5] M. A. Grechkoseeva, On element orders in covers of finite simple classical groups, J. Algebra, 339 (2011), 304-319.
[6] D. Gorenstein, Finite Groups, Harper and Row, New York, 1968.
[7] M. Hagie, The prime graph of a sporadic simple group, Comm. Algebra 31(9) (2003), 4405-4424.
[8] G. Higman, Finite groups in which every element has prime power order, J. London Math. Soc. 32 (1957), 335–342.
[9] M. Khatami, B. Khosravi and Z. Akhlaghi, NCF-distinguishability by prime graph of $PGL(2, p)$, where p is a prime, Rocky Mountain J. Math. (to appear).
[10] B. Khosravi, n-Recognition by prime graph of the simple group $PSL(2, q)$, J. Algebra Appl. 7(6) (2008), 735-748.
[11] B. Khosravi, B. Khosravi and B. Khosravi, 2-Recognizability of $PSL(2, p^2)$ by the prime graph, Siberian Math. J. 49(4) (2008), 749–757.
[12] B. Khosravi, B. Khosravi and B. Khosravi, On the prime graph of $PSL(2, p)$ where $p > 3$ is a prime number, Acta. Math. Hungarica 116(4) (2007), 295-307.
[13] R. Kogani-Moghadam and A. R. Moghaddamfar, Groups with the same order and degree pattern, Sci. China Math. 55 (4) (2012), 701-720.
[14] A. S. Kondrat'ev, Prime graph components of finite simple groups, Math. USSR-SB. 67(1) (1990), 235-247.
[15] A. Mahmoudifar and B. Khosravi, On quasirecognition by prime graph of the simple groups $A_n^+(p)$ and $A_n^-(p)$, J. Algebra Appl. 14(1) (2015), (12 pages).
[16] A. Mahmoudifar and B. Khosravi, On the characterization of alternating groups by order and prime graph, Sib. Math. J. 56(1) (2015), 125-131.
[17] A. Mahmoudifar, On finite groups with the same prime graph as the projective general linear group $\mathrm{PGL}(2, 81)$, (to appear).
[18] V. D. Mazurov, Characterizations of finite groups by sets of their element orders, Algebra Logic 36(1) (1997), 23-32.
[19] A. R. Moghaddamfar, W. J. Shi, The number of finite groups whose element orders is given, Beiträge Algebra Geom. 47(2) (2006), 463-479.
[20] D. S. Passman, Permutation groups, W. A. Bengamin, New York, 1968.
[21] A. V. Zavarnitsine, Fixed points of large prime-order elements in the equicharacteristic action of linear and unitary groups, Sib. Electron. Math. Rep. 8 (2011), 333-340.

Permissions

All chapters in this book were first published in JLTA, by IAU; hereby published with permission under the Creative Commons Attribution License or equivalent. Every chapter published in this book has been scrutinized by our experts. Their significance has been extensively debated. The topics covered herein carry significant findings which will fuel the growth of the discipline. They may even be implemented as practical applications or may be referred to as a beginning point for another development.

The contributors of this book come from diverse backgrounds, making this book a truly international effort. This book will bring forth new frontiers with its revolutionizing research information and detailed analysis of the nascent developments around the world.

We would like to thank all the contributing authors for lending their expertise to make the book truly unique. They have played a crucial role in the development of this book. Without their invaluable contributions this book wouldn't have been possible. They have made vital efforts to compile up to date information on the varied aspects of this subject to make this book a valuable addition to the collection of many professionals and students.

This book was conceptualized with the vision of imparting up-to-date information and advanced data in this field. To ensure the same, a matchless editorial board was set up. Every individual on the board went through rigorous rounds of assessment to prove their worth. After which they invested a large part of their time researching and compiling the most relevant data for our readers.

The editorial board has been involved in producing this book since its inception. They have spent rigorous hours researching and exploring the diverse topics which have resulted in the successful publishing of this book. They have passed on their knowledge of decades through this book. To expedite this challenging task, the publisher supported the team at every step. A small team of assistant editors was also appointed to further simplify the editing procedure and attain best results for the readers.

Apart from the editorial board, the designing team has also invested a significant amount of their time in understanding the subject and creating the most relevant covers. They scrutinized every image to scout for the most suitable representation of the subject and create an appropriate cover for the book.

The publishing team has been an ardent support to the editorial, designing and production team. Their endless efforts to recruit the best for this project, has resulted in the accomplishment of this book. They are a veteran in the field of academics and their pool of knowledge is as vast as their experience in printing. Their expertise and guidance has proved useful at every step. Their uncompromising quality standards have made this book an exceptional effort. Their encouragement from time to time has been an inspiration for everyone.

The publisher and the editorial board hope that this book will prove to be a valuable piece of knowledge for researchers, students, practitioners and scholars across the globe.

List of Contributors

M. Akdağ and F. Erol
Cumhuriyet University Science Faculty Department of Mathematics 58140 S_IVAS / TURKEY

M. Rashidi-Kouchi
Young Researchers and Elite Club Kahnooj Branch, Islamic Azad University, Kerman, Iran

R.J. Shahkoohi and A. Razania
Department of Mathematics, Science and Research Branch, Islamic Azad University, Tehran, Iran

A. Fallahzadeh and M. A. Fariborzi Araghi
Department of Mathematics, Islamic Azad University, Central Tehran Branch, Iran

V. Fallahzadeh
Department of Mathematics, Islamic Azad University, Arac Branch, Iran

M.Özkoç and B. S. Ayhana
Department of Mathematics, Faculty of Science Muğla Siki Koçman University, Menteşe-Muğla 48000 Turkey

M. Saleem Lone and D. Krishnaswamy
Department of Mathematics, Annamalai University, Chidambaram, Tamilnadu, India

M. Hakimi-Nezhaad and M. Ghorbani
Department of Mathematics, Faculty of Science, Shahid Rajaee Teacher Training University, Tehran, 16785-136, Iran

M. Amin khah
Department of Application Mathematics, Kerman Graduate University of High Technology, Iran

A. Askari Hemmat
Department of Mathematics, Shahid Bahonar University of Kerman, Iran

R. Raisi Tousi
Department of Mathematics, Ferdowsi University of Mashhad, Iran

Z. Sadati and Kh. Maleknejad
Department of Mathematics, Khomein Branch, Islamic Azad University, Khomein, Iran

S. Ebrahimi
Department of Mathematics, Payame Noor University, Tehran, Iran

Vijaya L. Gompa
Department of Mathematics, Troy University, Dothan, AL 36304

M. Ghorbani and Z. Gharavi-Alkhansari
Department of Mathematics, Faculty of Science, Shahid Rajaee Teacher Training University, Tehran, 16785-136, Iran

R. A. Rashwana
Department of Mathematics, Faculty of Science, Assuit University, Assuit 71516, Egypt

H. A. Hammad
Department of Mathematics, Faculty of Science, Sohag University, Sohag 82524, Egypt

S. P. Mondal
Department of Mathematics, National Institute of Technology, Agartala, Jirania-799046, Tripura, India

T. K. Roy
Department of Mathematics, Indian Institute of Engineering Science and Technology, Shibpur, Howrah-711103, West Bengal, India

F. M. Yaghoobi
Department of Mathemetics, College of Science, Hamedan Branch, Islamic Azad University, Hamedan, Iran

J. Shamshiri
Department of Mathematics, Mashhad Branch, Islamic Azad University, Mashhad, Iran

M. Hashemi and M. Polkouei
Faculty of Mathematical Sciences, University of Guilan, Rasht, Iran

Vijaya L. Gompa
Department Head and Professor of Mathematics, Jacksonville State University, Jacksonville, AL 36265

A. Mahmoudifar
Department of Mathematics, Tehran North Branch, Islamic Azad University,Tehran, Iran

Index

www.ingramcontent.com/pod-product-compliance
Lightning Source LLC
Chambersburg PA
CBHW082022190326
41458CB00010B/3239
* 9 7 8 1 6 8 2 8 5 6 3 4 5 *